Managing Your Biological Data with Python

T0132629

CHAPMAN & HALL/CRC
Mathematical and Computational Biology Series

Aims and scope:

This series aims to capture new developments and summarize what is known over the entire spectrum of mathematical and computational biology and medicine. It seeks to encourage the integration of mathematical, statistical, and computational methods into biology by publishing a broad range of textbooks, reference works, and handbooks. The titles included in the series are meant to appeal to students, researchers, and professionals in the mathematical, statistical and computational sciences, fundamental biology and bioengineering, as well as interdisciplinary researchers involved in the field. The inclusion of concrete examples and applications, and programming techniques and examples, is highly encouraged.

Series Editors

N. F. Britton
Department of Mathematical Sciences
University of Bath

Xihong Lin
Department of Biostatistics
Harvard University

Hershel M. Safer
School of Computer Science
Tel Aviv University

Maria Victoria Schneider
European Bioinformatics Institute

Mona Singh
Department of Computer Science
Princeton University

Anna Tramontano
Department of Physics
University of Rome La Sapienza

Proposals for the series should be submitted to one of the series editors above or directly to:
CRC Press, Taylor & Francis Group
3 Park Square, Milton Park
Abingdon, Oxfordshire OX14 4RN
UK

Published Titles

Algorithms in Bioinformatics: A Practical Introduction
Wing-Kin Sung

Bioinformatics: A Practical Approach
Shui Qing Ye

Biological Computation
Ehud Lamm and Ron Unger

Biological Sequence Analysis Using the SeqAn C++ Library
Andreas Gogol-Döring and Knut Reinert

Cancer Modelling and Simulation
Luigi Preziosi

Cancer Systems Biology
Edwin Wang

Cell Mechanics: From Single Scale-Based Models to Multiscale Modeling
Arnaud Chauvière, Luigi Preziosi, and Claude Verdier

Clustering in Bioinformatics and Drug Discovery
John D. MacCuish and Norah E. MacCuish

Combinatorial Pattern Matching Algorithms in Computational Biology Using Perl and R
Gabriel Valiente

Computational Biology: A Statistical Mechanics Perspective
Ralf Blossey

Computational Hydrodynamics of Capsules and Biological Cells
C. Pozrikidis

Computational Neuroscience: A Comprehensive Approach
Jianfeng Feng

Computational Systems Biology of Cancer
Emmanuel Barillot, Laurence Calzone, Philippe Hupé, Jean-Philippe Vert, and Andrei Zinovyev

Data Analysis Tools for DNA Microarrays
Sorin Draghici

Differential Equations and Mathematical Biology, Second Edition
D.S. Jones, M.J. Plank, and B.D. Sleeman

Dynamics of Biological Systems
Michael Small

Engineering Genetic Circuits
Chris J. Myers

Exactly Solvable Models of Biological Invasion
Sergei V. Petrovskii and Bai-Lian Li

Game-Theoretical Models in Biology
Mark Broom and Jan Rychtář

Gene Expression Studies Using Affymetrix Microarrays
Hinrich Göhlmann and Willem Talloen

Genome Annotation
Jung Soh, Paul M.K. Gordon, and Christoph W. Sensen

Glycome Informatics: Methods and Applications
Kiyoko F. Aoki-Kinoshita

Handbook of Hidden Markov Models in Bioinformatics
Martin Gollery

Introduction to Bioinformatics
Anna Tramontano

Introduction to Bio-Ontologies
Peter N. Robinson and Sebastian Bauer

Introduction to Computational Proteomics
Golan Yona

Introduction to Proteins: Structure, Function, and Motion
Amit Kessel and Nir Ben-Tal

An Introduction to Systems Biology: Design Principles of Biological Circuits
Uri Alon

Kinetic Modelling in Systems Biology
Oleg Demin and Igor Goryanin

Knowledge Discovery in Proteomics
Igor Jurisica and Dennis Wigle

Published Titles (continued)

Chapman & Hall/CRC Mathematical and Computational Biology Series

Managing Your Biological Data with Python

Allegra Via
Kristian Rother
Anna Tramontano

CRC Press
Taylor & Francis Group
Boca Raton London New York

CRC Press is an imprint of the
Taylor & Francis Group, an **informa** business

A CHAPMAN & HALL BOOK

CRC Press
Taylor & Francis Group
6000 Broken Sound Parkway NW, Suite 300
Boca Raton, FL 33487-2742

© 2014 by Taylor & Francis Group, LLC
CRC Press is an imprint of Taylor & Francis Group, an Informa business

No claim to original U.S. Government works

Printed on acid-free paper
Version Date: 20130808

International Standard Book Number-13: 978-1-4398-8093-7 (Paperback)

Library of Congress Cataloging-in-Publication Data

Via, Allegra.
    Managing your biological data with Python / Allegra Via, Kristian Rother, Anna Tramontano.
        pages cm. -- (Chapman & Hall/CRC mathematical and computational biology series)
    Includes bibliographical references and index.
    ISBN 978-1-4398-8093-7 (alk. paper)
    1. Biology--Data processing. 2. Python (Computer program language) I. Rother, Kristian. II. Tramontano, Anna. III. Title.

QH324.2.V526 2013
570.285--dc23   2013026177

Visit the Taylor & Francis Web site at
http://www.taylorandfrancis.com

and the CRC Press Web site at
http://www.crcpress.com

Table of Contents

PART IV **Summary**

PART V **Biopython**

Preface

Only a few years ago, programming was a prerogative of computational scientists. Notwithstanding this, programming is increasingly becoming a need of specialists in other fields such as biology. As a biologist, you are not necessarily interested in becoming an expert programmer, but you want to continue your scientific endeavors using programming as one of many tools. You may already have realized that programming techniques would dramatically speed up the management and analysis of your data. Maybe you want to deal with large amounts of data, repeat the same kind of analysis several times, or parse files with unusual formats. We can assure you that in all these cases programming is very useful. However, you may feel uncomfortable because you never had much interest in a "dry" and "conceptually hard" discipline such as computer science. In that case, this book is for you.

We wrote this book for life scientists who want to have more control of their data and, for this, need to learn some programming. It is aimed at empowering biologists without prior programming experience to work with biological data on their own using Python.

In the Preface, you will find a summary of what you can learn reading this book and an introduction of what a program is, followed by an overview of the Python programming language.

We hope that this book on programming is tailored to your needs as a biologist and will help you analyze your data and thus increase the likelihood to make better discoveries.

WHAT YOU CAN LEARN FROM THIS BOOK

In this book you will learn not only how to program but also how to manage your data, which means reading data from files, analyzing and manipulating them, and writing the results to a file or to the computer screen. Every single piece of code described in the book is aimed at solving biological problems; every example deals with biological questions. The book proposes as many different cases as possible; covers many strategies to organize, analyze, and present data; and solves biological problems in the form of "programming recipes." Exercises that you can use to test yourself or include in a programming course for biologists appear at the end of each chapter.

The book is organized in six parts and contains twenty-one chapters in total. Part I introduces the Python language and teaches you how to write your first programs. Part II introduces all the basic elements of the language, enabling you to write small programs independently. Part III is about creating bigger programs using techniques to write well-organized, efficient, and error-free code. Part IV is devoted to data visualization. You will learn how to plot your data, or draw a figure for an article or a slide presentation. It also introduces PyMOL, a program to visualize macromolecular structures. Part V introduces you to Biopython, a programming library that helps with reading and writing several biological file formats and facilitates querying the NCBI databases online and retrieving biological records from the web. Part VI is a cookbook containing twenty specific programming "recipes," ranging from secondary structure prediction and multiple sequence alignment analyses to superimposing protein three-dimensional structures.

Furthermore, the book has four appendices. Appendix A provides an overview of both Python and UNIX commands. Appendix B lists several links to Python resources freely available on the web. Appendix C contains sample file formats cited throughout the book, such as a sequence in FASTA format, a sequence in GenBank format, a PDB file, an MSA example, etc. Finally, Appendix D is a short UNIX tutorial.

WHAT IS PROGRAMMING?

This book will teach you how to write programs. What exactly is a program? A program is conceptually similar to a cooking recipe. Like a recipe lists ingredients and kitchenware at the beginning, a program needs to define what objects (data and functions) are necessary. For instance, you could define a given DNA sequence as your data and define a function that calculates the GC-content in it. A recipe also contains a list of actions that must be carried out to use ingredients and kitchenware to prepare a dish. Likewise, a program contains a written list of elementary instructions such as "read the DNA sequence from a file," "calculate the GC-content," or "print the GC-content to the screen." Creating a program means writing instructions in a suitable language (e.g., Python), typically to a text file. Running a program means executing the instructions (i.e., the lines of code) listed in the program.

There is one big difference between kitchen recipes and computer programs, though: a human cook can divert from the recipe and add ingredients creatively or react to unexpected mishaps, which is important to obtain a tasty meal! A computer, however, is never creative. It reads the instructions in the program one by one and executes them by the letter. On one hand, the lack of computer creativity makes it necessary for you to explicitly tell it every tiny step, which can sometimes be unnerving. Imagine you are talking to a cook who is intellectually disabled but incredibly fast. On the other hand, computer predictability makes it easy to precisely repeat instructions many times. Imagine what a cook would say to an order of 100,000 identical dishes! Programming means using the rigid logics of computers to your advantage.

You must be aware that most of programming happens in your head. When you struggle to write a program, it may be helpful to formulate small step-by-step instructions in human language first. When the overall structure of your program is ready and you know exactly what you want it to do, it is time to start writing instructions. To do this, you need a programming language. In fact, programming basically consists of writing instructions in a given language to a text file or to a special terminal shell and telling your computer to execute them. The lines containing instructions are commonly called source code. Accordingly, programming or coding means writing source code. Since computers do not understand English, Italian, or German, you need to use a programming language to write source code. Our favorite language for answering biological questions is Python.

WHY PYTHON?

Python is simple to learn. It is a high-level programming language that is interpreted and object oriented. Let's analyze these concepts one by one.

Python Is Simple to Learn

A program can be written in one of many programming languages: C, C++, Fortran, Perl, Java, Pascal, etc. Every programming language has formal rules and keywords (the syntax) and semantics (meaning). A key advantage of Python is that code is easy to read. Code can be more or less comprehensible to humans; for example, the Python instruction

```
print 'ACGT'
```

is quite intuitive (the computer will print the text ACGT to the screen), whereas the Perl instruction

```
$cmd = "imgcvt -i $intype -o $outtype $old.$num";
```

is less intuitive. Python is, compared to other programming languages, relatively similar to English and has a very simple syntax. We think this makes Python easy to learn for biologists.

Python Is a High-Level Programming Language

Python can also be used to do very complex things. You can represent complex data types like trees and networks, start other programs (e.g., bioinformatics applications) from Python, and download web pages. You also have tools to detect and handle errors in your programs. Finally, Python is not optimized for any particular purpose; it is therefore well apt to glue together other programs, web services, and databases in order to build customized scientific pipelines with a few lines of source code.

Python Is Interpreted

Some programming languages are interpreted, and some are compiled. For computers to execute a program, they need to translate the instructions to binary machine code, which is unreadable even for experienced programmers. In an interpreted language, each line is translated and executed one after another. In a compiled language, first the whole program is translated and only then executed. Execution of compiled languages is generally much faster than execution of interpreted ones. However, you need to compile the program each time you change something. With an interpreted language, you can see the effect of your changes immediately and, as a result, write programs faster. Therefore, we think that an interpreted language like Python is much easier to start with.

Python Is Object Oriented

In Python, everything is an *object*. Objects are independent program components representing data and instructions. They allow you to connect data with useful functionalities (e.g., you could have a sequence object that contains a DNA sequence and functions for transcribing and translating this sequence). Objects help to structure complex programs and make program components reusable.

Using Python, many developers have made reusable objects available in programming libraries. For instance, reading and parsing a FASTA sequence file using Biopython can be done in two lines of code. Without the library, you would have to write ten to thirty lines, depending on the programming language. Therefore, object orientation in Python helps you to write short programs.

In conclusion, we believe that Python is an ideal language for those who want to have fun with little or no pain and learn programming to pragmatically manage biological data, solve biological problems, and widen the horizon of their scientific discoveries. We hope you will enjoy using this book at least as much as we enjoyed writing it!

Code Downloads

All code examples presented in this book are available online at https://bitbucket.org/krother/python-for-biologists, following the "Source" link.

Acknowledgements

We would like to thank the students and trainees to whom we had the privilege to teach Python. Your questions, problems, and ideas during Python courses over the past seven years are the main source of inspiration for this book. We can't name all of you, but we want you to know that we learned much from your enthusiasm, cheerfulness, frustration, and success.

Special thanks go to Pedro Fernandes, a great course organizer, who provided us with the opportunity to condense existing material into a five-day course at the Gulbenkian Institute in Portugal. We learned many of the key questions of this book during these courses and during after-dinner discussions in Astrolabio.

Additional credit goes to Janusz M. Bujnicki, Artur Jarmolowski, Jakub Nowak, Edward Jenkins, Amelie Anglade, Janick Mathys, and Victoria Schneider for providing various Python training opportunities.

We are also grateful to Francesco Cicconardi for his help with the RNA-Seq output parser and the NGS pipeline on which Chapters 6 and 14 are respectively based. He not only suggested us a typical NGS pipeline but also provided code and verified that the biological and computational discussions of the problem were correct and exhaustive.

We would like to thank Justyna Wojtczak, Katarzyna Potrzebowska, Wojciech Potrzebowski, Kaja Milanowska, Tomasz Puton, Joanna Kasprzak, Anna Philips, Teresa Szczepinska, Peter Cock, Bartosz Telenczuk, Patrick Yannul, Gavin Huttley, Rob Knight, Barbara Uszczynska, Fabrizio Ferre', Markus Rother, and Magdalena Rother for providing examples and constructive feedback.

Finally, many thanks to Alba Lepore for discussions during the realization of the book and for key help in accomplishing the book's cover.

I

Getting Started

For the four brave Python apprentices who made it to the mountaintop during the Python and Friends Conference 2010 in Karpacz, Poland.

When you want to climb high mountains, what do you do? If you are good at mountain climbing, you gather your equipment, call up some fellow climbers, pick a mountain, and move out. In the stories written by professional mountain climbers, you will find that they use ropes, hooks, oxygen bottles, and sometimes nothing more than their bare hands. They fight with icy storms at altitudes of 4,000 meters and above, coordinate big teams distributed over several camps, and survive in the deadly zone near the mountaintop.

But what if you are a beginner interested in mountain climbing? Do you strap on the oxygen bottles and move out? No. Instead, you will probably start with an easy mountain. There are mountains with safe, clearly marked paths to the top. All you need is a map and a pair of boots. Still, the sight from the top of such a mountain can be breathtaking.

Programming is very similar to that. As a biologist learning to program, you do not need fancy equipment or tons of theoretical knowledge. Even simple programs can be powerful tools to master your data. A lot of programming can be done by collecting working fragments of code and then assembling and modifying them. The result may not be as elegant as a program written by a computer scientist, but it may solve a problem quickly. Your problem.

We want this book to be the map that helps you to climb the mountains of everyday data management. We want programming to make your life easier without you necessarily becoming a professional software developer first.

In the first part, we would like you to make your first steps in the Python programming language. You will see that commands in Python are very intuitive and close to the English language, so you won't need much effort to learn and remember most of the Python instructions. For example, if you want to calculate the length of a sequence, you just have to type `len('MALWMRLLPLLALLALWGPDPAA...')`. The aim of the two chapters of Part I is not only to show how simple the Python syntax is but also to make you scent the clever structure of the language. Python basically consists of a set of modules (typically files where programming instructions are written) that you can connect to each other.

When learning a new language, e.g., German, you may start by reading a text and analyzing the nature, role, and position in the text of each word. After reading and analyzing many texts, you will be able to extract language rules and write your own texts. Alternatively, you can first learn what kinds of object categories make up the language nouns, verbs, adjectives, etc. and the links between them (e.g., prepositions or German cases) and then use the structure of the language, associated with a good dictionary, to write your texts. In this part of the book, you will start grasping that Python is basically another language like English or German. In fact, it is made up of a limited number of object types (nouns, verbs, etc.) that you can connect to each other to form sentences. In this book, we blend the two approaches described previously to learn German, by alternating code examples that you can analyze and try with the explanation of language object categories. To indicate what specific objects belong to each category, Python provides a very good online dictionary, which is called the Python Standard Library (http://docs.python.org/2/library/), where you can look up the meaning of single words. Once the structure of the language is clear to you, and you are able to play with the various object categories, most will be done: at that stage you can basically improve your knowledge of the language by increasing your vocabulary or using the dictionary efficiently. The last step of learning is related to the good design of programs. This is in general good practice and may turn out to be very useful if you need to write big programs, efficiently collaborate with other programmers, maintain or extend your or other people's programs in the future, increase the performance of your work, or want to become an expert programmer, but it is not really indispensable in order to accomplish the tasks presented in this book. In any case, we will provide plenty of suggestions on how to write good programs in the second part of the book.

In Chapter 1, you will learn how to use the Python shell where you can enter simple commands. The simplest operations are similar to those on a pocket calculator. For instance, if you use the Python shell, you will see a prompt that looks like this: >>>. If you enter a simple mathematical operation at the right of the prompt and press the Enter key,

```
>>> 1 + 1
```

you get the result 2 immediately. You will also encounter variables as a way to store your data. You will learn how to perform calculations with numbers and to import and use a mathematical module that gives you extra functions like square roots and logarithms. In Chapter 2, you will write your first Python program. The program will be for counting amino acids in a protein sequence. For that you will need strings, a data structure for storing text. You will use a control flow structure for repeating instructions automatically, instead of writing the same line over and over. At the end of Part I, you will know most of the basic parts of the Python language.

## CHAPTER 1

The Python Shell

L EARNING GOAL: You can use Python as a scientific calculator.

1.1 IN THIS CHAPTER YOU WILL LEARN

- How to use the Python shell as a scientific calculator
- How to calculate the ΔG of ATP hydrolysis
- How to calculate the distance between two points
- How to create your own Python module

1.2 STORY: CALCULATING THE ΔG OF ATP HYDROLYSIS

1.2.1 Problem Description

$$ATP \rightarrow ADP + P_i$$

The hydrolysis of one phosphodiester bond from ATP results in a standard Gibbs energy (ΔG^0) of –30.5 kJ/mol. According to biochemistry textbooks, the real ΔG value depends on the concentration of the compounds. And these concentrations can differ quite a lot among tissues (see Table 1.1, according to Berg et al.[*]).

---

[*] Jeremy M. Berg, John L. Tymoczko, and Lubert Stryer, *Biochemistry*, 5th ed. (New York: W. H. Freeman, 2002).

TABLE 1.1   Compound Concentration in Different Tissues.

Tissue	[ATP] [mM]	[ADP] [mM]	[P_i][mM]
Liver	3.5	1.8	5.0
Muscle	8.0	0.9	8.0
Brain	2.6	0.7	2.7

How can the real ΔG value for ATP hydrolysis be calculated? The Gibbs energy as a function of the concentrations of the compounds can be written as

$$\Delta G = \Delta G^0 + RT * \ln ([ADP] * [P_i] / [ATP])$$

You can insert values from the table into this equation with many tools (e.g., a pocket calculator, the Windows calculator application, or your mobile phone). In this book, you are going to learn a much more efficient and powerful tool for calculations and data management: the Python programming language.

Using Python, you can do the calculation for *liver tissue* in the interactive Python interpreter (see Figure 1.1). The prompt >>> indicates the

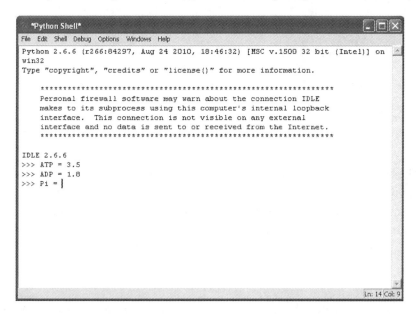

FIGURE 1.1   The Python shell. *Note:* To start it you have to type "python" at the prompt of the UNIX terminal shell (in UNIX/Linux or Mac OS X) or start 'Python (command line)' from the program menu (in Windows).

place where you can enter commands, and it appears when you start a Python interactive session (see Section 1.3.1). Python commands must be typed just at the right side of the prompt.

1.2.2 Example Python Session

```
>>> ATP = 3.5
>>> ADP = 1.8
>>> Pi = 5.0
>>> R = 0.00831
>>> T = 298
>>> deltaG0 = -30.5
>>>
>>> import math
>>> deltaG0 + R * T * math.log(ADP * Pi / ATP)
-28.161154161098693
```

Source: Adapted from code published by A.Via/K.Rother under the Python License.

1.3 WHAT DO THE COMMANDS MEAN?

In programming, most of what you do can be roughly summarized in five points: organize data, use other programs, calculate things, and read and write data. The previous example contains three. First, it organizes the parameters for the ΔG formula by storing them in variables. Variables are containers that help you not to write the same numbers repeatedly. Second, it uses an external program to calculate the logarithm: the math.log(x) function calculates the logarithm of x and is accessed through the import statement, which makes available extra Python functions by connecting a program to other *modules* (math in the example) where such functions are stored. Modules are programming units collecting variables, functions, and other useful objects. They are always stored in files. See Box 1.1 for more on the import statement and Python modules.

Finally, the example in Section 1.2.2 calculates the ΔG value. Simple arithmetical calculations work very similar to a pocket calculator. The second part of this book is dedicated to other ways in which you can manipulate your data. The first thing you can try to do yourself is to start the calculation in the previous section.

BOX 1.1 THE `import` STATEMENT AND THE CONCEPT OF MODULES

When you write

```
>>> import math
```

you are connecting to the `math` module. What exactly is `math`? `math` is a file on your computer; its actual name is `math.py`. The .py extension stands for Python, and the file contains Python instructions, i.e., definitions of variables and functions and instructions (for calculating things). The `math.py` file, in particular, contains instructions for the definition and calculation of mathematical functions (e.g., `sqrt()`, `log()`, etc.).

In Python, text files that contain Python instructions are called *modules*. The `import` instruction is needed to access an external module and read its content. This way, all the definitions present in a module will become available when you import it: effectively, the code is shared and can be used in many different programs.

How can you know which mathematical functions are defined in the `math` module? Either you can browse the Internet and find and open the file `math.py` and read its contents, or you can use the instruction

```
>>> import math
>>> dir(math)
```

You can use the `dir(math)` instruction only after having imported the `math` module, otherwise the `dir()` function does not know what its argument is. As a result, you will see a complete list of variables and functions present in the `math` module:

```
['__doc__', '__name__', '__package__', 'acos', 'acosh',
'asin', 'asinh', 'atan', 'atan2', 'atanh', 'ceil', 'copy-
sign', 'cos', 'cosh', 'degrees', 'e', 'exp', 'fabs',
'factorial', 'floor', 'fmod', 'frexp', 'fsum', 'hypot',
'isinf', 'isnan', 'ldexp', 'log', 'log10', 'log1p',
'modf', 'pi', 'pow', 'radians', 'sin', 'sinh', 'sqrt',
'tan', 'tanh', 'trunc']
```

You can get a short explanation of each function by typing, for instance,

```
>>> help(math.sqrt)
```

1.3.1 How to Run the Example on Your Computer

The Python programming language is a program that needs to be started before you can use it. On Linux (Ubuntu) and Mac OS X, Python is already installed and can be started from a text console by typing "python" at the command line prompt. See Appendix D to learn how to run a program from a text terminal. On Windows, you need to install Python first and then start a Python shell window in 'Start' → 'All programs' → 'Python' → 'IDLE or Python (command line)'. First, download Python 2.7 from www.python.org. Install it, then start 'Python (command line)' from the program menu. For more details, see Box 1.2. When you see the >>> sign in a text window on your screen, you have succeeded and are ready to write program code (see Figure 1.1). The Python shell can be exited by typing Ctrl+D.

BOX 1.2 HOW TO INSTALL PYTHON

On Linux and Mac OS X, Python is already installed. In rare cases where it is not, you can get the most recent version from the package manager or by typing at a command terminal:

```
sudo apt-get install python
```

On Windows you need to download the Python Windows installer from www.python.org. Make sure you download version 2.7 of Python. Versions 3.0 and above are at the time of writing experimental and not compatible with this book. The programming language can be installed like most programs by clicking and accepting the defaults.

To check whether your installation was successful, you should start Python. There are two ways to run Python code:

1. Using the interactive mode (Python shell). On Linux and Mac OS X, you type "python" from a text console and press Enter. On Windows you choose 'Start' → 'Programs' → 'Python 2.7' → 'Python (command line)'. The Python shell will start in a separate window. Alternatively, you can open a text console by entering "cmd" in the 'Start' → 'Execute' dialog, then change the directory to C:\Python27 and type "python" there. When you see the prompt >>>, your installation is successful.
2. Writing code into a script file having the .py extension (e.g., my_script.py) and executing the script by typing at the UNIX/Linux shell prompt:

```
python my_script.py
```

See also Section 2.3.1, "How to Execute the Program."

The Python Shell

The interactive mode is ideal for learning and for testing pieces of code. Each single instruction is written and directly executed. You can write instructions after the >>> sign and confirm them by pressing Enter. Each instruction is executed immediately.

```
>>> ATP = 3.5
>>> ATP
3.5
```

You can use the interactive mode for numerical calculations:

```
>>> 3 * 4
12
>>> 12.5 / 0.5
25.0
>>> (12.5 / 0.5) * 100
2500.0
>>> 3 ** 4
81
>>> 3 ** (4 + 2)
729
```

A disadvantage of the Python shell is that when you exit the session (by typing Ctrl+D), your code gets lost. Therefore, you can only save the code you have written by copying and pasting the instructions to a text editor. Text editors are described in Box 2.2 and Box D.2. To save your code, writing Python instructions to files directly is more convenient. See Example 1.1 or Chapter 2.

If something goes wrong, Python returns error messages, the content of which depends on the type of error. For example, if you mistype an instruction and write, for example,

```
>>> imprt math
```

instead of

```
>>> import math
```

you will get a message saying "SyntaxError: invalid syntax" plus some additional information to help you correct the error(s). The errors you can encounter and how you can manage them are described in Chapter 12. Making errors is normal in programming.

1.3.2 Variables

In Section 1.2.2, a number of variables are initially defined. That is, the values to be used in the calculation are put into named containers.

For example, when writing

```
>>> ATP = 3.5
```

the computer will remember the number 3.5 under the name ATP, so when you write later

```
>>> ATP
```

the computer will print the value 3.5.

In the same way, all numbers used (1.8, 5.0, 0.00831, 298, and –30.5) are recorded each in its own variable (ADP, Pi, R, T, and deltaG0, respectively). Note that none of the numbers have a unit. Like when using a pocket calculator, you need to take care to convert them properly. This is why for the gas constant R (8.31 J/kmol) the value

```
>>> R = 0.00831
```

is used, so that it fits to the unit of ΔG^0 (kJ/kmol). As with a pocket calculator, you are responsible for converting numbers to appropriate units.

Each kind of object can be stored in a variable. In other words, you can "label" a piece of data with a name and, instead of writing the whole data every time you need it, you can just use the name of the variable. The more complex and the more frequently used the data are (e.g., the nucleotide sequence of a whole gene), the more convenient it is to use a variable name in its place.

So, if you want to use the Gibbs energy value for ATP hydrolysis

$$\Delta G^0 = -30.5 \text{ kJ/mol}$$

several times, it would be better to put it into a variable and use the variable name instead of the whole number.

The operator used to assign an object to a variable name is the equal sign =:

```
>>> deltag = -30.5
```

Python distinguishes between integer and floating-point numbers:

```
>>> a = 3
>>> b = 3.0
```

In Python jargon, we say that the two variables a and b have different data *types*. The variable a is an *integer*; b is a *float*. Their difference can be seen when you divide these numbers by another integer:

```
>>> a / 2
1
>>> b / 2
1.5
```

You can enforce conversion of an integer to a float number by dividing the integer by a float:

```
>>> a / 2.0
1.5
```

You can assign numbers, text, and many other kinds of data to variables. More generally, you can refer to the data as Python *objects*. In the following example, you assign a floating-point number object to a variable:

```
>>> deltag = -30.5
```

If you assign a new value to an existing variable name, the second value will overwrite the first. In other words, by setting

```
>>> deltag = -28.16
```

deltag is now −28.16 and no longer −30.5. In later chapters, you will encounter more types of data.

There are some *rules* in the choice of variable names:

- Some words cannot be used for variable names because they have a meaning in Python. For instance, import cannot be used as a variable name. For a complete list of reserved words, see Box 1.3.

- The first character of a variable name cannot be a number.

- Variable names are case sensitive. Thus, var and Var are different names.

- Most special characters, i.e., all of $ % @ / \ . , [ ] ( ) { } # are not allowed.

BOX 1.3 RESERVED WORDS IN PYTHON

Python reserved words cannot be used for variables because they have a meaning in Python. Here are some examples: and, assert, break, class, continue, def, del, elif, else, except, exec, finally, for, from, global, if, import, in, is, lambda, not, or, pass, print, raise, return, try, while.

Q & A: DOES IT MATTER WHETHER I USE UPPERCASE OR LOWERCASE FOR VARIABLE NAMES?

Try the following code:

```
>>> ATP = 3.5
>>> atp = 8.0
>>> ATP
```

The result of the last command is 3.5, not 8.0. As a general rule in Python, it makes a difference whether uppercase or lowercase is used for naming variables.

Q & A: WHAT HAPPENS WHEN I USE A VARIABLE FOR THE FIRST TIME?

In some programming languages, you need to list all variables that you want to use and explicitly reserve memory for them. In Python, you don't have to do that. The Python interpreter treats everything as *objects*. This means every time you use a new variable name, Python recognizes the nature of the data (integer, float, text, etc.) and reserves sufficient memory for it. Python also automatically associates a list of *instruments* to the variable type. For example, the numerical variables a and b defined previously "know" that you can add, subtract, and multiply them and perform all numerical operations displayed in Table 1.2.

TABLE 1.2   Arithmetical Operations in Python.

Operator	Meaning
a + b	addition
a − b	subtraction
a * b	multiplication
a/b	division
a ** b	power (a^b)
a % b	modulo: the remainder of the division a / b
a // b	floor division, rounds down
a * (b + c)	parentheses, b + c will be done before the multiplication

1.3.3 Importing Modules

After defining variables, the next command in the Python session in Section 1.2.2 imports a module with mathematical functions. In Python, import is a command that activates installed extra libraries or single variables and functions. math is the name of a library module that is automatically installed with Python. It is activated by

```
>>> import math
```

In this chapter, the log function from the math module is being used to calculate a logarithm. For a complete list of available functions in math, see http://docs.python.org/2/library/math.html or type

```
>>> dir(math)
```

in the Python shell.

Every module can contain functions and variables. Modules are used to reuse code and to divide big programs into smaller parts and therefore organize them better. Every time you need, for example, a constant like the gas constant R, you can fetch it from its module without redefining it. Modules collected in the Python Standard Library are basically extra functions that somebody else wrote and optimized for you.

Python makes available hundreds of modules, i.e., sets of functions that become available through the import command. Moreover, you can create your own modules by writing Python instructions to a text file and saving the file with the .py extension (see Example 1.2). Modules will be discussed in more detail in Part III of the book.

To employ the logarithm function from the math module, we used the notation math.log. The *dot* between the module and function name has a very special role in Python. The dot is a "linker" between objects. We say that the object on the right of the dot is an attribute of the object on the left. So,

```
>>> math.log
```

means that the log object (a function) is an attribute of the math object (a module). In other words, log is a part of the math module, and if you want to use it after importing the module, you have to refer to it using the dot syntax. This is true for everything in Python. Whenever an object A contains another object B, the syntax to use it is A.B. If B contains C, and A contains B, you can write: A.B.C.

Objects can also be imported selectively from modules. In other words, you may want to import a single object or a few objects instead of the whole content of a module. To import only the logarithm function instead of the entire math module, you can write

```
>>> from math import log
```

To use the imported function now, instead of writing math.log, you need to directly write log. The question of which variable and function names are available at a given moment is best explained by the concept of Python namespaces (see Box 1.4).

BOX 1.4 NAMESPACES

The collection of object *names* (of variables, functions, etc.) defined in a module is called the *namespace* of that module. Each module has its own namespace. For instance, the namespace of the math module contains the names pi, sqrt, cos, and many others. The namespace of the random module contains none of the former but contains the names randomint and random instead. Even the Python shell has its own namespace, containing, for example, print.

The same name (e.g., pi) in two different modules may indicate two distinct objects, and the dot syntax makes it possible to avoid confusion between the namespaces of the two modules. What actually happens when the command import is executed? It happens that the code written in the imported module is entirely read and interpreted and its namespace is imported as well but kept separated from the namespace of the importing module. So, if you write

```
>>> import math
>>> sqrt(16)
```

```
Traceback (most recent call last):
        File "<stdin>", line 1, in <module>
NameError: name 'sqrt' is not defined
>>>
```

the name `sqrt` will not be recognized as an attribute of the `math` module unless you use the dot syntax

```
>>> math.sqrt(16)
4.0
>>>
```

But if you use

```
>>> from math import sqrt
```

you are actually merging the `math` namespace with the Python shell namespace. So, now, you can directly use

```
>>> sqrt(16)
4.0
>>>
```

You have to be careful when you merge the namespaces of two modules and be aware of how you are using variable names. In fact, if you import everything from the `math` module using the following instruction:

```
>>> from math import *
```

you will have

```
>>> pi
3.141592653589793
```

but by typing

```
>>> pi = 100
```

you are actually overwriting the `pi` variable imported from the `math` module, and `pi` will no longer have the π value. This may generate unexpected results in your calculations.

Q & A: WHY DO I HAVE TO IMPORT THE math LIBRARY WHEN IT IS INSTALLED ANYWAY?

In Python, there are about 100 different libraries in addition to `math`. Together, they have several thousand functions. Searching through all functions would make it easy to get lost even for experienced programmers. This is why they have been grouped into modules. Thus, you can add extra components to a Python program only if you need them.

1.3.4 Calculations

In the final part of the ΔG example, the calculation is done. The translation of the formula in Section 1.2.2 contains an addition (+), two multiplications (*), a division (/), and the natural logarithm (math.log(...)). The parentheses after the log are obligatory. Python also supports subtraction (−), power (**), floor division (//, rounding down), and modulo (%, resulting in the remainder of a division).

```
>>> deltaG0 + R * T * math.log(ADP * Pi / ATP)
```

Upon pressing Enter, you will see the result displayed immediately:

```
-28.161154161098693
```

Standard Arithmetical Operations
Most calculations will probably be simpler than calculating ΔG values. Arithmetical operations can be done right away from the command prompt

```
>>> a = 3
>>> b = 4
>>> a + b
7
```

Or you can leave the variables away and write numbers directly:

```
>>> 3 + 4
7
```

Table 1.2 gives an overview of the available arithmetical operations in Python.

Q & A: DO I NEED TO WRITE NUMBERS WITH DECIMAL PLACES?

There are two things to note: First, when you perform a calculation with integer numbers, the result is also an integer number. Second, when you calculate with floating-point numbers, the result will also be a floating-point number. For instance, if you execute the division

```
>>> 4 / 3
1
```

The result is `1` as an integer number, because it gets rounded down automatically. However, the result changes when you add one decimal place:

```
>>> 4.0 / 3.0
1.3333333333333333
```

The result of the second division is given with a precision of 16 decimal places. When you put together integer and floating-point numbers in a calculation, the result will also be a float.

Q & A: WHY DO WE USE VARIABLES AT ALL? WOULDN'T THE ΔG EXAMPLE BE SIMPLER IF WE JUST PUT THE NUMBERS INTO THE FORMULA DIRECTLY?

Yes and no. Yes, because it is fewer lines to write. No, because your code becomes much harder to read and not reusable. Consider the line for calculating the ΔG value:

```
>>> -30.5 + 0.000831 * 298 * math.log(1.8 * 5.0 / 3.5)
-30.26611541610987
```

How long would it take you to figure out that this result is actually wrong, although the calculation is mathematically correct? The problem becomes easier to spot if you have

```
>>> R = 0.000831
```

whereas it should be

```
>>> R = 0.00831
```

In the first line, one decimal place was forgotten while converting the units. This is a very common programming error. Often errors have nothing to do with the program itself but with misconceptions about the data. Ideas on how you can spot such problems more easily are explained in Chapter 12 and Chapter 15.

Mathematical Functions

When you issue the command

```
>>> import math
```

a set of mathematical functions from the `math` module is made available in the current Python interactive session. The most important functions from `math` are listed in Table 1.3.

TABLE 1.3 Some Important Functions Defined in the math Module.

Function	Meaning
log(x)	natural logarithm of x ($ln\ x$)
log10(x)	decadic logarithm of x ($log\ x$)
exp(x)	natural exponent of x (e^x)
sqrt(x)	square root of x
sin(x), cos(x)	sine and cosine of x (x given in radians)
asin(x), acos(x)	arcsin and arccos of x (result in radians)

When you are using functions in Python, the parentheses are mandatory:

```
>>> math.sqrt(49)
7.0
```

math also defines the constants math.pi (π = 3.14159) and math.e (e = 2.71828). They can be used just as any variable. For example, to calculate the volume of a 50 ml Falcon tube (a plastic cylinder used for centrifugation) that is 115 mm long and 30 mm wide, you can use math.pi:

```
>>> diameter = 30.0
>>> radius = diameter / 2.0
>>> length = 115.0
>>> math.pi * radius ** 2 * length / 1000.0
81.2887099116359
```

Source: Adapted from code published by A.Via/K.Rother under the Python License.

1.4 EXAMPLES

Example 1.1 How to Calculate the Distance between Two Points

A point in the three-dimensional space is defined by its Cartesian coordinates (x, y, z). The distance d between two points p_1 and p_2, the coordinates of which are (x_1, y_1, z_1) and (x_2, y_2, z_2), respectively, is given by the following equation:

$$d(p_1, p_2) = \sqrt{(x_1 - x_2)^2 + (y_1 - y_2)^2 + (z_1 - z_2)^2}$$

The coordinates of the two points can be stored in six variables: x1, y1, z1 and x2, y2, z2, respectively. You need two methods from the math module (pow() and sqrt()). In the following script, we actually import all functions (*) from the math module. The pow(i, j) method has two arguments: the number i you want to raise to the power of j, and j.

```
>>> from math import *
>>> x1, y1, z1 = 0.1, 0.0, -0.7
>>> x2, y2, z2 = 0.5, -1.0, 2.7
>>> dx = x1 - x2
>>> dy = y1 - y2
>>> dz = z1 - z2
>>> dsquare = pow(dx, 2) + pow(dy, 2) + pow(dz, 2)
>>> d = sqrt(dsquare)
>>> d
3.5665109000254018
```

Example 1.2 How to Create Your Own Modules

Technically, a Python module is a text file ending with .py (see Box 1.1). You can place variables and Python code, functions, etc., there. A short Python module can be written and used quickly. For instance, you could outsource the ATP constant to a module in four steps:

1. Create a new text file with a text editor.
2. Give it a name ending with .py (e.g., hydrolysis.py).
3. Add some code. For example, you could add the ATP constant

   ```
   ATP = -30.5
   ```

4. Finally, import the module from the Python shell:

   ```
   >>> import hydrolysis
   ```

   or

   ```
   >>> from hydrolysis import ATP
   ```

For the import to work, you need to store the module file in the same directory where you started the Python shell (on Linux and Mac) or in the Python library (C:/Python7/lib/site-packages/ on Windows). You may also save your modules to another directory (you may want to have a special directory where you collect all your modules) and add the directory path to a special Python variable (called sys. path, i.e., the variable path belonging to the module sys). Later in the book we will explain how to do it.

1.5 TESTING YOURSELF

Exercise 1.1 Calculate the ΔG Value for All Three Tissues

In which tissue does ATP hydrolysis set the most energy free? Use the code provided earlier to answer the question. (See Table 1.1.)

Exercise 1.2 Convert the Values to kcal

Calculate the three ΔG values for all three tissues to kcal/mol. The conversion factor is 1 kcal/mol = 4.184 kJ/mol.

Exercise 1.3 pH Calculation

In a solution you have a proton concentration of 0.003162 mM. What is the pH of the solution?

Exercise 1.4 Exponential Growth

Given optimal growth conditions, a single *E. coli* bacterium can divide within 20 minutes. If the conditions stay optimal, how many bacteria are there after 6 hours?

Exercise 1.5 Calculate the Volume of a Bacterial Cell

The average length of an *E. coli* cell is given as 2.0 μm, and its diameter as 0.5 μm. What would the volume of one bacterial cell be if it were a perfect cylinder? Use Python to do the calculation. Use variables for the parameters.

## CHAPTER 2

Your First Python Program

L EARNING GOAL: You can write programs that consist of input, output, and actions in between.

2.1 IN THIS CHAPTER YOU WILL LEARN

- How to compose a program of input, actions, and output

- How to repeat instructions

- How to write to the screen of your computer

- How to run a sliding window over a sequence

2.2 STORY: HOW TO CALCULATE THE AMINO ACID FREQUENCY IN THE SEQUENCE OF INSULIN

2.2.1 Problem Description

In this chapter, you will learn to analyze the protein sequence of insulin. Insulin was one of the first proteins discovered. Frederick Banting and John Macleod received the Nobel Prize in 1923 for discovering its function. Ninety years later, human insulin is of paramount medical and economical importance, mainly for the 285 million people affected by diabetes. The functional form of the protein itself is 51 amino acids long after proteolytic removal of two fragments from the translation product. The question this chapter deals with is, *How often does each of the 20 amino acids occur in the protein sequence?*

Analyzing amino acid frequencies of a protein helps to find out how many cysteines could form disulphide bonds, whether there is an unusual amount of nonpolar residues indicating a transmembrane domain or whether there are many positively charged residues that could be involved in nucleic acid binding. To determine these numbers for insulin, you have several possibilities:

- Count amino acids manually. This works fine as long as the protein is short and you have only one or a few proteins to analyze.

- Make clever usage of the "search–replace" functions in your favorite text editor for each amino acid. This works better than counting manually for long proteins, but if you want to analyze many proteins, this is not very convenient either.

- Write a computer program. In this chapter, you will use a program written in the Python language. Box 2.1 contains some reflections on how computers count residues.

BOX 2.1 HOW TO COUNT Cs

How many Cs are in the following sequence?

```
CCCHAJEAFIELAKJNFVLAIFEJLIEFJDCCCEFLEFJ
```

When looking at the sequence sharply, you will figure out that there are six Cs. Intuitively, you understand how the counting should be done and obtain the correct result. But how can you tell a computer to do the job for you?

The answer contains a lot about programming. You first need to fully understand what is to be done and then describe it precisely. Thus, how *exactly* did you count the Cs? Most probably you did one of the following:

- You looked at each character from the left to right and counted each C encountered.
- You looked at each character from the right to left and counted each C encountered.
- You made an estimate because you decided counting all characters would take too long.

Consider the first two options. In both, you essentially examine all characters and count all Cs. Why does it make a difference whether you start from the left or the right? Because for a computer it does! Computers have no intuition. They can't figure out by themselves which side to start from. They cannot conclude what you expect from them even though for you it may be obvious. So you must tell them what to do in the tiniest detail. A precise instruction that could be translated into program code easily would be as follows:

1. Set a counter to zero.
2. Look at the first character of the sequence.
3. If it is a C, add 1 to the counter.
4. If you have reached the last character, print the counter and then stop.
5. Otherwise, move to the next character and repeat from step 3.

Much of programming is chopping a task into very small operations. The third option uses an unexpected approach: estimates. For a huge sequence, it may be reasonable to make an educated guess from reference data. The conditions are that you tell the computer how to make guesses and the guesses are precise enough. In any case, when programming, be ready to be open to solutions that are counterintuitive. Whether it is counting or guessing, once you tell computers what to do, they do things incredibly quickly. No matter whether it is one short sequence, hundreds of them, or a whole genome you are interested in, we hope the sequence at the beginning of this box is the last one where you had to count all by yourself.

In the previous chapter, you learned how to store data in variables, to do calculations, and to import and use a module. In the Python session in Section 2.2.2, you are going to learn four more things. First, you will learn how to comment a line of code with a # symbol so that it will not be executed by the Python interpreter. Second, you will learn how to store text in a variable using a data type called *string*. For counting amino acids, a *method* (a function connected to a data object) of the protein string will be used. Third, you will learn how to repeat an action several times. For that a `for` loop will be used. Fourth and finally, you will see how to generate visible output on the screen using the `print` command (see Box 2.4).

Importantly, the following Python session is meant to be written to a file and executed (see Box 1.2 and Section 2.3.1). In the following, code lines *not* preceded by the Python shell prompt >>> are meant to be written to a text file and executed (see Figure 2.1).

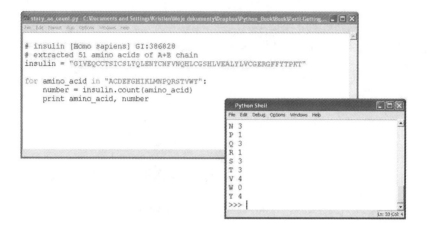

FIGURE 2.1   Text files and Python shell. *Note:* Left panel: A script written to a text file. Right panel: The execution of the script in the Python shell.

2.2.2 Example Python Session

```
# insulin [Homo sapiens] GI:386828
# extracted 51 amino acids of A+B chain
insulin = "GIVEQCCTSICSLYQLENYCNFVNQHLCGSHLVEALYLVCGERGFFYTPKT"
for amino_acid in "ACDEFGHIKLMNPQRSTVWY":
    number = insulin.count(amino_acid)
    print amino_acid, number
```

Source: Adapted from code published by A.Via/K.Rother under the Python License.

2.3 WHAT DO THE COMMANDS MEAN?

The program produces the following 20 × 2 table:

A	1
C	6
D	0
E	4
F	3
G	4
H	2
I	2
K	1
L	6
M	0

N	3
P	1
Q	3
R	1
S	3
T	3
V	4
W	0
Y	4

2.3.1 How to Execute the Program

When working on the Python shell, you had a window that was just for entering commands. Where can you enter a program? Of course, you could enter the above commands into the Python shell and they would work. But then you would have to retype the program each time you want to use it, including typing the insulin sequence!

A more convenient option is to store the program in a *text file*. Text files can be opened using a *text editor* (see Box 2.2 and Box D.2). Text files containing Python programs should have the ending .py (the suffix may not be visible). On Linux and Mac you can execute the Python program from a terminal window by typing

```
python aa_count.py
```

BOX 2.2   TEXT EDITORS FOR PROGRAMMING

A text editor for programming must allow you to create files, write to them, and save them on the hard disk. Examples of basic text editors are Notepad++, Vim (http://www.vim.org/), TextEdit, Pico, and Gedit (http://projects.gnome.org/gedit/). Most of them also highlight the syntax of Python code automatically. When using a normal text editor, make sure that tabs are automatically replaced by spaces. In Gedit you can configure this in Edit => Preferences. Go to the Editor tab and check the box "Insert spaces instead of tabs".

The IDLE editor (automatically installed with Python on Windows) can also recognize and manipulate code blocks (i.e., takes care of indentation). On Linux and Mac OS X, iPython is an improved Python shell that not only offers syntax highlighting of the code but also allows completion of many functions by pressing the TAB key (see http://ipython.org/).

> For Python there are several sophisticated editors, also called integrated development environments or IDEs, that help you with several aspects of programming. First, the editor helps to format the code consistently: adding spaces, highlighting keywords in different colors, highlighting missing parentheses. Some IDEs even highlight syntax errors. While writing code, you can look up names and documentation on the fly. While debugging you can go through the program step by step and trace the value of variables without adding print statements. Popular Python IDEs are Eric (Linux), SPE (Linux), and Sublime Text (Win/Mac/Linux, http://www.sublimetext.com).

On Windows, you can open the Python file in the IDLE editor (available from 'Start' → 'All Programs' → 'Python' → 'IDLE', create a new file) and press F5 to execute the program or select 'Run' → 'Run Module' from the menu.

2.3.2 How Does the Program Work?

In the Python programming language, a program is executed one line after the other. Each program line contains an instruction that tells the Python interpreter what it has to do. What happens in `aa_count.py` in each line?

- `# insulin`. The first line does nothing. The hash # indicates that this is a comment (see Section 2.3.3).

- `insulin = "MALWM..."`. Here, the protein sequence is stored in a *variable* called `insulin`. The protein sequence is defined as text, also called a string (see Section 2.3.4). The backslash \ at the end of the line indicates that the protein sequence continues on the next line.

- `for amino_acid`. This construction starts a loop going through the 20 characters A, C, D, E... and repeats for each of them the instructions in the next two lines. The name `amino_acid` is a variable that contains a single character in each round. The loop stops when the `amino_acid` variable reaches the last letter of the `"ACDEFGHIKLMNPQRSTVWY"` string of characters.

- `number = insulin.count(...)`. This calls a function that calculates how often a character occurs in a piece of text, in this case, the insulin protein sequence. The result is stored in a variable.

- `print amino_acid, number`. This writes both the amino acid and the result of the counting to the screen.

The last three lines form a *code block*. That means they are executed together, in this case, 20 times because the `for` command repeats a block of commands. The lines related to the `for` command are grouped by being indented by four spaces. The effect of this block of code is the same as if the program were written

```
number = insulin.count ("A")
print "A", number
number = insulin.count ("C")
print "C", number
number = insulin.count ("D")
print "D", number
. . .
```

Thus, combining the loop with a block of commands helps avoid writing a lot of redundant code.

In Python you normally have one instruction per line. However, long instructions may span several lines, like the insulin sequence. In that case, all but the last line need to end with a backslash character \:

```
>>> 3 + 5 + \
... 7
15
```

2.3.3 Comments

A *comment* is a portion of code that is addressed to other programmers (or yourself when you read your program after days or weeks) and not to the Python interpreter. In other words, it is a sort of documentation inside the program used to describe what the code does. The text you want the interpreter to ignore must be preceded by the # symbol.

```
>>> print "ACCTGGCACAA" # This is a DNA sequence
ACCTGGCACAA
```

2.3.4 String Variables

You saw in Chapter 1 that numbers can be stored in variables by simply assigning a number to a variable name:

```
x = 34
```

Text variables contain a data type called *string*. In contrast to numbers, strings in Python need to be enclosed by single ('abc'), double ("abc"), or triple ('''abc''' or """abc""") quotes. For instance, to store the protein sequence of insulin in a variable named insulin, you need to use the assignment operator =:

```
insulin = 'GIVEQCCTSICSLYQLENYCNFVNQHLCGSHLVEALYLVCGERGFFYTPKT'
>>> print insulin
GIVEQCCTSICSLYQLENYCNFVNQHLCGSHLVEALYLVCGERGFFYTPKT
```

Triple-quoted strings can span multiple lines (with no need for a backslash at the end of each line). This is useful to store longer pieces of text in variables:

```
>>> text = '''Insulin is a protein produced in the pancreas.
The protein is cut proteolytically.
Its deficiency causes diabetes.'''
>>> print text
Insulin is a protein produced in the pancreas.
The protein is cut proteolytically.
Its deficiency causes diabetes.
```

Strings have an intrinsic order. When you run a for loop on a string, the characters will always be processed in the same order. Strings are immutable objects. You cannot change single characters in an existing string or replace a substring with another one; you can only create a new string. Strings not only are useful for storing data but also provide a lot of powerful functions to manipulate and analyze text.

Indexing

By numerical indices in square brackets, you can extract characters in certain positions. The first character is treated as in position zero:

```
>>> 'Protein'[0]
'P'
```

```
>>> 'Protein'[1]
'r'
```

Negative indices address characters starting from the end:

```
>>> 'Protein'[-1]
'n'
>>> 'Protein'[-2]
'i'
```

Slicing

By introducing a colon in the square brackets, you can address parts of the string (*slices*). For instance [0:3] returns a substring starting at the beginning and stopping after the third character:

```
>>> 'Protein'[0:3]
'Pro'
```

The slice [1:] starts after the first character and stops at the end of the string.

```
>>> 'Protein'[1:]
'rotein'
```

String Arithmetics

Strings in Python can be added using the plus (+) operator. This simply results in a concatenation of both strings.

```
>>> 'Protein' + ' ' + 'degradation'
'Protein degradation'
```

Strings can also be multiplied with integer numbers:

```
>>> 'Protein' * 2
'ProteinProtein'
>>> '*' * 20
'********************'
```

Determining String Length

The len() function returns the length of a string in number of characters:

```
>>> len('Protein')
6
```

Counting Characters

The s.count() function counts how often a character or a short sequence occurs in a string:

```
>>> 'protein'.count('r')
1
```

More functions that work on strings can be found in Appendix A.

Q & A: WHY ARE SOME STRING FUNCTIONS LIKE count() ADDED AFTER THE STRING WITH A DOT, AND OTHERS LIKE len() TAKE THE STRING AS ARGUMENT?

Functions in Python are organized in several places. Some are so-called *built-in* functions (like len()); you can use them anywhere without constraints. They are widely applicable; e.g., len() works on other data types as well. Other functions are specific for certain data types; e.g., counting in the way used in the program in Section 2.2.2 works for strings only. To use count() you need to have a string first. Functions that are firmly connected to a certain type of object are also called *methods* of that object. The assignment of methods to data types helps keep complicated programs well organized. For more information on modular aspects of Python, see Box 2.3 and Part III.

BOX 2.3   EVERYTHING IS AN OBJECT

Whenever you use a new piece of data in your program (numbers, text, or more complex structures), the Python interpreter creates something called an *object*. Each object has a reserved piece of memory where the data is stored. A Python object also knows exactly what type of data it is. For instance, if you assign a number to a variable

```
a = 1
```

Python creates an *integer* object somewhere in the computer memory. The variable a points to that location in the memory. An object is also created when you create a string variable. When you have a for loop, each new value of the index variable is an object of its own as well. Each object contains some data, but it can also contain functions that work on data. The content of an object can be accessed by adding a dot to the variable name; e.g., sequence.count('A') calls a method of the sequence string variable.

Even a module that you import is an object. The math module, for instance, contains data (e.g., math.pi) and functions (e.g., math.log). Taken together, everything that you can give names to is an object, a container for both data and functions.

2.3.5 Loops with for

The control flow statement for is used for repeating actions. In particular, the for statement makes it possible to execute instructions a given number of times. The for loop needs a sequence-like object (see Box 2.3): a string it can go through character by character, a list of numbers that are processed one by one, or simply a pile of data items that the loop works through until the pile is empty. The elements of the sequence are made available as an index variable. The instructions that are repeated are the group of indented statements following the for instruction.

The general syntax of for loops is

```
for <index variable> in <sequence>:
    <command 1>
    <command 2>
    ...
    <command x>
```

<sequence> may be a string ("ACDEFGHIKLMNPQRSTVWY") or a collection of objects such as a list ([1, 2, 3], explained in Chapter 4). <index variable> is a variable name that takes the value of the elements of <sequence> as it runs over its contents. In the first round, the index variable has the first value of <sequence>; in the second round, it takes the second value; and so on. The instructions <command 1> and <command 2> are executed during each round of the loop. They are marked as belonging to the loop by being shifted right by four spaces (*indentation*). The last instruction to be executed in a for loop is to be shifted by four spaces. The instruction <command x> is executed as soon as the loop is exited, i.e., it has finished with the last element of <sequence>.

Running a Loop over a String
When the sequence used in a for loop is a string, the code inside the loop is repeated for each character:

```
for character in 'hemoglobin':
    print character,
```

Source: Adapted from code published by A.Via/K.Rother under the Python License.

For instance, the loop from the example program in Section 2.2.2 is run 20 times, one for each one-character amino acid code:

```
for amino_acid in "ACDEFGHIKLMNPQRSTVWY":
    number = insulin.count(amino_acid)
    print amino_acid, number
```

The `number = insulin.count(amino_acid)` and `print amino_acid, number` statements are each repeated 20 times. In each round of the loop, the `amino_acid` variable takes the value of the next character from the string `"ACDEFGHIKLMNPQRSTVWY"`. This way, each amino acid can be counted without duplicating code. The loop stops when the `amino_acid` variable has finished taking all the 20 values from the `"ACDEFGHIKLMNPQRSTVWY"` string; in other words, when `amino_acid` equals "Y", the `for` loop is executed one last time and then exits.

Running a Loop over a List of Numbers
A loop can simply print the contents of a list:

```
for i in [1, 2, 3, 4, 5]:
    print i,
```

which results in:

```
1 2 3 4 5
```

A loop can use a function to create a sequence on the fly. For instance, the built-in function `range(10)` creates a list of numbers from 0 to 9.

```
for number in range(10):
    print number,
```

will result in:

```
0 1 2 3 4 5 6 7 8 9
```

2.3.6 Indentation

In the Python session in Section 2.2.2, not all code lines start at the same point: two of them are preceded by a number of blank spaces. This is called *indentation* and is used in Python to mark blocks of code that are executed together.

Blocks of code occur in loops, conditional statements (see Chapter 4), functions (see Chapter 10), and classes (see Chapter 11). They all are identified by indentation. A block of code is initiated by a colon character and followed by the indented instructions of the block. The indentation length of the first instruction of a block should be four spaces at least, and all instructions of a block must be indented by the same number of spaces. Although it is possible to use tabs for indentation, it is advised to use spaces because tabs lead to problems in some text editors.

2.3.7 Printing to the Screen

The `print` command is a versatile way of displaying information on the screen. It writes numbers and text to the text console. You can influence the way the data are written in many ways. In the most straightforward way, you simply print the data:

```
print 7
```

will print 7 to the screen of your computer, whereas

```
print 'insulin sequence:'
```

will simply print `insulin sequence:`.

Instead of the plain data, you can print a variable:

```
sequence = 'MALWMRLLPLLALLALWGPDPAAA'
print sequence
```

will create the output

```
MALWMRLLPLLALLALWGPDPAAA
```

You can print multiple values in one command by separating them with commas:

```
print 7, 'insulin sequence:', sequence
```

will print

```
7 insulin sequence: MALWMRLLPLLALLALWGPDPAAA
```

By default, a line break (i.e., an end-of-line character) is added after each `print` statement, so that

```
print 7, 'insulin sequence:'
print sequence
```

will print 7 and `insulin sequence:` in one line and the sequence in a different line. Adding a comma at the end of the last item in the `print` statement suppresses the line break. The following commands produce the output in one line:

```
print 7, 'insulin sequence:',
print sequence
```

Escape Characters and Quotes
When printing text (or assigning it to a string variable) you cannot use every character in a Python string. Some characters need to be replaced by *escape characters*. For example, you need to write tabulators as \t (using a backslash), newline characters as \n, and backslashes as \\.

Of course, you cannot use the same quotes in a string as the ones you enclose the string in. If you enclose the string in single quotes, you can use double quotes inside the string and vice versa; triple-quoted strings may contain both:

```
print 'a single-quoted string may contain "".'
print "a double-quoted string may contain ''."
print '''a triple-quoted string may contain '' and "".'''
```

Both text and numbers can be formatted in more sophisticated ways, e.g., writing a fixed number of digits or right-justifying text (see Chapter 3). The main advantage of the `print` command is that it is a straightforward way to write any kind of information to the screen. Box 2.4 reports a list of the new concepts introduced in Chapter 2.

BOX 2.4 NEW PYTHON CONCEPTS

- Comments
- String variables
- `for` loops
- Indentation
- Print

2.4 EXAMPLES

Example 2.1 How to Create a Random Sequence

You can use a `for` loop for creating a random sequence (see also Recipe 2). The `range()` function allows you to repeat instructions a given number of times. The `random` module allows you to create random numbers. It provides a number of tools to manage random objects. For example, the `randint(i,j)` function generates a number between `i` and `j` with equal probabilities. In the following example program, a random number (ten) and string indexing (AGCT) are used to create a random sequence (of ten characters extracted from "AGCT"):

```python
import random
alphabet = "AGCT"
sequence = ""
for i in range(10):
    index = random.randint(0, 3)
    sequence = sequence + alphabet[index]
print sequence
```

Source: Adapted from code published by A.Via/K.Rother under the Python License.

In the first line, the `random` library is imported. The program defines a string with the characters of the four nucleotides and assigns it to the `alphabet` variable, then uses a `for` loop to randomly extract 10 times a character from `alphabet` and add it to the `sequence` string, which starts empty (`sequence = ""`). The random extraction is performed by taking letters from `alphabet` with random indices generated by `random.randint()`. The program output is a random sequence of 10 characters extracted from `'AGCT'`, for example,

```
GACTAAATAC
```

Notice that, being random, the output sequence will change each time you execute the program.

Example 2.2 How to Run a Sliding Window over a Sequence

When you look for sequence motifs, it is often necessary to consider all fragments of a sequence having a certain length. You can use a `for` loop to create all possible subsequences of a given length from a sequence:

```
seq = "PRQTEINSEQWENCE"
for i in range(len(seq)-4):
    print seq[i:i+5]
```

Source: Adapted from code published by A.Via/K.Rother under the Python License.

The variable i runs from zero to the length of the sequence minus four; len(seq) is 15, thus range(11) produces all numbers from 0 to 10. The instruction seq[i:i+5] extracts a subsequence of 5 characters in length from the sequence at position i. Python starts counting positions at zero; the first subsequence is from position 0 to 5, the last one from 10 to 14. Thus, the code produces

```
PRQTE
RQTEI
QTEIN
TEINS
...
WENCE
```

The range() function is described in detail in Chapter 4.

2.5 TESTING YOURSELF

Exercise 2.1 Frequency of Amino Acids in the Telomerase Protein Sequence

In 2009, Elizabeth H. Blackburn, Carol W. Greider, and Jack W. Szostak received the Noble Prize for discovering the function of the enzyme telomerase, which is responsible for extending the ends of chromosomes. Retrieve the 1,132-residue sequence of isoform 1 of human telomerase from the NCBI protein database. Which amino acid is the most frequent?

Exercise 2.2 Frequency of Nucleotide Bases in a DNA Sequence

Change the program from Exercise 2.1 to count the frequency of the four DNA bases instead. Test it first with a DNA sequence for which you know the result, for instance, "AAAACCCGGT". This approach makes it much easier to discover small program errors.

Exercise 2.3 Print an Amino Acid Sequence One Residue at a Time

Write a program that prints the first amino acid of the insulin sequence, then the first two, then the first three, and so on. You will need both the range() and the len() functions.

Exercise 2.4 Removing Indentation

In the example program in Section 2.2.2 for counting amino acids, replace the line

```
print amino_acid, number
```

with

```
print amino_acid, number,
```

Run the program again. Explain what happens.

Exercise 2.5 Twenty Commands or a for Loop?

The program for counting amino acids could simply consist of 20 commands for counting instead of the for loop:

```
number = insulin.count("A")
print "A", number
number = insulin.count("C")
print "C", number
number = insulin.count("D")
...
```

Would you prefer this implementation? Why or why not?

In Part I you learned all basic parts of the Python language. You can let your computer execute actions, such as *calculations*. You know what a *variable* is and how to use it comfortably; you also saw that data in Python can be of different types such as integer or floating numbers, or string, and that specific tools exist for working with different data types (e.g., the function count() for strings). You met *functions* and saw how some of them work. You also learned that some functions are built-in; i.e., they are "universal" and may act on many different *objects*, and some others can only apply to a specific object. For example, the function count() can only be applied to string objects, and this is expressed by the "dot" syntax (i.e., writing, for example, 'MALWMRLLPLLALLALWGPD'.count('L')). Importantly, you learned that a program does not necessarily need to be written in a single file but can be cleverly structured in several *modules*, which can be connected to one another through the import statement. In particular, Python provides literally hundreds of optimized modules that other programmers wrote for you, such as math. Apart from Python objects such as variables, functions, and modules, you also met a Python *control flow* structure, the for loop, which turns out to be incredibly useful when you have to repeat things several times. It is made of the for keyword, followed by a variable name (i, j, john, amino_acid, etc.) running over a sequence, followed by a colon and a block of indented code. So, you also learned what indentation is and what purpose it serves. Finally, you have seen how to use the print statement to display your result (or whatever text you want to display) on the computer screen.

II

Data Management

INTRODUCTION

In this part of the book, you will meet new *data structures* and *control flow structures*. The data structures are lists (already mentioned in Chapter 2), tuples, dictionaries, and sets. These structures make it possible to collect and conveniently manipulate data. Some of them are more appropriate for certain data and tasks than others, and it is up to you to choose the data structure that best fits the problem at hand. Strings, lists, and tuples are ordered collections of objects, whereas dictionaries and sets are unordered collections of objects. If keeping the order is a priority, strings, lists, or tuples are the appropriate structures to manage your data, but if you want to extract specific elements from a large collection quickly or to determine the intersection between two or more groups of objects, dictionaries and sets are the right choice for you.

Here, you will also learn about two more control flow structures: if conditions and while loops. The former makes it possible to execute a block of instructions only if one or more conditions are met. It is a structure that lets your program make decisions. The while loop combines the concepts of for loops and if conditions. In fact, it makes it possible to repeat actions until a given condition is verified. You can use it, for example, to read a nucleotide sequence until you meet something weird like a non-nucleotide character. All these new data and control flow structures are applied to solve typical problems in computational biology such as manipulating sets of numerical data, reading and writing sequence files, working with tables, parsing sequence data records, selectively extracting information from sets of data, sorting table columns, filtering and sorting data, or finding functional motifs in sequences or keywords in PubMed abstracts.

Chapter 3 is about data columns. It describes how to deal with sets of data: how to read them from a text file, how to collect them in a data structure and to manipulate them (e.g., convert them into floating numbers, add them, calculate their mean and standard deviation, etc.), and how to write them to a text file. In Chapter 4 you will learn how to extract information from sequence files, such as Uniprot, FASTA, or GenBank nucleotide files. Chapter 5 shows how to use dictionaries to store biological data and search them in a very quick manner. Chapter 6 shows ways to filter data by either combining loops with if statements or using the set data type. Chapter 7 teaches how to read, organize, manipulate, and write tabular data, for example, how to read a table in tab-separated format, delete one of its columns, and write it to a new file in comma-separated format. In Chapter 8, you will learn tricks for sorting your data, and in Chapter 9, you will see pattern matching tools, i.e., a syntax to encode a sequence consensus (e.g., extracted from a multiple alignment), and a set of functions to search hits of the consensus in a biological sequence.

The contents of Part I and Part II are the basics you need to repeat over and over in order to become fluent in Python.

## CHAPTER 3

Analyzing a Data Column

L EARNING GOAL: You can calculate the average and standard deviation from numbers in a text file.

3.1 IN THIS CHAPTER YOU WILL LEARN

- How to read numbers from a table

- How to read and write files

- How to calculate average values

- How to calculate a standard deviation

3.2 STORY: DENDRITIC LENGTHS

3.2.1 Problem Description

Neurobiological research studies, among other things, what conditions make neurons grow. The growth of neuron cells can be analyzed by fluorescent microscopy. You obtain images in which you can measure the dendritic arbor complexity. Software like Image J (Image Processing and Analysis in Java, http://rsb.info.nih.gov/ij/) makes it possible to calculate parameters like the dendritic length and to write the values to a text file. Simplified, the file neuron_data.txt is a text file containing the length of neurons in a single column of data:

```
16.38
139.90
441.46
```

```
29.03
40.93
202.07
142.30
346.00
300.00
```

If you need to work with such a set of numbers in a text file, there are several questions right at the beginning: How many measurements are there? What is the longest dendritic length? What is the shortest? What is the average length? What is the standard deviation? It would be great to see a quick summary before analyzing the data in more detail.

If you have many files with measurements of the same kind, loading all of them into Excel may be cumbersome. How can the data be read and analyzed using Python? In the Python session in Section 3.2.2, you will see a function that opens a file for reading or writing, open(*filename, option*). It generates a Python *file object*, which can be read or written, depending on whether *option* is `'r'` (= read) or `'w'` (= write), respectively. To write text to a file, you can use write(*some_text*). write() is a method of file objects, and as such, you have to use the dot syntax to write something to a file. In Section 3.2.2, a for loop is used to read line by line the file object returned by the open(*filename*) function.

3.2.2 Example Python Session

```python
data = []

for line in open('neuron_data.txt'):
    length = float(line.strip())
    data.append(length)

n_items = len(data)
total = sum(data)
shortest = min(data)
longest = max(data)

data.sort()

output = open("results.txt","w")
output.write("number of dendritic lengths : %4i \n"%(n_items))
output.write("total dendritic length       : %6.1f \n"%(total))
output.write("shortest dendritic length    : %7.2f \n"%(shortest))
output.write("longest dendritic length     : %7.2f \n"%(longest))
output.write("%37.2f\n%37.2f"%(data[-2], data[-3]))
output.close()
```

Source: Adapted from code published by A.Via/K.Rother under the Python License.

3.3 WHAT DO THE COMMANDS MEAN?

The output of the example in Section 3.2.2 is written to the `results.txt` output file. After running the example program, you will notice that a file named `results.txt` has appeared in the directory where you have started the program. If you open the output file, you will see that its content is

```
number of neuron lengths :     9
total length             : 1658.1
shortest neuron          :   16.38
three longest neurons    :  441.46
                            346.00
                            300.00
```

The program follows a straightforward input-processing-output pattern. It first reads, line by line, the dendritic lengths from the `neuron_data.txt` text file using a `for` loop. After having removed possible blank spaces and newline characters (`line.strip()`), it converts each line (i.e., each neuron length) into a floating-point number and adds it to the `data` list (`data.append(length)`). Then, it uses a couple of built-in functions to measure the number of dendritic lengths, the length of the longest dendrite, etc. Finally, it writes the result of the calculations to the text file `results.txt`. This chapter focuses on

- how to read and write files,

- how table columns are read to lists of numbers, and

- how lists of numbers can be evaluated.

3.3.1 Reading Text Files

A small number of data items can be written directly into the Python code. This is also called *hard-coding* information. For dozens, hundreds, or millions of entries, hard-coding data becomes increasingly impractical. Using text files for input data makes your program shorter, and often you can use the files you already have.

For instance, if your dendritic lengths are in a text file `neuron_data.txt`, you can read the entire body of data by three Python commands:

```
text_file = open('neuron_data.txt')
lines = text_file.readlines()
text_file.close()
```

Source: Adapted from code published by A.Via/K.Rother under the Python License.

The program does three things:

1. *Opens a text file.* The file is specified by a filename in the form of a string (between single or double quotation marks). It is assumed that the file is in the same directory as your Python program. Otherwise, you would need to add the directory *path* before the filename.

2. *Reads the dendritic lengths from the file.* The `readlines()` function simply reads everything that is in the file line by line and stores each line in a separate string. The strings are returned as a list of strings. In contrast, `read()` reads the entire file into a single string.

3. *Closes the text file.* It is good practice to close files that you open. Python closes files automatically as soon as your program terminates, but if you try to open the same file a second time without closing it before, you ask for trouble.

Many programs, like the Python session in Section 3.2.2, that read data from a text file will contain two lines similar to the following:

```
for line in open(filename):
    line = line.strip()
```

The first line opens the file for reading and runs a `for` loop over each line. This line is shorter than using the `readlines()` function on a file, but it does not create a list variable. The second line will be repeated for each line of the file. The `strip()` function removes blank spaces at the beginning and at the end of a line (if present) and the newline character from the end of the line. After that, the data are ready to be used.

3.3.2 Writing Text Files

Analogously, you can write the output of your program to a file instead of to the screen. This way, you can write the results to a spreadsheet or simply save them for later. In Python, a file is written by

```
output_file = open('counts.txt', 'w')
output_file.write('number of neuron lengths: 7\n')
output_file.close()
```

Source: Adapted from code published by A.Via/K.Rother under the Python License.

This code does the following:

1. *Opens a text file for writing.* This differs from the usage of open() you used for reading by the 'w' character (w = write). A file opened with the 'w' flag can be used only for writing.

2. *Writes a string to the file.* The write() function accepts only string data, so you need to convert to a string whatever you want to write. We will explain converting numbers to strings later. Note that the previous string ends with a newline character (the \n), because write() does not introduce line breaks automatically, so you have to add them explicitly if you need them. Alternatively, the writelines() function accepts a list of lines (each in the form of a string).

3. *Closes the file after usage.* Tidying up after finishing the work is good programming style. If you forget to close the file, nothing dangerous happens (your data are not lost). Python does that automatically when the program execution is over. However, it can lead to problems if you use the same file again:

```
>>> f = open('count.txt','w')
>>> f.write('this is just a dummy test')
>>> g = open('count.txt')
>>> g.read()
''
```

Why is the file empty? Where did the text go? It is still "floating" in the computer memory. In fact, strings are actually saved to files only when a certain number of characters has been reached (i.e., they are buffered). However, when you close the file, the data will be saved in any case, no matter how many characters you have written. Therefore, a better way to write the code is

```
>>> f = open('count.txt','w')
>>> f.write('this is just a dummy test')
>>> f.close()
>>> g = open('my_file')
>>> g.read()
'this is just a dummy test'
>>>
```

Source: Adapted from code published by A.Via/K.Rother under the Python License.

3.3.3 Collecting Data in a List

To work with the entire set of dendritic lengths, you need to put the data somewhere. The program uses a Python list for that. Lists in Python are containers for collections of data with arbitrary length. In the program, an empty list is created at the beginning:

```
data = []
```

In the `for` loop, the neuron lengths from the text file are converted to float numbers and then added to the list:

```
data.append(float(length))
```

The `for` loop in the previous program reads all neuron lengths from the text file into a single list variable. The content of the list after the loop is

```
data = [16.38, 139.90, 441.46, 29.03, 40.93, 202.07, 142.30, \
   346.00, 300.00]
```

Lists are one of the most powerful data structures in Python. The main advantage of using a list is that you can store an entire data set into a single variable. You don't even have to know the number of items in advance: the list grows automatically. You can use lists in a `for` loop and with many functions that evaluate data. Lists will be covered in more depth in the next chapter.

3.3.4 Converting Text to Numbers

When the program is reading the neuron lengths from the text file, they are string variables at first. To do calculations, however, the program must work with floating-point or integer numbers. In Python, you can convert a string to a floating-point number with `float()`:

```
number = float('100.12') + 100.0
```

which gives a totally different result (`200.12`) than

```
number = '100.34' + '100.0'
```

which results in `'100.34100'`. The conversion from strings to numbers results in an error if the text makes no sense numerically; for example,

```
float('hello')
```

returns the following error:

```
ValueError: invalid literal for float(): hello
```

In fact, `hello` is a string that cannot be converted to a number. It does not make any difference whether you convert a string that is hard-coded in the program (like above) or is assigned to a variable like in the line

```
length = float(line.strip())
```

You can convert floats to integer and vice versa.

```
>>> number = int(100.45)
>>> number
100
>>> f_number = float(number)
>>> f_number
100.0
```

When converting to integer, the number may contain decimal places, but they will be truncated.

3.3.5 Converting Numbers to Text

For information to be written to a text file, it must be in the form of a string. Both integer and float numbers (and every other type of data) can be converted to a string using the same function for the conversion:

```
>>> text = str(number)
>>> text
'100'
```

Using the `str()` function to convert numbers has a big disadvantage, however: the numbers in the text file will be unformatted. You cannot align numbers to occupy a given number of columns. Especially float numbers will have many decimal places that make your output less readable. An alternative to `str()` is to use string formatting. You can insert integer numbers into a string by using a *percent* character indicating the number of places you want to allocate for the integer:

```
>>> 'Result:%3i' % (17)
'Result:  17'
```

The %3i indicates that the string should contain an integer formatted to three positions. The actual value for the integer comes from the parentheses at the end. In the same way, you can insert floating-point numbers into a string with %x.yf, where x is the number of total characters (also counting the dot) and y is the number of decimal places.

```
>>> '%8.3f' % (12.3456)
'  12.346'
```

You can also format strings with %s:

```
>>> name = 'E.coli'
>>> 'Hello, %s' % (name)
'Hello, E.coli'
```

Normally, %s simply inserts the string, but you can also right-justify it by, e.g., %10s or left-justify it by %-10s. Multiple values can be inserted at the same time, and text between the formatting characters is allowed. Then, you need to take care to provide the exact number of values to insert at the end:

```
'text:%25s numbers:%4i%4i%5.2f' % ('right-justified', 1, 2, 3)
'text:      right-justified numbers:   1   2 3.00'
```

Summarizing, the string formatting in Python is a powerful option to create neatly formatted output from your data.

3.3.6 Writing a Data Column to a Text File

When your program has finished calculations, you may want to write the results to a text file. If your result is a list of numbers, you can format them to a list of strings and then pass the list to the writelines() method of file objects:

```
data = [16.38, 139.90, 441.46, 29.03, 40.93, 202.07, 142.30, \
    346.00, 300.00]
out = []
```

```
for value in data:
    out.append(str(value) + '\n')
open('results.txt', 'w').writelines(out)
```

Source: Adapted from code published by A.Via/K.Rother under the Python License.

The `for` loop goes through each value of the list, converts it to a string, and adds a *newline character* ('\n'). The values are collected in a list of strings that is written to a file at the end of the script. Note that, as mentioned earlier, for writing a list of strings into a file, you can use the `writelines()` function.

If you prefer to format the result as a long single string, the loop looks slightly different:

```
out = []
for value in data:
    out.append(str(value))
out = '\n'.join(out)
open('results.txt', 'w').write(out)
```

The '\n' `join()` function connects all values to a single string by newline characters, so that `write()` can be used. Notice that you can use `join()` to glue any number of strings together, connecting them by a linker string (a newline character '\n' in the previous example):

```
>>> L = ['1', '2', '3', '4']
>>> '+'.join(L)
'1+2+3+4'
```

The result is a string.

3.3.7 Calculations on a List of Numbers

A list of numbers is a powerful structure for your data. You can use a list as you would a column in a spreadsheet. Python has a set of built-in functions for working with lists of numbers and strings. Given the dendritic lengths in a list

```
>>> data = [16.38, 139.90, 441.46, 29.03, 40.93, 202.07, 142.30, \
    346.00, 300.00]
```

the `len()` function returns the length of the list, i.e., the number of items in the list:

```
>>> len(data)
9
```

With `max()` the largest element of the list is returned:

```
>>> max(data)
346.0
```

Analogously, `min()`, returns the smallest number:

```
>>> min(data)
16.38
```

Finally, `sum()` adds all elements to each other:

```
>>> sum(data)
1658.0
```

You can print the results of these operations or put them into variables for further calculations. The functions `min()` and `max()` generally work on lists of elements of any type. For instance it is possible to write:

```
>>> max(['a', 'b', 'c', 'd'])
'd'
```

or

```
>>> max(['Primary', 100.345])
'Primary'
```

however, in these cases, you will have less control on the result.

3.4 EXAMPLES

Example 3.1 How to Calculate a Mean Value

Assume you have measured five dendritic lengths and want to know their mean value:

```
data = [3.53, 3.47, 3.51, 3.72, 3.43]
average = sum(data) / len(data)
print average
```

Source: Adapted from code published by A.Via/K.Rother under the Python License.

In the first line, all five measurements are put into a list and stored in the variable data. In the second line, the arithmetic mean is calculated using the formula

$$\mu = \frac{1}{N} \sum_{i=1}^{N} x_i$$

The sum() function adds up all values in the list data.

```
>>> sum(data)
17.66
```

The len() function gives the number of items in the list (its length).

```
>>> len(data)
5
```

By using the combination of len() and sum(), you can calculate the arithmetic mean of any list that is not empty. You don't have to know how many items are inside at the time you write the program. If your data consist of integer numbers, the result will be rounded down by default. If you prefer to see an average with decimal places, you need to convert the sum or length to a float:

```
data = [1, 2, 3, 4]
average = float(sum(data)) / len(data)
print average
```

Source: Adapted from code published by A.Via/K.Rother under the Python License.

When one of the numbers of the division is a float, the result will be a float as well.

Example 3.2 How to Calculate a Standard Deviation

Calculating a standard deviation is a little more complicated, because you need a for loop to calculate the square difference for each value. It is necessary to have the average calculated previously. Then, for each value, you have to subtract the average and square the result ((value – average) ** 2). All squared differences must be added together and the result divided by the total number of values. Finally, you have to calculate the square root of the result. To add together the squared differences, you can set a variable to 0.0 and add the squared difference for each value to it.

The formula for the standard deviation is

$$\sigma = \sqrt{\frac{1}{N} \sum_{i=1}^{N} (x_i - \mu)^2}$$

and the script that calculates it is

```
import math
data = [3.53, 3.47, 3.51, 3.72, 3.43]
average = sum(data) / len(data)
total = 0.0
for value in data:
    total += (value - average) ** 2
stddev = math.sqrt(total / len(data))
print stddev
```

Source: Adapted from code published by A.Via/K.Rother under the Python License.

Example 3.3 How to Calculate a Median Value

Another useful measure is the median, the value dividing a data set in two equal halves. To calculate the median from a list of numbers, it is necessary to sort the data. Depending on whether the number of elements is odd or even, the calculation differs slightly:

```
data = [3.53, 3.47, 3.51, 3.72, 3.43]
data.sort()
mid = len(data) / 2
if len(data) % 2 == 0:
    median = (data[mid - 1] + data[mid]) / 2.0
else:
    median = data[mid]
print median
```

Source: Adapted from code published by A.Via/K.Rother under the Python License.

The function `data.sort()` sorts the data in ascending order (see Chapter 8 for details). The `if` statement (explained in Chapter 4) is used to distinguish between lists where the length can be divided by two and odd lengths. Finally, the square brackets, e.g., `data[mid]`, are used to access individual elements of the list (also explained in Chapter 4).

3.5 TESTING YOURSELF

Exercise 3.1 Read and Write a File

Write a program that reads the file with neuron lengths and saves an identical copy of the file.

Exercise 3.2 Calculate Average and Standard Deviation

Extend the example in Section 3.2.2 so that it calculates the average neuron length and standard deviation.

Exercise 3.3 Frequency of Nucleotides

Write a program that reads a DNA sequence from a plain text file. Count the frequency of each base. The program has to determine how often the most frequent base occurs.

Hint: You don't have to identify which base it is.

Exercise 3.4 GC-Content from a DNA Sequence

Write a program that calculates the GC-content of a DNA sequence from a plain text file.

Exercise 3.5

Write the results of Exercises 3.3 and 3.4 to a text file.

## CHAPTER 4

Parsing Data Records

L EARNING GOAL: You can extract information from text files.

4.1 IN THIS CHAPTER YOU WILL LEARN

- How to integrate mass spectrometry data into metabolic pathways

- How to parse sequence FASTA files

- How to parse GenBank sequence records

4.2 STORY: INTEGRATING MASS SPECTROMETRY DATA INTO METABOLIC PATHWAYS

4.2.1 Problem Description

To parse data files, you need to know two things: how to use lists to collect data, and how to make decisions in your program in order to extract the data you want. In this chapter, you will first meet these two central elements of the Python language in a simple example. Then, you will be ready to parse a couple of real biological data files in Section 4.4.

Suppose you want to identify proteins belonging to a given metabolic or regulatory pathway (e.g., cell cycle) that are expressed in a given cancer cell. Your initial data set may be represented by (1) a list (list_a) of proteins (in the form of, e.g., Uniprot ACs) participating in the cell cycle that you read from a text file (file_a) and (2) a second list (list_b) of proteins that have been detected in a given cancer cell by means of, e.g., a mass spectrometry experiment and that you read from a second text file (`file_b`). Such text files can be downloaded from resources like Reactome (see Box 4.1) or can be the result of your experiments. In both cases, you need to read the data into your program before you can compare the two lists. File formats easily readable by a script are `csv` (comma-separated values) or `tsv` (tab-separated values), but a simple text file (as those in Section 4.2.2) with protein identifiers would be appropriate as well (see Box 4.1).

BOX 4.1   MASS SPECTROMETRY

Mass spectrometry (MS) is a technique used to determine the elemental composition of a sample of molecules. This technique can be used for the characterization and sequencing of proteins, and, as such, MS-based proteomics can be applied to obtain comprehensive pictures of gene expression. In this sense, the very final output of an MS experiment basically consists of a list of peptides that have been detected (i.e., that are expressed) in a sample under study. Thanks to the availability of specific data analysis software (e.g., Mascot, http://www.matrixscience.com/), MS peptides can be matched against Uniprot sequences so that the output list can be available in the form of Uniprot IDs and typically stored to a `csv` (comma-separated values) text file:

```
protein_hit_num,prot_acc,prot_score,prot_matches
1," P43686",194,15
2," P62333",41,4
...
```

As for metabolic pathways, several resources are freely available on the Internet. Reactome (http://www.reactome.org/) is one of these. If you click on the Browse Pathways link on the Reactome website, you can select an organism and a pathway and download a list of proteins participating in that pathway in a format of your choice, e.g., textual:

```
Uniprot ID
P62258
P61981
P62191
P17980
P43686
P35998
P62333
Q99460
O75832
...
```

Once MS data have been read, the problem of integrating the data into a pathway comes down to the problem of printing all proteins from list_a that are also in list_b, which is a pretty simple task.

4.2.2 Example Python Session

```
# proteins participating in cell cycle
list_a = []
for line in open("cell_cycle_proteins.txt"):
    list_a.append(line.strip())
print list_a
# proteins expressed in a given cancer cell
list_b = []
for line in open("cancer_cell_proteins.txt"):
    list_b.append(line.strip())
print list_b

for protein in list_a:
    if protein in list_b:
        print protein, 'detected in the cancer cell'
    else:
        print protein, 'not observed'
```

Source: Adapted from code published by A.Via/K.Rother under the Python License.

The output of the script is

```
P62258 not observed
P61981 not observed
P62191 not observed
P17980 not observed
P43686 detected in the cancer cell
P35998 not observed
P62333 detected in the cancer cell
Q99460 not observed
O75832 not observed
```

4.3 WHAT DO THE COMMANDS MEAN?

In the Python session in Section 4.2.2, two small lists of proteins are read from text files and compared. In the program, you can recognize a for loop and a list data structure, which were introduced in Chapters 2 and 3. The two lists both contain Uniprot ACs and have been assigned to the list_a and list_b variables. Finally, there is a new language construct, the if ... else ... clause. It is used to make decisions inside the program. Both if ... else clauses and lists are essential to parse files more complicated than a simple list of protein identifiers. If you are planning to write your own parsers, it is worth it to get comfortable with both of these Python structures first. This will make it a lot easier to combine them in the right way in your own programs.

4.3.1 The if/elif/else Statements

Parsing a data record essentially consists of reading the record line by line ("for line in file_a:"), as you learned in the previous chapter, selecting from each line the information relevant to you and loading it into a data structure (list_a.append(line.strip())) if you need to further manipulate it or copying it to a new file if you want to store it for future use. To extract specific parts from a line of text, you need to execute a set of statements only on certain conditions (e.g., you may want to print a Uniprot AC from a list only if it is also present in another list). Here, we will explain how to do this, i.e., how to make choices using Python.

The structure to be used is the following:

```
if <condition 1>:
        <statements 1>
[elif <condition 2>]:
        <statements 2>]
[elif <condition 3>]:
        pass]
...
[else:
        <statements N>]
```

The elif (else + if) and else statements and the corresponding blocks of instructions are optional.

In the previous example, the following instruction

```
for item in list_a:
```

makes the index item run over the objects listed in list_a (i.e., Uniprot ACs in cell cycle), and for each value taken by the index item, the following commands are carried out:

```
if item in list_a:
    print item, 'detected in the cancer cell'
else:
    print item, 'not observed'
```

In other words, if a Uniprot AC from `list_a` is also found in `list_b`, it will be printed followed by the string `'detected in the cancer cell'`; otherwise it will be printed followed by the string `'not observed'`.

This can be restated as follows: if the condition `item in list_b` is true, print the `item` followed by the string `'detected in the cancer cell'`; otherwise print the `item` followed by the string `'not observed'`.

To formulate conditions in `if` clauses, you need special operators to compare numbers and objects and, for example, to verify if they are equal (`==`) or different (`!=` or `<>`), if one is greater than the other or vice versa (`<`, `<=`, `>=`, `>`), if one is contained in the other or not (`in`, `not in`), and if one is the other or not (`is`, `is not`). For comparison operators, also see Box 4.2.

What all expressions used in `if` clauses have in common is that they will result in one of two possible values: `True` or `False`. These two possible logical values are a data type called Boolean (named after George Boole). If the returned value is `True`, the corresponding block of statements will be executed; otherwise it will not.

You can also verify if two or more conditions are met together or alternatively. The three Boolean operators `and`, `not`, `or` make it possible to combine conditions:

```
seq = "MGSNKSKPKDASQRRRSLEPAENVHGAGGGA\
    FPASQTPSKPASADGHRGPSAAFAPAAAE"
if 'GGG' in seq and 'RRR'in seq:
    print 'GGG is at position: ', seq.find('GGG')
    print 'f'RRR is at position: ', seq.find('RRR')
if 'WWW' in seq or 'AAA' in seq:
    print 'Either WWW or AAA occur in the sequence'
if 'AAA' in seq and not 'PPP' in seq:
    print 'AAA occurs in the sequence but not PPP'
```

Source: Adapted from code published by A.Via/K.Rother under the Python License.

Q & A: HOW CAN I USE THE `elif` STATEMENT?

`elif` stands for `else if`, and it is used to test a condition only if all conditions of the preceding `if`/`elif` statements are not satisfied. Here, we identify numbers in the range 0–30 that are not divisible by 2, 3, 5 and therefore are prime numbers (this is true for numbers in the range 4–30). The order of the `if`/`elif` statements matters; the first condition that applies is taken. This is why we first need to check whether a number is less than 4, because 2 and 3 are divisible by 2 and 3 yet they are prime numbers.

BOX 4.2   OPERATORS USED IN `if` CONDITIONS

If a condition returns the Boolean value `True`, the statements in the corresponding block will be executed. If it returns the Boolean value `False`, the statements in the corresponding block will be ignored.

Condition	Meaning	Example	Boolean Value
A < B	A lower than B	3 < 5	True
		5 < 3	False
A <= B	A lower than or equal to B	(1 + 3) <= 4	True
		4 <= 3	False
A > B	A greater than B	3*4 > 2*5	True
		10 > 12	False
A >= B	A greater than or equal to B	10/2 >= 5	True
		5 >= 2	False
A == B	A equal to B	'ALA' == 'ALA'	True
		'ALA' == 'CYS'	False
A != B	A different from B	'ALA' != 'CYS'	True
		'ALA' != 'ALA'	False
A <> B	A different from B	'ALA' <> 'CYS'	True
		'ALA' <> 'ALA'	False
A is B	A is the same thing as B	'ALA' is 'ALA'	True
		'ALA' is 'CYS'	False
A is not B	A is not the same thing as B	'A' is not 'C'	True
		'A' is not 'A'	False
A in B	A is present in the sequence B	'A' in 'ACTTG'	True
		'U' in "ACTTG"	False
A not in B	A is not present in the sequence B	"U" not in "ACTTG"	True
		"A" not in "ACTTG"	False

In conditions, the number 1 and nonempty objects (e.g., a nonempty string) correspond to the Boolean value `True`, and 0 and empty objects (e.g., an empty string `''` or an empty list `[]`) correspond to the Boolean value `False`. For example,

```
>>> if 1:
... print 'This is True'
...
This is True
>>> if '':
... print 'Nothing will be printed'
...
>>>
```

```
for i in range(30):
    if i < 4:
        print "prime number:", i
    elif i % 2 == 0:
        print "multiple of two:", i
    elif i % 3 == 0:
        print "multiple of three:", i
    elif i % 5 == 0:
        print "multiple of five:", i
    else:
        print "prime number:", i
```

Source: Adapted from code published by A.Via/K.Rother under the Python License.

Q & A: IS THERE A LIMIT ON THE NUMBER OF `ELIF` STATEMENTS THAT I CAN USE?

No, but keep in mind that as soon as one condition is met, all subsequent `elif` statements in the same `if/elif/else` block will be ignored by the Python interpreter. Moreover, you can use the `else` statement only once in an `if/elif/else` block. It will be executed provided that none of the preceding conditions have been met.

4.3.2 List Data Structures

Python provides three data structures to work with sequences of items (i.e., ordered collections of objects): strings, tuples, and lists. You met strings in Chapter 2, and in Chapter 3 lists were introduced to store a set of strings and a set of numbers. However, lists can do a lot more. They are very powerful and versatile tools to manipulate any kind of data.

A list is an *ordered mutable* set of objects (strings, numbers, lists, etc.) enclosed in square brackets. The fact that lists are mutable means that they can be modified at any moment, i.e., new elements can be added, and elements can be replaced or removed altogether. The elements of a list can be any kind of object (numbers, strings, tuples, other lists, dictionaries, sets, or even functions [to be introduced in later chapters]) or a blend of different objects.

For instance,

```
>>> list_b = ['P43686', 'P62333']
```

is a list of strings, whereas

```
>>> list_c = [1, 2.2, 'P43686', [1.0, 2]]
```

is a list made up of an integer, a floating number, a string, and a list. Other examples of lists include the following:

```
>>> d1 = []
>>> d2 = [1, 2, 5, -9]
>>> d3 = [1, "hello", 12.1, [1, 2, "three"], "seq", (1, 2)]
```

The first one is an empty list, which is useful for collecting data later in a program. The last list d3 also includes a tuple, which is defined by round brackets (see Box 4.3).

BOX 4.3 TUPLES

Tuples are *immutable ordered* sequences of objects and are indicated with round brackets, (a, b, c), or by simply listing the sequence of items separated by commas: a, b, c. This implies that once you have defined a tuple, you cannot change or replace its elements.

```
data = (item1, item2, item3,…)
```

Notice that brackets are optional; i.e., you can use either data = (1,2,3) or data = 1,2,3

A tuple of a single item must be written either

```
data = (1,) or data = 1,
```

The operations of indexing and slicing are allowed also with tuples:

```
>>> my_tuple = (1,2,3)
>>> my_tuple[0]              #indexing
1
>>> my_tuple[:]             #slicing
(1, 2, 3)
>>> my_tuple[2:]            #slicing
(3,)
>>> my_tuple[0] = 0   #reassigning (Forbidden)
Traceback (most recent call last):
      File "<stdin>", line 1, in <module>
TypeError: 'tuple' object does not support item assignment
```

Like strings, lists support indexing and slicing operations (see Chapter 2). With reference to the list d3, we can use indexing to extract the first element of the list:

```
>>> d3[0]
1
```

Or we can use slicing to generate a list containing the third-to-the-last element of the original list:

```
>>> d3[2:]
[12.1, [1, 2, "three"], "seq", (1, 2)]]
```

Notice that, similarly to strings, the index of the first element of a list is 0. You can also count from the last element using negative integers, so that, for example,

```
>>> d2[-1]
-9
```

Sometimes you will see two indices with square brackets:

```
>>> d3[3][2]
"three"
```

What is happening here? The instruction selectively fished the third element ("three") of the fourth element ([1,2,"three"]) of the list (this is possible because lists can contain other lists).

Notice that

```
L[i][j]
```

is actually the element of a table (or matrix) corresponding to the ith row and jth column. Lists of lists are called *nested lists* and are the Python way to represent and manipulate tables. Nested lists are discussed in more detail in Chapter 7.

Even more brackets are possible:

```
>>> d3[3][2][0]
"t"
```

This command selected the first element ("t") of the third element ("three") of the fourth element ([1,2,"three"]) of the list. The first two pairs of square brackets address elements of lists; the last picks a single character ("t") from a string ("three").

A complete list of operators and methods common to strings, tuples, and lists is reported in Appendix A.

Lists Are Mutable Objects
In contrast to strings and tuples, which are immutable objects, lists can be modified (see Appendix A, Section A.2.7, "Lists," subsection "Modifying Lists"). This makes them very flexible objects. For example, you can reassign any element of a list:

```
>>> data = [0,1,2,3,4]
>>> data[0] = 'A'
>>> data
['A', 1, 2, 3, 4]
```

In this case, the first element of the list is replaced. When you reassign one of its items, you can see that the original list changes. In other words, you did not create a new list but rather modified the original one, which does not exist anymore in its original form.

Some of the actions that can be performed on lists are carried out by specific methods:

```
>>> data = [0,1,2,3,4]
>>> data.append(5)
>>> data
[0, 1, 2, 3, 4, 5]
```

The method append() adds the item in brackets (the number 5 in this case) to the end of the list. It is a method of the list object (i.e., a function acting only on lists), and as such, you have to use the dot (see also Chapter 1 and Box 1.4) to link it to "its" object. The list (data) contains the append() method in a similar way as the math module contains the sqrt() function (see Chapter 1).

The dot is a "linker" between two objects. In this case the first of the two objects is a list (data), and the second object is a function (append()). When a function refers to a specific object, it is called a *method* of that object. So, we can say "the method append() of the list data." This concept, as you will also see later in this book, applies to all kinds of objects, so we can talk about methods of modules (e.g., math.sqrt()) or methods of classes (see Chapter 11).

4.3.3 Concise Ways to Create Lists

Creating a List of Consecutive Numbers

If you want to generate an ordered list of integers, you can use the function range(i). i is the number of integers generated by the function.

```
>>> range(3)
[0, 1, 2]
```

This may turn out to be very useful when you want to execute a for loop running over a very long list of ordered integers. The built-in function range() will be described in detail in Chapter 10 (in particular, see Box 10.3).

Creating a List of Zeroes

Sometimes it is useful to have a list that contains just a given number of zeroes (or some other numbers). Instead of writing a for loop, you can use the multiplication operator:

```
data = [0.0] * 10
```

List Comprehension

Another way to concisely generate lists is the *list comprehension*. It basically consists of putting between square brackets (1) a variable (e.g., x), (2) the set of values taken by the variable (e.g., x in range(5)), and (3) an expression (e.g., x**2) defining the values of the list elements.

```
>>> data = [x**2 for x in range(5)]
>>> data
[0, 1, 4, 9, 16]
```

In the following example, a new list data is created by running the variable base in the seq string and including an element in data only if it is also present in the bases list:

```
bases = ['A', 'C', 'T', 'G']
seq = 'GGACXCAGXXGATT'
seqlist = [base for base in seq if base in bases]
print seqlist
```

Source: Adapted from code published by A.Via/K.Rother under the Python License.

The code results in:

```
['G', 'G', 'A', 'C', 'C', 'A', 'G', 'G', 'A', 'T', 'T']
```

4.4 EXAMPLES

With the if ... elif construct and the list data type explained, it is time to look at some real file parsers. In the following examples, you will see how to read the content of a protein sequence FASTA file line by line (see Appendix C, Section C.1, "A Single Protein Sequence File in FASTA Format," and Section C.2, "A Single Nucleotide Sequence File in FASTA Format") and make choices by means of the if/elif/else construct. You can download the sample files by going to the Uniprot website (http://www.uniprot.org/), access the entry of a protein of your choice, and click on the "fasta" link. Then, you can copy and paste the FASTA-formatted sequence to a local text file. To create a multiple sequence file, you can repeat this procedure several times (or download from the Uniprot website the entire SwissProt data set in FASTA format).

Example 4.1 Read a Sequence File in FASTA Format and Write Only the Sequence Header to a New File

Each sequence record in a FASTA file is composed of two parts: a header line and the 64-character-long sequence line(s) (the number of which depends on the sequence length). The header line is marked by a greater than symbol (>) at the first position of the line, and you can use it to distinguish this line from the ones with the sequence.

```
fasta_file = open('SwissProt.fasta','r')
out_file = open('SwissProt.header','w')
for line in fasta_file:
    if line[0:1] == '>':
        out_file.write(line)
out_file.close()
```

Source: Adapted from code published by A.Via/K.Rother under the Python License.

Example 4.2 How to Extract a List of Accession Codes from a Multiple Sequence FASTA File

How were the input data files for the Python script in Section 4.2.2 created? Consider SwissProtSeq.fasta, a FASTA file of the form reported in Appendix C, Section C.4, "A Multiple Sequence File in FASTA Format." The accession codes may originate from the header lines like the following, enclosed by the pipe symbol '|':

```
>sp|P03372|ESR1_HUMAN Estrogen receptor OS = Homo sapiens
    GN = ESR1 PE = 1 SV = 2
```

The following simple script extracts the SwissProt accession code from each header line in the file, adds it to a list, and prints the list:

```
input_file = open("SwissProtSeq.fasta","r")
ac_list = []
for line in input_file:
    if line[0] == '>':
        fields = line.split('|')
        ac_list.append(fields[1])
print ac_list
```

The output of this program is

```
['P31946', 'P62258', 'Q04917', 'P61981', 'P31947',...]
```

You will have noticed that the list, before being filled using the append() method, has been initialized to an empty list. Moreover, to extract the AC from each header line, it is first identified through an if condition (which makes use of the fact that header lines in FASTA files are marked by an initial '>' character). Then the split() method of string objects is used to cut the string into pieces. It returns a Python list, the elements of which are substrings delimited by the argument in brackets ('|'in this case), which acts as delimiter. As usual, the method is linked to its object (i.e., the string variable line) through a dot.

This example teaches you one important thing: if you want to parse a record file (any record file) programmatically, you have to analyze the file format and structure first and find out a "trick" (e.g., a recurring element that marks specific lines or specific records, a symbol associated with specific words, a suitable delimiter to split a line in columns, etc.) to selectively extract the information relevant to you.

Example 4.3 How to Parse GenBank Sequence Records

If you want to extract the accession codes from a GenBank entry (see Section C.5, "A GenBank Entry," for the full entry), you first need to locate that information in the file:

```
LOCUS           AY810830   705 bp   mRNA   linear   HTC
22-JUN-2006
DEFINITION      Schistosoma japonicum SJCHGC07869 protein mRNA,
partial cds.
ACCESSION       AY810830
VERSION         AY810830.1 GI:60600350
KEYWORDS        HTC.
SOURCE          Schistosoma japonicum
...
```

Now you can (1) select the line where the ACCESSION keyword appears, (2) split the line using the blank space as the delimiter, and (3) collect the last element of the list returned by the split() method.

The following script reads the text file shown in Example 4.3 and in Appendix C, Section C.5, "A GenBank Entry," and writes the nucleotide sequence to a new file in FASTA format (using the ACCESSION number as a header):

```python
InputFile = open("AY810830.gbk")
OutputFile = open("AY810830.fasta","w")
flag = 0
for line in InputFile:
    if line[0:9] == 'ACCESSION':
        AC = line.split()[1].strip()
        OutputFile.write('>' + AC + '\n')
    elif line[0:6] == 'ORIGIN':
        flag = 1
    elif flag == 1:
        fields = line.split()
        if fields != []:
            seq = ''.join(fields[1:])
            OutputFile.write(seq.upper() + '\n')
InputFile.close()
OutputFile.close()
```

Source: Adapted from code published by A.Via/K.Rother under the Python License.

The strip() method of strings erases blank spaces before and after a string of characters:

```
>>> " ACTG ".strip()
'ACTG'
```

To find the nucleotide sequence, we used a small trick: a flag variable was set when the ORIGIN keyword was found (a flag is a normal variable used internally to indicate that something has happened instead of storing data). If the flag is already set, the line must contain nucleotides. In the nucleotide lines, the blanks between the 10 nucleotide portions are removed (fields = line.split()) and then joined together again by empty strings, skipping the numbers at the beginning of the lines (seq = ''.join(fields[1:])).
Try to run the script. Once everything is clear to you, try to add the organism name in the header line, using a '|' as a delimiter to separate it from the ACCESSION number.

Example 4.4 Read a Multiple Sequence File in FASTA Format and Write Records from *Homo sapiens* to a New File

In this example, entire FASTA records, not just accession codes, will be extracted. To make this example work, you need a multiple sequence FASTA file that contains at least one sequence from *Homo sapiens*. The whole SwissProt data set would be appropriate.

This example script is in principle very similar to the previous one. Here, you have to identify header lines and check if the keyword "homo sapiens" occurs in them. The main difference is that now you have to write to the output file entire records (header + sequence line(s), not only the header line) that fulfill the condition. This makes things a bit more complicated.

```
fasta_file = open('SwissProt.fasta','r')
out_file = open('SwissProtHuman.fasta','w')
seq = ''
for line in fasta_file:
    if line[0] == '>' and seq == '':
# process the first line of the input file
        header = line
    elif line [0] != '>':
```

```
# join the lines with sequence
            seq = seq + line
        elif line[0] == '>' and seq != '':
# in subsequent lines starting with '>',
# write the previous header and sequence
# to the output file. Then re-initialize
# the header and seq variables for the next record
            if "Homo sapiens" in header:
                    out_file.write(header + seq)
            seq = ''
            header = line
# take care of the very last record of the input file
if "Homo sapiens" in header:
    out_file.write(header + seq)
out_file.close()
```

Source: Adapted from code published by A.Via/K.Rother under the Python License.

See Figure 4.1 for a graphical scheme of the program. By running this script, you will observe that the sequence in the output file is formatted as in the input file (i.e., including new lines). This is because the `line` variable contains, besides the visible characters (amino acids), an invisible one called *newline character,* which in Python programming is encoded by "\n". For example, see

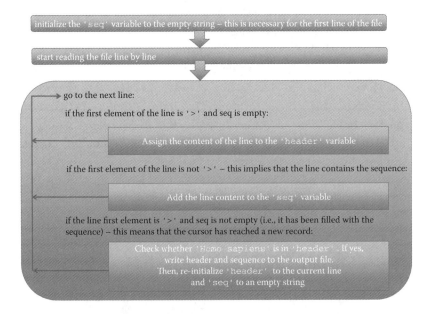

FIGURE 4.1   Flowchart describing Example 4.4.

what happens if you remove the newline characters by replacing the line

```
seq = seq + line
```

with

```
seq = seq + line.strip()
```

4.5 TESTING YOURSELF

Exercise 4.1 Read and Write Multiple Sequence FASTA Files

Read a multiple sequence file in FASTA format and write each record (header + sequence) to a different file.

Hint: The instruction to open the output file must be inserted in the `for` loop (a new file must be opened for each sequence record).

Hint: You can select the AC number from each header line (using the `split()` method), collect it into a variable (e.g., `AC = line.split()[1].strip()`), and use it to assign a name to the output file (e.g., `outfile = open(AC, "w")`).

Exercise 4.2 Read and Filter FASTA Files

Read a multiple sequence file in FASTA format and write to a new file only the sequences starting with a methionine and containing at least two tryptophanes.

Hint: This exercise is very similar to Example 4.4. In this case, you have to apply conditions (first character must be `'M'` and `seq.count('W') > 1`) to the `seq` variable instead of to the `header` variable.

Exercise 4.3 Nucleotide Frequency in a Single DNA
Sequence in FASTA Format

Read a nucleotide sequence file in FASTA format and calculate the frequency of each of the four nucleotides in the sequence.

Hint: You have to count the number of occurrences of each nucleotide (e.g., `seq.count("A")`) and divide it by the length of the sequence (`len(seq)`), after having converted it to a floating-point number (`float(len(seq))`), otherwise the division will return an unexpected value.

**Exercise 4.4 Nucleotide Frequency in Several DNA
            Sequences in FASTA Format**

Redo Exercise 4.3 using a nucleotide *multiple* sequence file in FASTA format.
Print, for each record, the AC and the four (A, C, T, G) frequencies. Is there
a sequence in your file with an anomalous frequency of some nucleotides?

**Exercise 4.5 Nucleotide Frequency in Several DNA
            Sequences in GenBank Format**

Redo Exercise 4.4 using a nucleotide multiple record file in GenBank format.

## CHAPTER 5

Searching Data

L EARNING GOAL: You can use dictionaries to store and search your data.

5.1 IN THIS CHAPTER YOU WILL LEARN

- How to convert an RNA sequence into a protein sequence

- How to use dictionaries to store and search your data

- How to search through a list of data

5.2 STORY: TRANSLATING AN RNA SEQUENCE INTO THE CORRESPONDING PROTEIN SEQUENCE

5.2.1 Problem Description

Suppose you have one or more RNA sequences and you want to translate them into the corresponding protein sequences using the codon table representing the genetic code. This means that you have to read an RNA sequence from a file (in the following called 'A06662-RNA.fasta'), e.g., in FASTA format:

```
>A06662.1 Synthetic nucleotide sequence of the human GSH
transferase pi gene
UGGGACCAGUCAGCAGAGGCAGCGUGUGUGCGCGUGCGUGUGCGUGUGUGUGCGUGUGUGUG
UGUACGCUUGCAUUUGUGUCGGGUGGGUAAGGAGAUAGAGAUGGGCGGGCAGUAGGCCCAGG
UCCCGAAGGCCUUGAACCCACUGGUUUGGAGUCUCCUAAGGGCAAUGGGGGCCAUUGAGAAG
UCUGAA...
```

The RNA sequence must be read three characters at a time, and for each group of three characters (i.e., a codon), the corresponding amino acid must be found in the genetic code. Special characters for STOP and truncated codons (e.g., '*' and '-', respectively) need to be taken into account. This procedure should be repeated for each reading frame, i.e., one starting at the first nucleotide of the RNA sequence, then at the second one, and finally at the third one. In practice, you need to search for a lot of codon:amino acid correspondences. Intuitively, you could use a `for` loop for searching the codon and the corresponding amino acid in a list:

```
genetic_code = [('GCU', 'A'), ('GCC', 'A'), ...]
for codon, amino_acid in genetic_code:
    if codon == triplet:
        seq = seq + amino_acid
```

This search pattern works, but it is highly inefficient. If your sequences are long, the program quickly becomes very slow. It would be better to have a data structure where a base triplet from the genetic code could be used to look up the corresponding amino acid directly, so that you could specifically extract the codon:amino acid pair without having to scan the entire data structure each time.

This kind of data structure exists in Python and is called a *dictionary*. Dictionaries are useful to store and quickly extract data selectively.

The program in Section 5.2.2 stores the genetic code as codon:amino acid pairs in a dictionary, reads an RNA sequence from a FASTA file as a string, and translates the sequence. The program goes through the RNA string in steps of three characters (nucleotide triplets) and replaces each nucleotide triplet with the corresponding amino acid and adds it to a new protein string, which is finally printed to the screen. This is repeated for each reading frame. For printing the output in blocks of 48 amino acids, a `while` loop is used instead of a `for` loop. A `while` loop executes a set of statements until a given condition is fulfilled. In Section 5.2.2, the condition is `i < len(prot)` (which is `True` until the entire sequence has been written; also see Box 4.2 and Box 5.1). Once the index variable `i` exceeds the length of the sequence, the condition corresponds to `False`, and the loop exits.

5.2.2 Example Python Session

```
codon_table = {'GCU':'A','GCC':'A','GCA':'A','GCG':'A','CGU':'R',
               'CGC':'R','CGA':'R','CGG':'R','AGA':'R','AGG':'R',
               'UCU':'S','UCC':'S','UCA':'S','UCG':'S','AGU':'S',
```

```
                    'AGC':'S','AUU':'I','AUC':'I','AUA':'I','AUU':'I',
                    'AUC':'I','AUA':'I','UUA':'L','UUG':'L','CUU':'L',
                    'CUC':'L','CUA':'L','CUG':'L','GGU':'G','GGC':'G',
                    'GGA':'G','GGG':'G','GUU':'V','GUC':'V','GUA':'V',
                    'GUG':'V','ACU':'T','ACC':'T','ACA':'T','ACG':'T',
                    'CCU':'P','CCC':'P','CCA':'P','CCG':'P','AAU':'N',
                    'AAC':'N','GAU':'D','GAC':'D','UGU':'C','UGC':'C',
                    'CAA':'Q','CAG':'Q','GAA':'E','GAG':'E','CAU':'H',
                    'CAC':'H','AAA':'K','AAG':'K','UUU':'F','UUC':'F',
                    'UAU':'Y','UAC':'Y','AUG':'M','UGG':'W',
                    'UAG':'STOP','UGA':'STOP','UAA':'STOP' }
rna = ''
for line in open('A06662-RNA.fasta'):
    if not line.startswith('>'):
        rna = rna + line.strip()
# translate one frame at a time
for frame in range(3):
    prot = ''
    print 'Reading frame ' + str(frame + 1)
    for i in range(frame, len(rna), 3):
        codon = rna[i:i + 3]
        if codon in codon_table:
            if codon_table[codon] == 'STOP':
                prot = prot + '*'
            else:
                prot = prot + codon_table[codon]
        else:
            # handle too short codons
            prot = prot + '-'
# format to blocks of 48 columns
i = 0
while i < len(prot):
    print prot[i:i + 48]
    i = i + 48
```

Source: Adapted from code published by A.Via/K.Rother under the Python License.

The output will contain a translated sequence for each reading frame:

```
Reading frame 1
WDQSAEAACVRVRVRVCACVCVRLHLCRVGKEIEMGGQ*AQVPKALNP
LVWSLLRAMGAIEKSEQGCV*M*GLEGSSREASSKAFAIIW*ENPARM
DRQNGIEMSWQLKWTGFGTSLVVGSKQRRIWDSGGLAWGRRGCLRGWE
G*E*DDTWWCLAGGGQG*LCEGTARATEAF*DPAVPEPGRQDLHCGRP
GEHLA
Reading frame 2
GTSQQRQRVCACVCVCVRVCVYACICVGWVRR*RWAGSRPRSRRP*TH
WFGVS*GQWGPLRSLNRAVSECEV*KDPPEKPALKLLQSSGERTQQGW
TGRME*R*VGS*SGQDLVLAWLWGASRGESGTLVVWPGADGGVSGAGR
DESRMIHGGVWQEAGKDDYVKALPGQLKPFETLLSQNQGGKTFIVGDQ
```

BOX 5.1 TRUE AND FALSE BOOLEAN VALUES

The Boolean values True and False are important especially for if conditions and while loops. In particular, if and while statements return a False value when they are applied to 0, None, or empty objects such as empty data structures ('', (), [], {}) and return a True value if applied to a number different from 0 or to a nonempty data structure.

The statements following

```
while 1:
```

will be repeated to infinity, unless you insert a break statement. On the other hand, the following loop will never be executed even once, because the condition is False:

```
while []:
```

The statements belonging to the block of an if or a while statement are executed only if the statement returned value is True. For instance, the following loop will be repeated four times:

```
>>> n = 0
>>> while n < 4:
...     n = n + 1
...     print n
...
1
2
3
```

VSIW-
Reading frame 3
GPVSRGSVCARACACVCVCVCTLAFVSGG*GDRDGRAVGPGPEGLEPT
GLESPKGNGGH*EV*TGLCLNVRSRRILQRSQL*SFCNHLVREPSKDG
QAEWNRDELAAEVDRIWY*PGCGEQAEENLGLWWSGLGQTGVSQGLGG
MRVG*YMVVSGRRRARMTM*RHCPGN*SLLRPCCPRTREARPSLWETR
*ASG-

5.3 WHAT DO THE COMMANDS MEAN?

5.3.1 Dictionaries

The codon_table object defined at the beginning of Section 5.2.2 is a dictionary. A dictionary is a nonordered collection of objects in the form key:value pairs and is enclosed by curly brackets: {'GCU': 'A', 'GCC': 'A'}.

More specifically, dictionaries are structures for mapping immutable objects (*keys*) to arbitrary objects (*values*). Immutable objects are numbers, strings, and tuples. This means that lists and dictionaries themselves cannot be used as dictionary keys but can be used only as values. A key and its value are separated by a colon, and key:value pairs are separated by a comma.

Dictionaries are useful for searching information quickly. The codon_ table dictionary can be used to retrieve the amino acid for any given codon:

```
>>> print codon_table['GCU']
'A'
```

Other examples of dictionaries are as follows:

1. A dictionary where each key is a Uniprot AC of a sequence belonging to the organism listed in the corresponding value:

```
UniprotAC_Organism = {
        'P034388': 'D.melanogaster',
        'O42785': 'C.trifolii',
        'P01119': 'S.cerevisiae'
        }
```

2. A dictionary where keys are single-letter amino acid codes and values are the propensity of each amino acid for being in a loop (i.e., the propensity of *not* being in either an alpha-helical conformation or a beta-sheet conformation):

```
propensities = {
    'N': 0.2299, 'P': 0.5523, 'Q': -0.1877,
    'A': -0.2615, 'R': -0.1766, 'S': 0.1429,
    'C': -0.01515, 'T': 0.0089, 'D': 0.2276,
    'E': -0.2047, 'V': -0.3862, 'F': -0.2256,
    'W': -0.2434, 'G': 0.4332, 'H': -0.0012,
    'Y': -0.2075, 'I': -0.4222, 'K': -0.100092,
    'L': 0.33793, 'M': -0.22590
    }
```

3. A dictionary where keys are tuples collecting single-letter codes of amino acids sharing a physicochemical property and values are keywords pointing at the corresponding properties:

```
aa_properties = {
  ('A', 'C', 'G', 'I', 'L', 'M', 'P', 'V'): 'hydrophobic',
  ('N', 'S', 'Q', 'T'): 'hydrophilic',
  ('H', 'K', 'R') = 'pos_charged',
```

```
('D', 'E') = 'neg_charged',
('F', 'W', 'Y') = 'aromatic'
}
```

Keys must be unique; i.e., the same key cannot be associated to more than one value. If you try to insert two identical keys in a dictionary, the newer one will overwrite the other.

A dictionary can be defined as shown in Section 5.2.2 by defining all key:value pairs, separated by a comma between curly brackets. However, you can also assign elements one by one. First, you need to create an empty dictionary by empty curly brackets. Then, you can assign the value 'A' to the key 'GCU'.

```
>>> codon_table = {}
>>> codon_table['GCU'] = 'A'
>>> codon_table
{'GCU': 'A'}
>>> codon_table['CGA'] = 'R'
>>> codon_table
{'GCU': 'A', 'CGA': 'R'}
```

If you want to look up the value corresponding to a given key, you can use square brackets. In other words, you can extract specific values from a dictionary; for instance, from the previous example:

```
>>> codon_table['GCU']
'A'
```

As with other data structures, many operations and methods are available for dictionaries. For example, you can selectively delete one key:value pair by del codon_table['GCU']; get a list of all keys or values (codon_table.keys(), codon_table.values()); check if a dictionary contains a given key (if 'GCU' in codon_table:, which is True if 'GCU' is a key; otherwise it is False); calculate the number of elements of the dictionary (len(codon_table)); etc. For a complete list of methods and operations available for dictionaries, see Appendix A, Section A.2.9, "Dictionaries."

5.3.2 The while Statement

In the Python session in Section 5.2.2, the translated protein sequence is collected in a string variable (prot). To produce an output that is pleasant to read, the protein sequence is printed in blocks of 48 symbols per line. This is achieved by a new instruction:

```
i = 0
while i < len(prot):
    print prot[i:i + 48]
    i = i + 48
```

What is the `while` statement? It repeats a block of statements until some condition is met. The `while` loop is practically a combination of `for` and `if` statements.

The general `while` loop syntax is

```
while <condition>:
        <statements>
```

The `while` instruction implements a loop that executes `<statements>` until the expression contained in `<condition>` returns the value `True` (see Box 4.2 and Box 5.1). A `while` block ends with the last indented line. When `<condition>` is met (i.e., in this case, when `line` is no more a nonempty string, which happens at the end of the file), the loop is interrupted. The loop in Section 5.2.2 first defines an index variable that runs over `prot` in steps of 48 characters each. The `while i < len(prot):` checks whether the end of the sequence has been reached already. If not, the next two lines are executed. The first of them prints the next part of the protein sequence, the secod increases the index variable by 48.

The main thing to keep in mind when writing `while` loops is the exit condition. Consider what happens when you remove the line

```
i = i + 48
```

Q & A: WHAT IF <CONDITION> IS NEVER MET?

The condition in the following `while` loop is always the Boolean `True`:

```
while 1:
    print 'while loop still running'
```

You may safely execute this code. Python will not stop executing it by itself. To stop the program, press Ctrl-C. As explained in Box 4.2 and Box 5.1,

0 corresponds to a condition returning the Boolean `False`, whereas any number greater than 0 corresponds to a Boolean that returns `True`. Therefore, the statement starts a never-ending loop. The same happens when you forget to increment the index variable in the loop in Section 5.2.2. Therefore, you need to design and test your `while` conditions carefully.

5.3.3 Searching with `while` Loops

In general, `while` loops are good for searching data structures or in files because you can stop them when your search has been successful. This avoids waste of computing resources. Suppose that you need to extract a specific record from the Uniprot database. The optimized choice would consist of interrupting your search when you find the record you are looking for, e.g., P01308, which is the Uniprot AC for the human insulin. This can be easily done with a `while` loop. If you have downloaded the whole SwissProt database in FASTA format, you can write:

```
swissprot = open("SwissProt.fasta")
insulin_ac = 'P61981'
result = None
while result == None:
    line = swissprot.next()
    if line.startswith('>'):
        ac = line.split('|')[1]
        if ac == insulin_ac:
            result = line.strip()
            print result
```

Source: Adapted from code published by A.Via/K.Rother under the Python License.

Here, the `while` condition returns `True` until `result` is not empty. This applies as soon as the insulin record is found (`if ac == insulin_ac:`), and the corresponding header (i.e., the line starting with `>`) is printed. However, if the insulin accession code is not found, either you insert a `break` statement (see Box 5.2) or the end of the file will be reached and the program will exit with a `StopIteration` error. You can use a `try ... except` block to prevent that (see Chapter 12).

5.3.4 Searching in a Dictionary

In Section 5.2.2, the RNA sequence is scanned three times: once starting at the first sequence position, once at the second position, and once at the third.

> **BOX 5.2 break AND continue STATEMENTS**
>
> When the interpreter encounters the `break` statement, it exits the loop—without executing the rest of the loop statements, including the `else` statement (if present)—and goes to the first statement following the loop.
>
> The `continue` statement skips the rest of the loop statements and goes to the loop's next step.

The code to search the `codon_table` dictionary is contained in the following lines:

```
codon = rna[i:i + 3]
if codon in codon_table:
    if codon_table[codon] == 'STOP':
        prot = prot + '*'
    else:
        prot = prot + codon_table[codon]
else:
    # handle too short codons
    prot = prot + '-'
```

The scanning is carried out in steps of three characters, and for each triplet (`rna[i:i+3]`) a one-letter code amino acid extracted from the `codon_table` dictionary is added to the translated sequence `prot`. The instruction

```
codon_table[codon]
```

will return the one-letter code for the amino acid corresponding to the codon. If the value of a codon is STOP, a `'*'` symbol will be added instead; if the codon is not among the keys of the `codon_table` dictionary (this may happen if the codon is truncated, if it erroneously contains non-nucleotide characters, or if you forgot to insert it in the dictionary definition), a dash `'-'` will be inserted into the protein sequence.

Notice that the task of translating an RNA sequence into a protein sequence can be easily achieved for multiple RNA sequences by inserting the code from Section 5.2.2 into a `for` loop running over a list of records, as shown in Chapter 4.

5.3.5 Searching in a List

To locate the RNA sequence in the input FASTA file, a combination of a `for` loop and an `if` statement is used:

```
rna = ''
for line in open('A06662-RNA.fasta'):
    if not line.startswith('>'):
      rna = rna + line.strip()
```

Source: Adapted from code published by A.Via/K.Rother under the Python License.

The loop collects the RNA sequence from all lines not starting with a '>' (i.e., all but the header line). This combination of for and if is a very simple search pattern. A general scheme for searching in lists and files is the following: run over the list using a for loop, use if/else conditions to verify if what you are looking for is in the current element, and store the data you want in a variable.

An alternative way to search something in a list is the in and not in operators that check if an element is (or is not) in a list and return True or False:

```
>>> bases = ['A','C','T','G']
>>> seq = 'CAGGCCATTRKGL'
>>> for i in seq:
...     if i not in bases:
...         print i, "is not a nucleotide"
...
R is not a nucleotide
K is not a nucleotide
L is not a nucleotide
```

5.4 EXAMPLES

In the following examples you will see how to build and use dictionaries.

Example 5.1 How to Fill a Dictionary from a FASTA File Where the Uniprot ACs Are the Keys and the Corresponding Sequences Are Their Values

In this example, a multiple sequence FASTA file downloaded from UniProt is read (see the sample reported in Appendix C, Section C.4, "A Multiple Sequence File in FASTA Format"), and for each record the AC number is extracted and the corresponding sequence is placed in a variable (seq). Then the AC:seq pairs are used to fill a dictionary for further use. The dictionary is finally printed.

```
sequences = {}
ac = ''
seq = ''
for line in open("SwissProt.fasta"):
    if line.startswith('>') and seq != '':
        sequences[ac] = seq
        seq = ''
    if line.startswith('>'):
        ac = line.split('|')[1]
    else:
        seq = seq + line.strip()
sequences[ac] = seq
print sequences.keys()
print sequences['P62258']
```

Source: Adapted from code published by A.Via/K.Rother under the Python License.

Notice that the procedure to parse the multiple sequence FASTA file is basically the same as the one described in Chapter 4 and, in particular, in Example 4.4. The sixth line contains the instruction to assign a value (the content of the variable seq) to a key (the content of the variable AC) in the sequences dictionary, which has been initialized in the first line of the example.

Example 5.2 How to Write a Simple Protein Sequence Loop Predictor

This example program predicts the disordered (loop) regions in a protein sequence. Even though the definition of *protein disorder* is rather controversial, one of the most accepted definitions is that a protein region is "disordered" if it is neither in alpha-helical conformation nor in beta-sheet conformation. The idea of this predictor is that each amino acid has a specific propensity to be in a secondary structure element. You can approximate a propensity by the frequency f of finding an amino acid type in a secondary structure element in a large collection of protein structures (the Protein Data Bank). The propensity of an amino acid to be "disordered" (i.e., to reside in a loop) can be calculated as $1 - f$. The propensity values for the 20 amino acids have been normalized (this also introduces some negative values) and reported in the propensities dictionary (at the beginning of Section 5.3).

To establish whether a given amino acid is disordered, you have to set a threshold (e.g., 0.3) and then sum up the propensities for all amino acids in the sequence.

The program prints the protein sequence using lowercase characters for disordered (loop) residues (propensity ≥ 0.3) and uppercase characters for "ordered" residues (i.e., residues that have the tendency of occurring in secondary structure elements).

```
propensities = {
 'N': 0.2299, 'P': 0.5523, 'Q':-0.18770, 'A':-0.2615,
 'R':-0.1766, 'S': 0.1429, 'C':-0.01515, 'T': 0.0089,
 'D': 0.2276, 'E':-0.2047, 'V':-0.38620, 'F':-0.2256,
 'W':-0.2434, 'G': 0.4332, 'H':-0.00120, 'Y':-0.2075,
 'I':-0.4222, 'K':-0.1001, 'L': 0.33793, 'M':-0.2259
 }
threshold = 0.3

input_seq = "IVGGYTCGANTVPYQVSLNSGYHFCGGSLINS\
    QWVVSAAHCYKSGIQVRLGEDNINVVEGNEQFISASKSIVH\
    PSYNSNTLNNDIMLIKLKSAASLNSRVASISLPTSCASAGTQ\
    CLISGWGNTKSSGTSYPDVLKCLKAPILSDSSCKSAYPGQI\
    TSNMFCAGYLEGGKDSCQGDSGGPVVCSGKLQGIVSWG\
    SGCAQKNKPGVYTKVCNYVSWIKQTIASN"
output_seq = ""
# Cycle over every amino acid in input_seq
for res in input_seq:
    if res in propensities:
        if propensities[res] >= threshold:
            output_seq += res.upper()
        else:
            output_seq += res.lower()
    else:
        print 'unrecognized character:', res
        break
print output_seq
```

Source: Adapted from code published by A.Via/K.Rother under the Python License.

The output of this program is:

```
ivGGytcGantvPyqvsLnsGyhfcGGsLinsqwvvsaahcyks
GiqvrLGedninvveGneqfisasksivhPsynsntLnndim
LikLksaasLnsrvasisLPtscasaGtqcLisGwGntkssGtsyP
dvLkcLkaPiLsdsscksayPGqitsnmfcaGyLeGGkdscqGdsGG
PvvcsGkLqGivswGsGcaqknkPGvytkvcnyvswikqtiasn
```

Example 5.3 How to Extract the Amino Acid Sequence from a PDB File

This example uses a dictionary to convert three-letter amino acids to one-letter code. The keys of the dictionary are the amino acid three-letter codes, and the values are the corresponding amino acid one-letter codes. The program uses that dictionary to read the residue names (in the form of three-letter codes) from the SEQRES lines of a PDB file (see Appendix C, Section C.8, "An Example of the SEQRES Lines of a PDB File (from File 1TDL)"), convert them in one-letter codes, and concatenate these later in a string in order to obtain the protein sequence. Finally, the sequence is printed in FASTA format.

To use the program, you need to go to the PDB archive (http://www.rcsb.org/) and download and save a PDB file. In this example, the PDB file 1TLD.pdb is used, which contains the crystal structure of the bovine beta-tripsin at 1.5 Å resolution. The SEQRES lines of a PDB file list the sequence of the protein used in the experiment. Each SEQRES line can be divided in 17 columns. The first one contains the keyword "SEQRES," the second one contains the sequence line number (starting from 1), the third contains the chain ID, and the fourth contains the number of residues composing the sequence. Residues (in three-letter code) are reported from the fifth column to the last. Columns are separated by a blank space.

```
aa_codes = {
  'ALA':'A', 'CYS':'C', 'ASP':'D', 'GLU':'E',
  'PHE':'F', 'GLY':'G', 'HIS':'H', 'LYS':'K',
  'ILE':'I', 'LEU':'L', 'MET':'M', 'ASN':'N',
  'PRO':'P', 'GLN':'Q', 'ARG':'R', 'SER':'S',
  'THR':'T', 'VAL':'V', 'TYR':'Y', 'TRP':'W'}
seq = ''
for line in open("1TLD.pdb"):
    if line[0:6] == "SEQRES":
        columns = line.split()
        for resname in columns[4:]:
            seq = seq + aa_codes[resname]
i = 0
print ">1TLD"
while i < len(seq):
    print seq[i:i + 64]
    i = i + 64
```

Source: Adapted from code published by A.Via/K.Rother under the Python License.

5.5 TESTING YOURSELF

Exercise 5.1 A Simple Dictionary

Create a dictionary where the following five codons are associated with their corresponding values:

```
'UAA':'Stop'
'UAG':'Stop'
'UGA':'Stop'
'AUG':'Start'
'GGG':'Glycin'
```

Exercise 5.2 Counting START and STOP Codons in a Nucleotide Sequence

Write a program that counts the number of STOP codons and the number of START codons in an input nucleotide sequence. The program must print two elements: the number of START codons and the number of STOP codons.

Hint: Download an RNA sequence from NCBI in FASTA format and use what you learned from Section 5.2.2.

Exercise 5.3 Search Keywords in a PubMed Abstract

Copy and paste to a text file the title and abstract of a paper of your choice (e.g., you can download it from PubMed: http://www.ncbi.nlm.nih.gov/pubmed). Check if two (or more) keywords of your choice (e.g., "calmodulin" or "CALM2") are (both or alternatively) present in the abstract, and if yes, print that you found them or otherwise that you didn't.

Exercise 5.4 Secondary Structure Predictor

Write a Sequence-Based Predictor for Secondary Structure Elements.

Hint: Use the following table of preferences, where the second column is for helices and the third is for beta sheets (http://www.bmrb.wisc.edu/referenc/choufas.html):

	pref_H	pref_E
A	1.45	0.97
C	0.77	1.30
D	0.98	0.80
E	1.53	0.26
F	1.12	1.28
G	0.53	0.81
H	1.24	0.71
I	1.00	1.60
K	1.07	0.74
L	1.34	1.22
M	1.20	1.67
N	0.73	0.65
P	0.59	0.62
Q	1.17	1.23
R	0.79	0.90
S	0.79	0.72
T	0.82	1.20
V	1.14	1.65
W	1.14	1.19
Y	0.61	1.29

Scan the input sequence residue by residue and replace each residue with H (helix) if its pref_H ≥ 1 and its pref_E < pref_H, with E (sheet) if its pref_E ≥ 1 and its pref_H < pref_E, and with L (loop) otherwise. Print (or write to a file) the input and output sequences, one on top of the other.

Exercise 5.5 Write a Predictor for the Solvent Accessibility of Amino Acidic Residues in a Protein Sequence

The input of the predictor must be a protein sequence file in FASTA format. The output must be the same sequence with residues in uppercase if they are predicted to be accessible to the solvent and in lowercase otherwise. You can find the solvent-exposed area of PDB residues (in Appendix C, Section C.10, "Solvent Accessibility of Amino Acids in Known Protein Structures") and consider that a residue has a propensity to be accessible if it has >70% in the >30 Å^2 column. Try to change the propensity threshold and see what happens to your output.

## CHAPTER 6

Filtering Data

LEARNING GOAL: You can find common, unique, and redundant items in data sets.

6.1 IN THIS CHAPTER YOU WILL LEARN

- How to find common items in two or more data sets

- How to merge data sets

- How to remove duplicates from a data set

- How to detect data set overlaps, intersections, and differences with the set data structure

- How to remove noise from NGS raw data

6.2 STORY: WORKING WITH RNA-SEQ OUTPUT DATA

6.2.1 Problem Description

The output of the NGS data analysis program Cuffcompare is the `transcripts.tracking` file described in the caption of Figure 6.1 and shown in Appendix C, Section C.9, "An Example of the Cuffcompare Output for Three Samples (q1, q2, and q3)" (for three biological samples). The first row of the file for three samples (q1, q2, and q3) looks like the following:

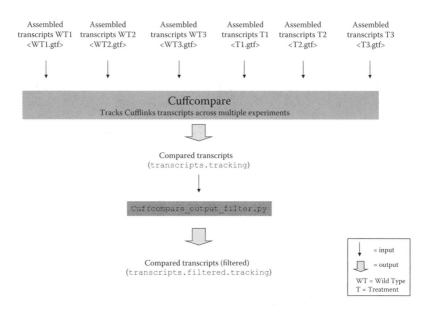

FIGURE 6.1 Input and output of the Cuffcompare program for transcript comparison. *Note:* The pipeline shown in Figure 15.1 can be applied to different sample cells to determine the transcriptome (the assembled transcripts) of each of them. For example, it can be applied to three replicas of a wild type cell (WT1, WT2, WT3) and to three replicas of a treated cancer cell (T1, T2, T3). Replicas are necessary to obtain more reliable results. If you want to compare the six transcriptomes obtained from different sample cells, you can use the Cuffcompare program from the Cufflinks package. Cuffcompare takes Cufflinks .gtf output files (containing assembled transcripts) as input and tracks transcripts across multiple experiments (i.e., samples). The output file (transcripts.tracking) consists of a table where each row corresponds to a single transcript and the (tab-separated) columns contain information about different samples. For each transcript, the information for a given sample (e.g., the one labeled with the symbol q1) looks like:

q1:NSC.P419.228|uc007afh.1|100|35.109496|34.188903|36.030089|397.404732|2433

where q1 is the sample label and all the other fields (separated by a pipe) provide details about the transcript in that sample, including the transcript ID (NSC. P419.228), the gene ID (uc007afh.1), the fmi (fraction of the major isoform, 100), the fpkm (expression mean value, 35.109496), min and max expression values (34.188903 and 36.030089, respectively), the transcript coverage (397.404732), and the transcript length (2433). When a given transcript has not been detected in a sample, the corresponding cell in the table contains a dash ("-"). The transcripts.tracking file can be filtered to retain only transcripts that appear in at least two out of the three replicas at hand (or in whatever fraction of replicas). This is what the program in Section 6.2.2 does.

```
Medullo-Diff_00000001  XLOC_000001  Lyplal|uc007afh.1
q1:NSC.P419.228|uc007afh.1|100| 35.109496| 34.188903|
    36.030089|397.404732|2433
q2:NSC.P429.18|uc007afh.1|100|15.885823|15.240240|
    16.531407|171.011325|2433
q3:NSC.P437.15|uc007afh.1|100|18.338541|17.704857|
    18.972224|181.643949|2433
```

The file is a tab-separated table reporting the results of a comparison among the transcriptomes obtained from different DNA sequence samples. In particular, six samples are taken into account: WT1, WT2, and WT3, which are three replicas of a wild type cell type (denoted with q1, q2, and q3 in the file, respectively), and T1, T2, and T3 (denoted with q4, q5, and q6 in the file, respectively), which are three replicas of the same cell type after pharmacological treatment (T stands for treated). Replicas are necessary to ensure robustness of data, and to this aim, you may want to retain only transcripts that have been observed in the transcriptome of all replicas or at least in two out of three of them.

This corresponds to removing from transcripts.tracking the transcripts (i.e., file rows) occurring in only one out of the three WT1, WT2, and WT3 (or T1, T2, and T3) samples. When a transcript is absent in a given sample, the corresponding cell in the table (see the transcripts.tracking file content description reported in the Figure 6.1 caption and Figure 6.2) is filled in with a dash ("-").

Here are two lines of the complete file (with six replicas). The first one is the example of a line (starting with Medullo-Diff_00000001) that

WT1	WT2	WT3		T1	T2	T3	
q1	q2	q3		q4	q5	q6	save
q1	-	q3		q4	q5	q6	save
q1	q2	q3		-	q2	-	skip
-	q2	q3		q4	-	q6	save
q1	-	-		q4	q5	q6	skip
q1	q2	-		-	-	-	skip

FIGURE 6.2 Saved and skipped rows in the transcripts.tracking file of Figure 6.1. *Note:* The presence of "qi" in a cell means that the information about the replica is available. "-" means that the information is not available. Each row corresponds to a different transcript.

will be saved in the output file, whereas the second line (starting with
Medullo-Diff_00000002) will be skipped:

```
Medullo-Diff_00000001 XLOC_000001    Lypla1|uc007afh.1
q1:NSC.P419.228|uc007afh.1|100|35.109496| 34.188903|
    36.030089 |397.404732|2433
q2:NSC.P429.18|uc007afh.1|100|15.885823|15.240240|
    16.531407|171.011325|2433
q3:NSC.P437.15|uc007afh.1|100|18.338541|17.704857|
    18.972224|181.643949|2433
q4:CSC.Mmb8.236|uc007afh.1|100|22.594194|21.925964|
    23.262424|225.248080|2433
q5:CSC.Mmb10.251|uc007afh.1|100|22.778360|22.025125|
    23.531595|255.416281|2433
q6:CSC.Mmb21.221|uc007afh.1|100|17.288114|16.675834|
    17.900395|184.487708|2433
Medullo-Diff_00000002   XLOC_000002   Tcea1|uc007afi.2=
q1:NSC.P419.228|uc007afi.2|18|1.653393|1.409591|
    1.897195|18.587029|2671
                    -
q3:NSC.P437.108|uc007afi.2|100|4.624079|4.258801|
    4.989356|45.379750|2671
                    -
                    -
                    -
```

The script to evaluate the transcripts.tracking file needs to identify lines where at least two replicas are present for the wild type and for
the treated cell lines. All lines where more than one replica out of (wt1,
wt2, wt3) or (t1, t2, t3) is missing, are skipped; otherwise, the lines
will be written to the output file (see Figure 6.2).

6.2.2 Example Python Session

```python
tracking = open('transcripts.tracking', 'r')
out_file = open('transcripts-filtered.tracking', 'w')
for track in tracking:
    # split tab-separated columns
    columns = track.strip().split('\t')
    wildtype = columns[4:7].count('-')
    treatment = columns[7:10].count('-')
    if wildtype < 2 and treatment < 2:
        out_file.write(track)
tracking.close()
out_file.close()
```

Source: Adapted from code published by A.Via/K.Rother under the
Python License.

The output file looks like the input file except for the absence of lines (i.e., transcripts) containing more than one dash in wild type or in treated replicas (i.e., for which less than two replicas are expressed).

6.3 WHAT DO THE COMMANDS MEAN?

6.3.1 Filtering with a Simple for...if Combination

The Python session in Section 6.2.2 shows how to filter out data in a file using a simple for...if combination. The for loop is needed to run through all lines in the file, and the if condition is needed to skip lines depending on the number of dashes. Actually, the program (1) splits each line into fields separated by a tab ('\t') and stores the result in a list called columns, (2) uses the function count() to count the number of dashes in each list of three samples (columns[4:7] and columns[7:10]), and (3) checks through the if condition whether the number of dashes for each set of three samples is lower than two (i.e., at most one).

The same result could be achieved with a more explicit code, replacing the count() function with a series of if statements:

```
output_file = open('transcripts-filtered.tracking', 'w')
for track in open('transcripts.tracking'):
    columns = track.strip().split('\t')
    wt = 0
    t = 0
    if columns[4] != '-': wt += 1
    if columns[5] != '-': wt += 1
    if columns[6] != '-': wt += 1
    if columns[7] != '-': t += 1
    if columns[8] != '-': t += 1
    if columns[9] != '-': t += 1
    if wt > 1 and t > 1:
        output_file.write(track)
output_file.close()
```

Source: Adapted from code published by A.Via/K.Rother under the Python License.

Here, the program checks column by column to determine if there is a dash. If yes, it increases a counter (wt or t for wild type and treated, respectively) by one. There are two counters, one for each group of samples (wt1, wt2, wt3 and t1, t2, t3) to treat them separately. If the counter value is greater than one, it means that the transcript is present in at least two samples, and the line is copied to the output file. Otherwise, it is ignored.

6.3.2 Combining Two Data Sets

Another situation you may want to be able to tackle is to retain items from a data set A only if they are in a data set B. This can be done with a simple for...if combination working on Python lists. In the following example, we consider two lists of integer numbers (they also could be lists of Uniprot IDs; see the example shown in Section 6.3.5) and write all the elements belonging to both lists to a new list (a_and_b):

```
data_a = [1, 2, 3, 4, 5, 6]
data_b = [1, 5, 7, 8, 9]
a_and_b = []
for num in data_a:
    if num in data_b:
        a_and_b.append(num)
print a_and_b
```

This combination of a for loop and if statement preserves the order of elements in data_a. If the order is not relevant, you can make the code shorter by using the set data type (explained in more detail in Section 6.3.6):

```
data_a = set([1, 2, 3, 4, 5, 6])
data_b = set([1, 5, 7, 8, 9])
a_and_b = data_a.intersection(data_b)
print a_and_b
```

Source: Adapted from code published by A.Via/K.Rother under the Python License.

6.3.3 Differences between Two Data Sets

A related problem is to find which elements are different in two lists. The next example collects elements of data_a that are not in data_b (a_not_b) and elements of data_b that are not in data_a (b_not_a).

```
data_a = [1, 2, 3, 4, 5, 6]
data_b = [1, 5, 7, 8, 9]
a_not_b = []
b_not_a = []
for num in data_a:
    if num not in data_b:
        a_not_b.append(num)
for num in data_b:
    if num not in data_a:
        b_not_a.append(num)
print a_not_b
print b_not_a
```

Again, the program becomes shorter with sets, but you lose the order of elements:

```
data_a = set([1, 2, 3, 4, 5, 6])
data_b = set([1, 5, 7, 8, 9])
a_not_b = data_a.difference(data_b)
b_not_a = data_b.difference(data_a)
print a_not_b
print b_not_a
```

Source: Adapted from code published by A.Via/K.Rother under the Python License.

6.3.4 Removing from Lists, Dictionaries, and Files

Python has functions to remove items from data structure objects, such as lists and dictionaries.

Removing Elements from Lists

There are several methods to remove items from a list. If you want to remove the last element of a list, you can use the pop() method of lists without passing an argument to the method:

```
>>> data = [1,2,3,6,2,3,5,7]
>>> data.pop()
7
>>> data
[1, 2, 3, 6, 2, 3, 5]
```

Notice that pop() returns the value of the element before deleting it. If you want to remove an element of a list data in a given position i, you can use the pop() method, passing the index of the position as an argument:

```
>>> data = [1,2,3,6,2,3,5,7]
>>> data.pop(0)
1
>>> data
[2, 3, 6, 2, 3, 5, 7]
```

Alternatively, you can use the built-in function del(data[i]):

```
>>> data = [1,2,3,6,2,3,5,7]
>>> del(data[0])
>>> data
[2, 3, 6, 2, 3, 5, 7]
```

But if you want to remove an element with a certain value (e.g., the number 3 in the list `data` above), you have to use the list method `remove()`:

```
>>> data = [1, 2, 3, 6, 2, 3, 5, 7]
>>> data.remove(3)
>>> data
[1, 2, 6, 2, 3, 5, 7]
```

You may have noticed that the `remove()` method has only removed the element 3 the first time it occurred in the list. If you want to remove all elements with the value of 3, you can use a list comprehension for that (see Chapter 4, Section 4.3.3):

```
>>> data = [1, 2, 3, 6, 2, 3, 5, 7]
>>> data = [x for x in data if x != 3]
>>> data
[1, 2, 6, 2, 5, 7]
```

In the second row of this example, the list `data` is redefined as a list that has the same elements of the original list except for the elements that equal 3.

Notice that all these functions permanently modify the original list. If you want to keep the original list, you have to create a copy beforehand or use another variable name for the list generated using list comprehension.

An additional trick consists of using the slicing property of lists to remove one or more portions of a list:

```
data2 = data[:2] + data[3:]
```

Here, the element `data[2]` will not appear in list `data2`.

Removing Elements from Dictionaries
The method `pop()` also exists for dictionaries, where it works as follows:

```
>>> d = {'a':1, 'b':2, 'c':3}
>>> d.pop('a')
1
>>> d
{'c': 3, 'b': 2}
```

It is slightly different from the corresponding method for lists. In fact, you cannot use it without an argument, and the argument must be the key of the (*key, value*) pair you want to remove.

The del() built-in function can be used for the same purpose as well:

```
>>> d = {'a':1, 'b':2, 'c':3}
>>> del d['a']
>>> d
{'c': 3, 'b': 2}
```

Deleting Particular Lines from a Text File
There are several simple ways to filter lines in certain positions from a text file. Here, we suggest two of them. Suppose you have the input file text. txt and you want to remove the first and second lines and the fifth and sixth lines. You can remove them by slicing the list of lines:

```
lines = open('text.txt').readlines()
open('new.txt','w').writelines(lines[2:4]+lines[6:])
```

Notice that, in this example, the lines of the input file have been stored in a list through the file object method readlines(). This is not very convenient if you are using very big files. In this case, the for...if combination is a better solution. To remove the right lines, use a counter variable to keep track of the line number. In the next example, the number of lines to be removed (first, second, fifth, and sixth) is stored in a list ([1,2,5,6]). Then a counter is initialized to 0 and increased by 1 for each new line. When the counter is 1, 2, 5, or 6 (i.e., it is in the [1,2,5,6] list), the line is skipped (pass); otherwise the line is written to the output file:

```
in_file = open('text.txt')
out_file = open('new.txt', 'w')
index = 0
indices_to_remove = [1, 2, 5, 6]
for line in in_file:
    index = index + 1
    if index not in indices_to_remove:
        out_file.write(line)
out_file.close()
```

Source: Adapted from code published by A.Via/K.Rother under the Python License.

If you do not want to introduce a counter, you can use the enumerate() built-in function:

```
out_file = open('new.txt', 'w')
indices_to_remove = [1, 2, 5, 6]
```

```
for index, line in enumerate(open('text.txt')):
    if (index + 1) not in indices_to_remove:
        out_file.write(line)
out_file.close()
```

Given a list x, enumerate(x) returns tuples (i, x[i]) of indexes i and values x[i] of the list:

```
>>> x = [1,2,5,6]
>>> for i,j in enumerate(x):
...     print i,j
...
0 1
1 2
2 5
3 6
>>>
```

In this example, enumerate() returns, for each line of the file, a tuple composed of the line number (starting from 0) and the corresponding line content.

6.3.5 Removing Duplicates Preserving and Not Preserving Order

Being able to remove duplicates from redundant data is very useful in many situations. To remove the duplicates, you need to identify unique objects. This can be done preserving the order of elements (which may be important) or without preserving it (which is faster).

Selectively Remove Duplicate Records from a Text File Preserving Order
This is something that may happen relatively often to you: the need to remove duplicate lines in a text file and create a new file containing only unique elements. Suppose you have the following input file with Uniprot ACs

```
P04637
P02340
P10361
Q29537
P04637
P10361
P10361
P02340
```

and you want the output to contain only unique Uniprot ACs:

```
P04637
P02340
P10361
Q29537
```

Here is how to remove duplicates using a list:

```
input_file = open('UniprotID.txt')
output_file = open('UniprotID-unique.txt','w')
unique = []
for line in input_file:
    if line not in unique:
        output_file.write(line)
        unique.append(line)
output_file.close()
```

Source: Adapted from code published by A.Via/K.Rother under the Python License.

Writing unique entries to the output file directly ensures that the order of the lines is preserved.

Selectively Remove Duplicate Records from a Text File without Preserving Order

If you do not care about the order of your records, you can read the entries to a set (described in Section 6.3.6).

```
input_file = open('UniprotID.txt')
output_file = open('UniprotID-unique.txt','w')
unique = set(input_file)
for line in unique:
    output_file.write(line)
```

Source: Adapted from code published by A.Via/K.Rother under the Python License.

In this example, the lines of the input file are added to a set `unique` by reading the lines into the set (`unique = set(input_file)`). Sets are unordered collections of unique elements, so that a file line identical to one that is already present in the set won't be added to it. Finally, a `for` loop is used to read and write the set items to the output file.

How to Remove Sequences with More Than 90% Identity

A quite common task in bioinformatics is to remove redundancy from a set of sequences. More precisely, you want to generate another set of sequences with a level of sequence identity not greater than a given cutoff (e.g., 90%). This is not as easy as it sounds, as you not only need to group the similar sequences together, but you also need a rule establishing which of the similar sequences in a group to select. In the past decade, several algorithms have been developed for fast sequence redundancy removal. A well-optimized and easy to use tool is CD-HIT, which is briefly described in Box 6.1.

BOX 6.1 CD-HIT (CLUSTER DATABASE AT HIGH IDENTITY WITH TOLERANCE)

This is a very fast program for clustering protein sequences based on a user-defined similarity threshold. The program takes as input a set of sequences in FASTA format (Appendix C, Section C.4, "A Multiple Sequence File in FASTA Format") and returns two files: one with a list of clusters, and one with the sequence of the cluster representatives. The program can be downloaded at http://bioinformatics.org/cd-hit/. A manual with installation instructions is also available. Once the program has been installed, the command line to run it looks like

```
cd-hit -i redundant_set -o nr-90 -c 0.9 -n 5
```

where `redundant_set` is the input filename, `nr-90` is the output, `0.9` means 90% identity, and `5` is the size of word (instructions on how to choose the size of word are provided in the manual). Many further options are available.

6.3.6 Sets

Sets are unordered collections of *unique* objects. This means that they are not sequential objects like lists and that they cannot contain identical elements. Sets are an ideal data structure to remove duplicates and to calculate the intersection, the union, and the difference between two or more groups of objects as long as the order is not important. Sets do not support indexing and slicing operations, but the 'in' and 'not in' operators can be used to test an element for membership in a set.

Creating a Set
To create a set, you have to use the method set(x), where x is a sequence-like object (string, tuple, list).

```
>>> set('MGSNKSKPKDASQ')
set(['A', 'D', 'G', 'K', 'M', 'N', 'Q', 'P', 'S'])
>>> set((1, 2, 3, 4))
set([1, 2, 3, 4])
>>> set([1, 2, 3, 'a', 'b', 'c'])
set(['a', 1, 2, 3, 'c', 'b'])
```

The elements of a set are the same as those of the input data even if they are listed in a different order. The elements of a set must be immutable objects such as numbers, strings, or tuples; thus lists, dictionaries, and other sets cannot be elements of a set.

Since sets are collections of unique elements, redundant elements will be removed automatically when you create a set:

```
>>> id_list = ['P04637', 'P02340', 'P10361', 'Q29537',
'P04637', 'P10361', 'P10361']
>>> id_set = set(id_list)
>>> id_set
set(['Q29537', 'P10361', 'P04637'])
```

This is a compact way of finding unique identifiers.

Methods of Sets

The method add() is used to add an element to a set. If you add an element that already exists in the set, add() has no effect. The method update() is used to add several elements to a set, unless they are present in the set already. pop(), remove(), and discard() make it possible to remove elements from a set.

```
>>> s1 = set([1, 2, 3, 4, 5])
>>> s1.add(10)
>>> s1
set([1, 2, 3, 4, 5, 10])
>>> s1.update(['a', 'b', 'c'])
>>> s1
set(['a', 1, 2, 3, 4, 5, 10, 'c', 'b'])
```

Checking Set Membership

The operator in allows you to check whether an element is contained in a set or not.

```
>>> 5 in s1
True
>>> 6 in s1
False
>>> 6 not in s1          #Test 6 for non-membership in s1
True
>>> s2 = set([10, 4, 5])
>>> s1.issubset(s2)      #Test if s1 is a subset of s2
False
>>> s1.issuperset(s2)    #Test if s1 is a superset of s2
True
```

Use Sets to Determine Data Overlap/Differences

The *union* between two sets (s1 and s2) creates a new set containing all elements from both s1 and s2.

```
>>> s1 = set(['a','b','c'])
>>> s2 = set (['c','d','e'])
>>> s1.union(s2)
set(['a', 'c', 'b', 'e', 'd'])
```

The *intersection* of two sets s1 and s2 creates a new set with the elements common to both s1 and s2.

```
>>> s1 = set(['a', 'b', 'c'])
>>> s2 = set (['c', 'd', 'e'])
>>> s1.intersection(s2)
set(['c'])
```

The *symmetric difference* of two sets s1 and s2 creates a new set with elements in either s1 or s2 but not in both.

```
>>> s1 = set(['a', 'b', 'c'])
>>> s2 = set (['c', 'd', 'e'])
>>> s1.symmetric_difference(s2)
set(['a', 'b', 'e', 'd'])
```

The *difference* of two sets s1 and s2 creates a new set with elements in s1 but not in s2.

```
>>> s1 = set(['a', 'b', 'c'])
>>> s2 = set (['c', 'd', 'e'])
>>> s1.difference(s2)
set(['a', 'b'])
>>> s2.difference(s1)
set(['e', 'd'])
```

6.4 EXAMPLES

Example 6.1 Comparing More Than Two Sets of Data

If you have more than two sets of data and you want, e.g., to find the elements common to all of them, you can compare them as follows:

```
a = set((1, 2, 3, 4, 5))
b = set((2, 4, 6, 7, 1))
c = set((1, 4, 5, 9))
```

```
triple_set = [a, b, c]
common = reduce(set.intersection, triple_set)
print common
```

Source: Adapted from code published by A.Via/K.Rother under the Python License.

The built-in function reduce() takes two arguments: the first one is a function taking two variables (f(x,y)) (see Chapter 10 for more on functions), and the second one is an iterable object i (a tuple or a list). reduce(f, i) passes the first two elements of the iterable (i[0] and i[1]) to the function f and calculates the value returned by the function, then passes that value to f as the first argument, and the third element of the iterable (i[2]) as the second argument, and so on. For example if you define a multiplication function

```
def multiply(x,y):
    return x * y
print reduce(multiply, (1, 2, 3, 4))
```

you obtain 24. In fact, reduce() calculates 1*2 = 2, then 2*3 = 6, and finally 6*4 = 24. Thus, at each step, the new first argument x of the multiply(x) function is the accumulated result of the function calls, and the second argument is the subsequent element of the tuple (1,2,3,4).

In the previous example with the sets, the first argument is a function that calculates the intersection of two set variables x, y, and the second argument is a list of sets (triple_set = [a, b, c]). reduce() applies the intersection function to the first two elements of the list of sets (i.e., it intersects a and b) and calculates the returned value, i.e., the intersection of a and b, which is a new set, then applies the intersection function to that new set and to the third element of triple_set, i.e., to c, thus obtaining the intersection of the new set and c. In summary, the reduce() function finds the intersection (i.e., the common values) between the first set (a) and the second (b) and then the intersection of that set and the third set (c).

Example 6.2 Compare/Update Different Releases of a Database (e.g., Uniprot)

Sets can be easily applied to detect differences between two releases of a database. Suppose you have two Uniprot releases, in the form of lists of Uniprot ACs (or FASTA files from which you can easily extract Uniprot ACs), and want to see which entries are new, which entries disappeared (deprecated entries), and what is absent in either the old or the new release (unique entries). This can be done using sets:

```
# read old database release
old_db = set()
for line in open("list_old.txt"):
        accession = line.strip()
        old_db.add(accession)
# read new database release
new_db = set()
for line in open("list_new.txt"):
        accession = line.strip()
        new_db.add(accession)
# report differences
new_entries = new_db.difference(old_db)
print "new entries", list(new_entries)
old_entries = old_db.difference(new_db)
print "deprecated entries", list(old_entries)
unique_entries = new_db.symmetric_difference(old_db)
print "unique entries", list(unique_entries)
```

Source: Adapted from code published by A.Via/K.Rother under the Python License.

Notice that another way to do the same thing is to use a for...if combination, as explained in Section 6.3.5.

6.5 TESTING YOURSELF

Exercise 6.1 Copy Only Selected FASTA Records to a File

Read a multiple sequence FASTA file and copy to a new file the ID (one per line) of the sequences starting with a methionine.

Hint: You can use what you learned in Chapter 4 about FASTA file parsing.

Hint: Since you have to check the type of the first residue, you need to collect not the whole sequence of each record but just its first character.

Exercise 6.2

Remove the even (or the odd) lines from a text file of your choice.

Hint: You can use the % operator, which returns the remainder of a division:

```
>>> 7%2
1
```

If you use a line counter, the remainder of the division by 2 will be 0 for even lines (and 1 for odd lines).

Exercise 6.3 Finding Differences between Files Having the Same Number of Lines

Write a program that reads two text files and prints their differences (line by line).

Hint: Use the file method `readlines()` to put the lines of each file into a list. If you do this separately for the two files you want to compare, you will end up with two lists. Use counters to count how many lines (i.e., list elements) are identical in the two files (i.e., in the two lists), how many are present in the first file and absent in the second, and vice versa.

Exercise 6.4 A More Sophisticated Way of Printing Differences between Files

Implement the program of Exercise 6.3 in order to print lines present in the first file and absent in the second preceded by a ">", lines present in the second and absent in the first preceded by a "<", and lines present in both files preceded by a "#".

Exercise 6.5 A Further Filter for Transcripts

Modify the Python session in Section 6.2.2 to retain only transcripts that are expressed in at least three samples, regardless if WT or T.

## CHAPTER 7

Managing Tabular Data

L EARNING GOAL: You can organize and edit data in tables.

7.1 IN THIS CHAPTER YOU WILL LEARN

- How to store data in two-dimensional tables using nested lists

- How to insert and delete table rows and columns

- How to create an empty table

- How to represent a table using dictionaries

7.2 STORY: DETERMINING PROTEIN CONCENTRATIONS

7.2.1 Problem Description

In 1951, Oliver Lowry described a general procedure using Folin phenol reagent to measure protein concentrations.[*] The test has the advantage that multiple samples can be measured quickly using a photometer (in contrast to, e.g., the Kjeldahl method[†]). Its ubiquitous applicability has made Lowry's paper the most cited article of all time.

In the Lowry assay, you record the extinction of a series of samples with known protein concentration, from which you construct a line of best fit. Then, the concentration of an unknown sample can be determined by finding the concentration corresponding to an extinction value on that

---

[*] O.H. Lowry, N.J. Rosebrough, A.L. Farr, and R.J. Randall "Protein measurement with the Folin phenol reagent." *Journal of Biological Chemistry* 193 (1951): 265–275.

[†] Dr. D. Julian McClements (http://people.umass.edu/~mcclemen/581Proteins.html).

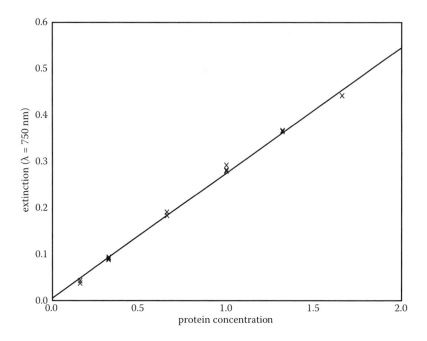

FIGURE 7.1   Best-fit line according to Lowry data.

line. Back in 1951, the line of best fit was drawn on paper (see Figure 7.1), and the unknown sample was determined manually. Most of this chapter is about handling a lot of sample data in Python using Lowry's extinction values as an example. An example of a standard series was given by Lowry in Table IV (Measurement of Small Amount of Protein from Rabbit Brain) of his publication, where he determined the protein concentrations of 18 rabbit brain samples. He examined six concentrations three times each, obtaining the values reported in Table 7.1.

TABLE 7.1   Lowry's Measurement of Small Amount of Protein from Rabbit Brain.

Protein (%)	Extinction 1 (Optical density at 750 nm)	Extinction 2 (Optical density at 750 nm)	Extinction 3 (Optical density at 750 nm)
0.16	0.038	0.044	0.040
0.33	0.089	0.095	0.091
0.66	0.184	0.191	0.191
1.00	0.280	0.292	0.283
1.32	0.365	0.367	0.365
1.66	0.441	0.443	0.444

TABLE 7.2   Same Data as Shown in Table 7.1 but
Distributed into Two Columns Instead of Four.

Protein	Extinction
0.16	0.038
0.16	0.044
0.16	0.040
0.33	0.089
...	...

This table can be parsed easily from a file. But the formula to calculate a best-fit line requires just one set of x/y values, i.e., a table with two columns. Thus, for calculating the concentration of an unknown protein sample, you need a single list of protein concentration–extinction pairs, as in Table 7.2.

How can the table with the original data (Table 7.1) be converted to the simpler one (Table 7.2)? In the following Python session, a number of manipulation steps are performed on the initial table, which is represented as a list of lists. A new function (zip(), explained in detail in Section 7.3.4) is used twice: the first time, to turn a table around by 90 degrees and, the second time, to combine two columns of a table to obtain a new two-dimensional table.

7.2.2 Example Python Session

```python
table = [
    ['protein', 'ext1', 'ext2', 'ext3'],
    [0.16, 0.038, 0.044, 0.040],
    [0.33, 0.089, 0.095, 0.091],
    [0.66, 0.184, 0.191, 0.191],
    [1.00, 0.280, 0.292, 0.283],
    [1.32, 0.365, 0.367, 0.365],
    [1.66, 0.441, 0.443, 0.444]
    ]
table = table[1:]
protein, ext1, ext2, ext3 = zip(*table)

extinction = ext1 + ext2 + ext3
protein = protein * 3

table = zip(protein, extinction)

for prot, ext in table:
    print prot, ext
```

Source: Adapted from code published by A.Via/K.Rother under the Python License.

The output will look like

```
0.16 0.038
0.33 0.089
0.66 0.184
1.0 0.28
1.32 0.365
1.66 0.441
0.16 0.044
0.33 0.095
...
```

7.3 WHAT DO THE COMMANDS MEAN?

When you start the program, it prints the table transformed to a two-column format. The program is divided into two parts. At the top, data are written as a list containing other lists. Using this *nested list*, the program can unambiguously represent two-dimensional data. At the bottom, the conversion is done in five steps. First, the row with the labels is removed (table = table[1:]). Second, four tuples containing one column each are created (protein, ext1, ext2, ext3 = zip(*table)). Third, the extinction columns are copied into a single tuple (extinction = ext1 + ext2 + ext3), and the protein column is extended accordingly (protein = protein * 3). Here, the protein tuple has been multiplied by 3, resulting in the same values repeating three times over, because in the previous table there are three extinction values for one protein concentration. Fourth, both columns are combined to a new two-dimensional table (table = zip(protein, extinction)). Finally, the contents of the table are printed line by line. The result is a nested list containing pairs of protein concentrations and corresponding extinction values.

7.3.1 Representing a Two-Dimensional Table

We start with a table containing one column of protein concentrations and three columns of extinction values (see Table 7.1). The type of the table variable is a list that contains other lists. Any table can be encoded as a list of lists, also called a *nested list*. For example, the table

1	2	3
4	5	6
7	8	9

can be encoded as the nested list

```
square = [[1, 2, 3], [4, 5, 6], [7, 8, 9]]
```

or as a list of nested tuples

```
square = [(1, 2, 3), (4, 5, 6), (7, 8, 9)]
```

As shown in Chapter 4, Python lists can keep all kinds of data, including other lists. In a table represented by a nested list, there is a single outer list (containing the rows) and inner lists (one for each row). The outer list contains the inner lists. This way the table has a clear structure. This nested list structure is also called a *2D array*.

Q & A: CAN I HAVE THREE-DIMENSIONAL TABLES IN PYTHON?

There is no limit to how many lists you can store within each other. For instance, you can create a three-dimensional list with $2 \times 2 \times 2$ elements, such as the following:

```
cube = [[[0, 1], [2, 3]], [[4, 5], [6, 7]]]
```

With a three-dimensional table, the size of your data may become huge very quickly (e.g., if your table has 100 positions in each direction, you will have 100^3 or 10^6 cells in the table). With such numbers, programs get slow easily unless you employ sophisticated algorithms and have lots of memory and/or powerful computers. As a rule of thumb, the more data you have, the more you have to plan in advance (see also Chapter 16). When your data becomes more complex, accessing a table by indices like [1][2][3] can become cumbersome. The use of classes for a more human-readable representation becomes useful then (see Chapter 11).

7.3.2 Accessing Rows and Single Cells

When a table is represented as a nested list, you can access each row by an index, in the same way as with any list. For instance, the second row would be as follows (indices starting from zero):

```
second_row = table[1]
```

By adding a second index for the column, you can access individual cells. For instance, the cell from the second row in the third column could be accessed by

```
second_row_third_column = table[1][2]
```

Or if you want to manipulate the data, you can assign to a given cell:

```
table[1][2] = 0.123
```

With a for loop, you can do something to all rows in a table:

```
for row in table:
    print row
```

With a double for loop, you access each cell in each row one by one:

```
for row in table:
    for cell in row:
        print cell
```

Accessing rows and single cells is the most straightforward operation on nested lists in Python. Accessing columns is slightly more complicated and will be discussed in Section 7.3.4.

7.3.3 Inserting and Removing Rows

When doing calculations on the Lowry data, you must get rid of the row with the labels first. Since the table is stored as a list, you can use all operations lists provide. The entire first row is removed by a slicing operation that takes all but the first element:

```
table = table[1:]
```

Alternatively, you could use the pop() method:

```
table.pop(0)
```

Remember that the indices start with 0. Similarly, you can remove any other row, e.g., the third one, with

```
table.pop(2)
```

or alternatively with slicing:

```
table = table[:2] + table[3:]
```

Analogously, you can use a list function to insert a new row in a given position:

```
table.insert(2, [0.55, 0.123, 0.122, 0.145])
```

Or you can add a new row at the end:

```
table.append([0.55, 0.123, 0.122, 0.145])
```

Adding and removing rows of a table can be done with a single line of code.

7.3.4 Accessing Columns

The disadvantage of the nested list approach is that accessing columns is less straightforward, because the data for one column are distributed over all rows. Of course, you could run a loop over your table to collect all data from one column:

```
protein = []
for row in table:
    protein.append(row[0])
```

If you want to extract many columns this way or access the same column many times, your program will grow long and hard to read. There is, however, a more efficient abbreviation in Python:

```
protein, ext1, ext2, ext3 = zip(*table)
```

The `zip(*table)` command puts each column into a separate variable as a tuple thus effectively turning the table around by 90 degrees. This operation, although syntactically short, takes time to complete. Thus, for large data sets, you may be better off with a `for()` loop.

Combining Multiple Columns

After all four columns are stored in separate variables (containing lists or tuples), they need to be combined into a two-column table (see Table 7.2).

The plus (+) and multiplication (*) operators in Python can be applied to combine and multiply, respectively, lists and tuples. Multiplication extends a list by copying it:

```
protein = protein * 3
```

This results in three subsequent copies of the same data:

```
>>> [1, 2, 3] * 3
[1, 2, 3, 1, 2, 3, 1, 2, 3]
```

Addition concatenates two or more lists or tuples into one:

```
>>> [1, 2, 3] + [4, 5, 6]
[1, 2, 3, 4, 5, 6]
```

The result is a single list or tuple containing all data items, one after another:

```
extinction = ext1 + ext2 + ext3
```

In the previous program, the result of these lines is that the three columns with extinctions get merged into a single one, and the information in the protein column is tripled to contain the corresponding values.

Q & A: What If I Combine Lists That Contain Different Kinds of Data?

Python does not care what is inside the lists you multiply with the "*" operator or concatenate with the "+" operator. You can easily create lists that first contain numbers and then strings, for instance. But when you want to use a `for` loop or a function like `sum()` on all elements of the list, having different data leads to problems. If you feel that a nested list is not well structured enough for your purposes, consider using nested dictionaries (see Example 7.2) or classes (see Chapter 11).

The zip() *Function*

How exactly does the built-in zip() function and the asterisk work? The zip() command allows you to combine elements from two or more lists one by one, e.g.,

```
>>> zip([1, 2, 3], [4, 5, 6])
[(1, 4), (2, 5), (3, 6)]
```

In the result, the first elements from each input list are paired together, then the second elements, etc. The argument(s) of the zip() function must be iterable (lists, tuples, strings). It returns a list of tuples, where the ith tuple contains the ith element from each of the arguments. In the example in Section 7.2.2, the asterisk tells the zip function to use all lists from the nested lists as arguments, so that you can write

```
zip(*table)
```

instead of

```
zip(table[0], table[1], table[2], table[3])
```

Thinking of zip() arguments as the rows of a table, the zip(*table) notation rotates the table by 90 degrees:

```
>>> data = [[1, 2, 3], [4, 5, 6]]
>>> zip(*data)
[(1, 4), (2, 5), (3, 6)]
```

Summarizing, the zip() function pairs items from lists like a zipper. The * symbol interprets the given variable like a list of arguments for each row (also see Chapter 10, Section 10.3.2). A very common usage of zip() is to rotate (or transpose) tables. This makes accessing table columns much easier.

7.3.5 Inserting and Removing Columns

With the zip() function you can turn a table by 90 degrees:

```
table = zip(*table)
```

With this trick in mind, you can access, insert, and remove columns in the same way as rows. For instance, to insert a column, you need to first turn the table, insert a row, and then turn the table back:

```
table = zip(*table)
table.append(['ext4', 0, 0, 0, 0, 0, 0])
table = zip(*table)
```

This code adds an extra column full of zeroes with a label in the first row.

If you want to delete a column from a table, you can do it using the same pattern:

```
table = zip(*table)
table.pop(1)
table = zip(*table)
```

This removes the entire second column. This method has a slight disadvantage, however: the zip(*table) operation converts the inner lists to tuples. As mentioned in Chapter 5, tuples are immutable. This means that after manipulating columns using the zip() pattern, you cannot manipulate individual cells anymore. You need to convert the row to a list again:

```
table[1] = list(table[1])
table[1][2] = 0.123
```

These two lines change the value of an individual cell after a zip(*table) command. Figure 7.2 summarizes the actions that can be carried out on tables.

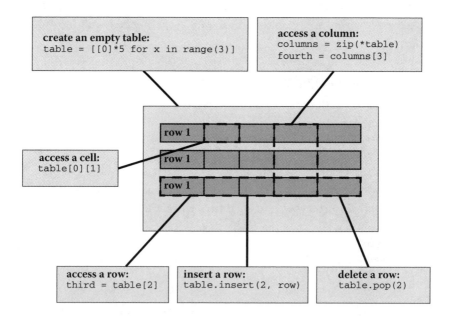

FIGURE 7.2 Actions that can be carried out on tables.

7.4 EXAMPLES

Example 7.1 Creating an Empty Table

For some calculations it is useful to create an empty table first. For example, when you would like to count features in a two-dimensional matrix, you can fill the table for counting with zeroes first. Creating a one-dimensional list of zeroes can be done as follows:

```
>>> row = [0] * 6
[0, 0, 0, 0, 0, 0]
```

Many rows can be created by repeating the above line in a loop:

```
table = []
for i in range(6):
    table.append([0] * 6)
```

Source: Adapted from code published by A.Via/K.Rother under the Python License.

Or you can use a list comprehension (introduced in Chapter 4, Section 4.3.3):

```
>>> table = [[0] * 6 for i in range(6)]
```

Beware not to write:

```
>>> row = [0] * 3
>>> table = [row] * 3
>>> table
[[0, 0, 0], [0, 0, 0], [0, 0, 0]]
```

In the resulting table, the rows will not be copies of the empty row, only references to the same row. So, the resulting table contains the same list object three times. Every time you change a cell in one row of the resulting table, all other rows will change simultaneously. For clarification, try the following:

```
>>> table[0][1] = 5
>>> table
[[0, 5, 0], [0, 5, 0], [0, 5, 0]]
```

Example 7.2 Representing Tables with Dictionaries

When using tables that are nested lists, you need to know the numerical indices of the cells. Sometimes this can make the code hard to read. Also, lists are good for sorting but bad for searching. Are there any alternatives?

Using Dictionaries Instead of Lists

Dictionaries are good for searching or looking up information. The protein and extinction data can be represented as a list of dictionaries:

```
table = [
    {'protein': 0.16, 'ext1': 0.038, 'ext2': 0.044, 'ext3': 0.040},
    {'protein': 0.33, 'ext1': 0.089, 'ext2': 0.095, 'ext3': 0.091},
    {'protein': 0.66, 'ext1': 0.184, 'ext2': 0.191, 'ext3': 0.191},
    {'protein': 1.00, 'ext1': 0.280, 'ext2': 0.292, 'ext3': 0.283},
    {'protein': 1.32, 'ext1': 0.365, 'ext2': 0.367, 'ext3': 0.365},
    {'protein': 1.66, 'ext1': 0.441, 'ext2': 0.443, 'ext3': 0.444}]
```

Source: Adapted from code published by A.Via/K.Rother under the Python License.

The outer list contains all rows, and a dictionary with the column labels contains the data for each row. Accessing cells becomes more readable this way:

```
cell = table[1]['ext2']
```

Dictionaries in Dictionaries

If you have unambiguous labels for the rows as well, a table can be represented by dictionaries in a dictionary. With Lowry's data, you would have to add an ID number for each row:

```
table = {
    'row1': {'protein': 0.16, 'ext1': 0.038, 'ext2': 0.044, 'ext3': 0.040},
    'row2': {'protein': 0.33, 'ext1': 0.089, 'ext2': 0.095, 'ext3': 0.091},
    'row3': {'protein': 0.66, 'ext1': 0.184, 'ext2': 0.191, 'ext3': 0.191},
    'row4': {'protein': 1.00, 'ext1': 0.280, 'ext2': 0.292, 'ext3': 0.283},
    'row5': {'protein': 1.32, 'ext1': 0.365, 'ext2': 0.367, 'ext3': 0.365},
    'row6': {'protein': 1.66, 'ext1': 0.441, 'ext2': 0.443, 'ext3': 0.444}
    }
```

With such a nested dictionary, looking up particular cells is fast and easy:

```
cell = table['row1']['ext2']
```

Alternatively, you could also do the following:

- *Put lists into a dictionary.* You still can search for a given item easily, but the data for each row are in a simpler format.

- *Have each column as a separate list.* Practically, the table is turned around by 90 degrees. This is good if you are doing calculations, e.g., calculating the average for several columns. However, with this representation, sorting the rows becomes impossible.

Q & A: When I Want to Both Search and Sort a Table, What Can I Do?

You can store your table as a nested list and create an extra dictionary that you can use for searching (see Chapter 5). A search dictionary used to speed up your program is also called an *index*. Using a list for the data and an extra dictionary with the same data for searching creates redundancy. As soon as your list changes, the dictionary gets out of date and needs to be updated. This is a potential source for very nasty program bugs. One possible way is to use a dedicated class for storing your data (see Chapter 11). Another way to use indices, especially if you have lots of data, is to use an SQL database.

Example 7.3 How to Convert Table Representations

All methods for storing tables, *nested lists*, *nested dictionaries*, and *mixed* (see Figure 7.3), have their particular advantages and

list in a list
`[`
`[1, 2],`
`[3, 4]`
`]`

dictionary in a list
`[`
`{'x':1, 'y':2},`
`{'x':3, 'y':4}`
`]`

	x	y
a	1	2
b	3	4

list in a dictionary
`{`
`'a':[1, 2],`
`'b':[3, 4]`
`}`

dictionary in a dictionary
`{`
`'a':{'x':1, 'y':2},`
`'b':{'x':3, 'y':4}`
`}`

FIGURE 7.3 All methods for storing tables: nested lists, nested dictionaries, and mixed.

disadvantages. Pros and cons of the different table representations are discussed in Box 7.1. Often, none of them will perfectly fit to everything you want to do. Then, it is most likely that you will need to convert one representation of the table into the other.

BOX 7.1   PROS AND CONS OF TABLE REPRESENTATIONS

Lists in Lists

- *Pros:* Adding and deleting rows to a table is easy. The list can be sorted with a single command.
- *Cons:* To find a certain protein by its name, you need to run a for loop over the entire table, which is slow. To address individual elements, you need to use numerical indices, which makes the code harder to read.

Dictionaries in Dictionaries

- *Pros:* Finding any entry in the table by a protein ID is easy and fast. The explicit labeling of cells by the column names makes the code easier to read.
- *Cons:* A dictionary is by definition unsorted, so it is not possible to sort the data in this representation.

Mixed Lists and Dictionaries

- *Pros:* This combines the advantages of both types. You can choose to use lists for the rows and dictionaries for the columns, or vice versa.
- *Cons:* Using the table becomes a little less straightforward, and you need to remember in which way to access rows and in which columns. The code will be a little harder to read.

Taken together, there is no clear answer to which type is better. Representing a table as lists in lists is good if you want to sort and edit the data a lot. Representing a table as dictionaries in dictionaries is good if you want to search the data and keep your code transparent.

To convert a nested list into a nested dictionary, you will find that a single for loop is sufficient. In each round of the loop, a new dictionary for each row is added to an overall dictionary that stores the rows. A counter is introduced to name the rows.

```
table = [
['protein', 'ext1', 'ext2', 'ext3'],
[0.16, 0.038, 0.044, 0.040],
[0.33, 0.089, 0.095, 0.091],
[0.66, 0.184, 0.191, 0.191],
[1.00, 0.280, 0.292, 0.283],
[1.32, 0.365, 0.367, 0.365],
[1.66, 0.441, 0.443, 0.444]
]

nested_dict = {}
n = 0
key = table[0]
# to include the header, run the for loop over
# ALL table elements (including the first one)
for row in table[1:]:
    n = n + 1
    entry = {key[0]: row[0], key[1]: row[1], key[2] : row[2],
    key[3] : row[3]}
    nested_dict['row'+str(n)] = entry

print nested_dict
```

Source: Adapted from code published by A.Via/K.Rother under the Python License.

Converting a list of dictionaries to a nested list can be done with a single for loop as well. The code looks similar to the previous example, only the indexing is replaced by the dictionary keys. In the example below, nested_dict is the dictionary created in the previous example.

```
nested_list = []
for entry in nested_dict:
    key = nested_dict[entry]
    nested_list.append([key['protein'],key['ext1'],
    key['ext2'], key['ext3']])

print nested_list
```

Source: Adapted from code published by A.Via/K.Rother under the Python License.

Example 7.4 How to Read Files with Tabular Data

Tables can be read from tab-separated text files using a simple pattern:

```
table = []
for line in open('lowry_data.txt'):
    table.append(line.strip().split('\t'))
```

Source: Adapted from code published by A.Via/K.Rother under the Python License.

The first line creates an empty table. The second line defines a `for` loop over all lines in the file with the given name. The `lowry_data.txt` input file contains tab-separated columns with the data shown in Table 7.1. This expression is an abbreviation for using first `open()`, then `readlines()`, then `for`. The third line first removes the new-line character from a single line of text from the file (using `strip()`), then dissects it into columns at tabulator symbols (using `split()`), and finally adds the row to the nested list. This piece of code is worth exercising, because it is useful in many programs.

Example 7.5 How to Write Files with Tabular Data

For writing a tab-separated `table` to a file from a nested list, a similar pattern exists (here, `table` is the nested list generated in Example 7.4):

```
out = ''
for row in table:
    line = [str(cell) for cell in row]
    out = out + '\t'.join(line) + '\n'
open('lowry_data.txt', 'w').write(out)
```

Source: Adapted from code published by A.Via/K.Rother under the Python License.

The first line creates an empty string where the contents of the output file are collected. The second line loops through the table. The third line converts all cells of a row to strings and puts them into a new list (this step can be omitted if they are strings already, or it can be changed if you need more sophisticated string formatting). The fourth line connects all cells of a row by tabulators (using `join()`) and adds a newline character (`'\n'`). Finally, the output string is written to a new file. This pattern occurs with modifications in many programs as well.

A remarkable number of scientific programs work by reading data from a tab-separated text file, doing some manipulation to the data, and storing the result in another tab-separated text file. Knowing these read–write patterns inside out is certainly a good investment. If you want to know more about how to write tab-separated or comma-separated files, also consider the Python `csv` module.

7.5 TESTING YOURSELF

Exercise 7.1

Add a row with average concentrations or extinctions to the table in the code example in Section 7.2.2 and print it.

Exercise 7.2

Convert the table from the code example in Section 7.2.2 to a list of dictionaries.

Exercise 7.3 Reading Matrices from Text Files

You have a similarity matrix of RNA bases:

	A	G	C	U
A	1.0	0.5	0.0	0.0
G	0.5	1.0	0.0	0.0
C	0.0	0.0	1.0	0.5
U	0.0	0.0	0.5	1.0

Write the matrix to a text file. Write a program that reads the matrix from the file to a table and prints it to the screen.

Exercise 7.4: Similarity of RNA Sequences

Write a program that calculates the similarity of the two RNA sequences:

```
AGCAUCUA
ACCGUUCU
```

Hint: To calculate the similarity, you need to extract similarity values from the matrix from Exercise 7.3. You will need a `for` loop that runs over both sequences simultaneously. This can be achieved using the instruction

```
for base1, base2 in zip(seq1, seq2):
```

Hint: The sequence similarity of the two sequences is the sum of all base–base similarities.

Exercise 7.5 Printing Table Columns and Rows Selectively

Write a program that prints the entire second row of the Lowry table (Table 7.1). Then print the entire protein concentration column. Do this for the nested list and for the nested dictionary you obtained in Exercise 7.2. What advantages and disadvantages do you observe in both approaches?

## CHAPTER 8

Sorting Data

L EARNING GOAL: You can sort your data according to a custom parameter or column.

8.1 IN THIS CHAPTER YOU WILL LEARN

- How to sort iterable objects (lists, tuples, dictionaries)

- How to sort tables

- How to sort in ascending and descending order

- How to sort tables by a given column

- How to sort the output of BLAST according to a given parameter

8.2 STORY: SORT A DATA TABLE

8.2.1 Problem Description

A common need in dealing with tables consists of sorting them. This
task can be accomplished using MS Excel or similar Office programs.
However, this becomes complicated if you have to manage big amounts
of data. Moreover, you may want to sort the table based on given conditions or to automatize the sorting procedure. Sorting data can be easily done using a Python script. The following Python session sorts an
input table by the values of its second column in descending order. To
do this, the built-in function `sorted()` is used in combination with the
`itemgetter` function, which is imported from the `operator` module.
A list comprehension (see Chapter 4, Section 4.3.3) is used to convert the

string elements of a list into floating numbers (at the beginning) and vice versa (at the end). The program works as follows: the input file containing a table of numbers is read to a Python table (a nested list of floats). Then the table is sorted, and the elements are converted back to strings. The elements belonging to the table columns are joined by a tabulator ('\t') and finally printed to the screen.

8.2.2 Example Python Session

```
from operator import itemgetter
# read table to a nested list of floats
table = []
for line in open("random_distribution.tsv"):
    columns = line.split()
    columns = [float(x) for x in columns]
    table.append(columns)

# sort the table by second column (index 1)
column = 1
table_sorted = sorted(table, key = itemgetter(column))

# format table as strings
for row in table_sorted:
    row = [str(x) for x in row]
    print "\t".join(row)
```

Source: Adapted from code published by A.Via/K.Rother under the Python License.

8.3 WHAT DO THE COMMANDS MEAN?

Most of the code lines in Section 8.2.2 are needed for opening a file, reading, converting strings into floating-point numbers and saving them into a list, and then converting floating-point numbers back into strings and printing them. Only one row is related to sorting. The sorted() function appears in the following instruction:

```
table = sorted(table, key = itemgetter(column))
```

Here table is a list of lists and sorted() is a built-in function. This chapter will focus on that single line and possible variations.

8.3.1 Python Lists Are Good for Sorting

In the program, table is a nested list that has been initialized to an empty list and then filled with floating-point numbers extracted from

space	8	P	h
!	9	Q	i
"	:	R	j
#	;	S	k
$	<	T	l
%	=	U	m
&	>	V	n
' (apostrophe)	?	W	o
(	@	X	p
)	A	Y	q
*	B	Z	r
+	C	[	s
, (comma)	D	\	t
- (dash)	E	]	u
. (period)	F	^	v
/	G	_ (underline)	w
0	H	` (ticmark)	x
1	I	a	y
2	J	b	z
3	K	c	}
4	L	d	|
5	M	e	{
6	N	f	~
7	O	g	DEL

FIGURE 8.1   ASCII sort order chart. *Note:* Notice that numbers precede the alphabet characters and that uppercase characters precede lowercase ones.

a tab-separated input file (random_distribution.tsv, a portion of which is shown in Figure 13.1). The conversion to floating-point numbers is necessary if you want to sort the table based on numerical values. Strings are sorted based on the order of the ASCII chart (see Figure 8.1).

In Python, there are two techniques to sort data: the sort() method for sorting lists, and the built-in function sorted(), which can sort any iterable data structure. The sort() method modifies the list in-place, whereas sorted() builds a new sorted list from any iterable (lists, tuples, dictionary keys).

To sort a list of numbers or strings, you can use the sort() method of lists: it sorts the contents of the list. Notice that, as mentioned earlier, this method does not return a new list but modifies the list in-place. This is why the sort() method returns the value None. The default behavior of the sort() method is to sort in ascending order, i.e., from the lowest value to the highest value of the list:

```
>>> data = [1, 5, 7, 8, 9, 2, 3, 6, 6, 10]
>>> data.sort()
>>> data
[1, 2, 3, 5, 6, 6, 7, 8, 9, 10]
```

If you want to sort in a descending order, you may first sort in ascending order and then reverse the list:

```
>>> data.reverse()
>>> data
[10, 9, 8, 7, 6, 6, 5, 3, 2, 1]
```

Alternatively, you may pass an optional comparison function as an argument to the sort() method. The default comparison function cmp(a,b) takes two arguments a,b to be compared and returns a negative value if a < b, zero if a = b, or a positive value if a > b. The function can be customized to return negative values, zero, or positive values depending on different conditions. If you are interested in the built-in function cmp() and how to customize it, look at Box 8.1.

BOX 8.1   THE cmp() BUILT-IN FUNCTION

The cmp(a,b) built-in function compares two objects a and b and returns an integer depending on the outcome. In particular, it returns –1 if a < b, 0 if a = b, and 1 if a > b. The sort() method of lists compares all pairs of elements of a list and implicitly uses the cmp() function to determine which is the smaller or bigger one. If you pass a modified cmp() function to sort() (e.g., my_cmp(a,b)), where you invert the returned values (i.e., –1 if a > b and 1 if a < b), the list will be sorted in reverse order. This function can be further modified in order to obtain a more fine-tuned sorting. For example, you may sort by the first column of a table and, when two or more values in the first column are equal, by the second column.

For example, if you want to sort a list in descending order, you can write

```
>>> def my_cmp(a,b):
...     if a > b: return -1
...     if a == b: return 0
...     if a < b: return 1
...
>>> L = [1, 2, 3, 4, 5, 6, 8, 8, 9, 9, 30]
>>> L.sort(my_cmp)
>>> L
[30, 9, 9, 8, 8, 6, 5, 4, 3, 2, 1]
```

If you have a table and you want to sort it by the first column, then by the second, then by the third, you can read the table and put it into a list of lists (see Chapter 7) and then sort it using the sort() function in combination with the following customized cmp() function:

```
def my_cmp(a,b):
        if a[x] < b[x]: return 1
        if a[x] == b[x]:
                if a[x + 1] < b[x + 1]: return 1
                if a[x + 1] == b[x + 1]:
                        if a[x + 2] < b[x + 2]: return 1
                        if a[x + 2] == b[x + 2]: return 0
                        if a[x + 2] > b[x + 2]: return -1
                if a[x + 1] > b[x + 1]: return -1
        if a[x] > b[x]: return -1
```

Notice that in Section 8.2.2, `table` is not a simple list but a list of lists. Therefore, the `sort()` method has to sort a list of lists. By default, the sorting is carried out based on the first element of each list:

```
>>> data = [[1, 2], [4, 2], [9, 1], [2, 7]]
>>> data.sort()
>>> data
[[1, 2], [2, 7], [4, 2], [9, 1]]
```

But if you want to sort by the second element, you have to use a different approach. Either you customize the `sort()` method as described in Box 8.1 or you use the `sorted()` built-in function. In the Python session in Section 8.2.2, the nested list is sorted according to the second element of each sublist (`column = 1`).

8.3.2 The `sorted()` Built-in Function

Alternatively to the `sort()` method of lists, you can use the `sorted()` built-in function to sort data. The advantage of `sorted()` is that it can sort many kinds of data, such as lists, tuples, or dictionary keys, whereas the method `sort()` only applies to lists. The `sorted()` built-in function returns a new sorted list from any iterable:

```
>>> data = [1, 5, 7, 8, 9, 2, 3, 6, 6, 10]
>>> newdata = sorted(data)
>>> newdata
[1, 2, 3, 5, 6, 6, 7, 8, 9, 10]
```

8.3.3 Sorting with `itemgetter`

Importantly, the `sorted()` built-in function can be used to sort by custom parameters (e.g., the values of a given column in a table). This result can be achieved using the `itemgetter` function from the `operator` module, as shown in Section 8.2.2. `operator.itemgetter(i)(T)` returns the ith element of T, which can be a string, a list, a tuple, or a dictionary. In the case of a dictionary, it returns the value associated to key i. If you use two or more indices, the function returns a tuple:

```
>>> from operator import itemgetter
>>> data = ['ACCTGGCCA', 'ACTG', 'TACGGCAGGAGACG', 'TTGGATC']
>>> itemgetter(1)(data)
'ACTG'
>>> itemgetter(1, -1)(data)
('ACTG', 'TTGGATC')
```

In Section 8.2.2, `itemgetter` is used to sort `table` by the second column (column index 1).

If you want to sort `table` first by the second column and then, e.g., by the fourth, you can write the column indices (1 and 3 in this case) into the `itemgetter()` function:

```
new_table = sorted(table, key = itemgetter(1, 3))
```

Q & A: WHAT DO THE DOUBLE PARENTHESES ()() MEAN?

The `itemgetter` function returns a function. The first pair of parentheses is used to call `itemgetter`, which creates an intermediate function for extracting columns. The second pair calls that function on `data` to produce the actual column(s). Using intermediate functions allows one to access data in tables in a very flexible way.

8.3.4 Sorting in Ascending/Descending Order

To sort in descending order, the additional argument `reverse = True` can be passed to the `sorted()` function:

```
>>> sorted(data, reverse = True)
[30, 9, 9, 8, 8, 6, 5, 4, 3, 2, 1]
```

And in Section 8.2.2, you can do as follows:

```
table = sorted(table, key = itemgetter(1), reverse = True)
```

8.3.5 Sorting Data Structures (Tuples, Dictionaries)

Sort a Dictionary According to Its Keys Passing through a List

To sort a dictionary, you can extract all keys into a list and sort that list:

```
data = {1: 'a', 2: 'b', 4: 'd', 3: 'c',
  5: 't', 6: 'm', 36: 'z'}
keys = list(data)
keys.sort()
for key in keys:
    print key, data[key]
```

Source: Adapted from code published by A.Via/K.Rother under the Python License.

The output of this program is:

```
1 a
2 b
3 c
4 d
5 t
6 m
36 z
```

Sort a Dictionary Using the `sorted()` *Built-in Function*

If the sorted keys are used only once, the `sorted()` function is shorter to write:

```
for key in sorted(data):
    print key, data[key]
```

Source: Adapted from code published by A.Via/K.Rother under the Python License.

Sort a Tuple Passing through a List

Tuples are immutable and therefore cannot be sorted themselves. To sort a tuple, you need to convert it to a list, sort the list, and convert the list back to a tuple:

```
data = (1, 4, 5, 3, 8, 9, 2, 6, 8, 9, 30)
list_data = list(data)
list_data.sort()
new_tup = tuple(list_data)
print new_tup
```

Source: Adapted from code published by A.Via/K.Rother under the Python License.

Sort a List Using the `sorted()` *Built-in Function*
Again, the function `sorted()` provides a shortcut:

```
data = (1, 4, 5, 3, 8, 9, 2, 6, 8, 9, 30)
new_tup = tuple(sorted(data))
print new_tup
```

Notice that tables can also be sorted using the UNIX/Linux `sort` command (see Box 8.2).

BOX 8.2   SORT FILES FROM THE COMMAND SHELL

A table saved in a text file can be sorted using the `sort` UNIX/Linux command in the shell terminal.

Sort in Alphabetical Order

```
%sort myfile.txt
```

Sort in Numerical Order

```
%sort -n myfile.txt
```

Sort on Multiple Columns

```
%sort myfile.txt -k2n -k1
```

This will first sort on column 2 (in numerical order: –k2n) then on column 1 in alphabetical order.

Sort a Comma-Separated Table

```
%sort -k2 -k3, -k1 -t ',' myfile.txt
```

Sort in Reversed Order

```
%sort -r myfile.txt
```

The command

```
%sort myfile.txt -nrk 2 -st '|'
```

will sort a pipe-separated file in reversed order by the second column.

8.3.6 Sorting Strings by Their Length

You can use the `sorted()` built-in function with a lambda function (Chapter 10, Box 10.6) as a custom parameter instead of `itemgetter`. For instance, you can sort a list of strings by their length. To do so, you need to provide a function with a single argument returning the "key" to be used for sorting as a parameter to `sorted()`:

```
>>> data = ['ACCTGGCCA', 'ACTG', 'TACGGCAGGAGACG', 'TTGGATC']
>>> bylength = sorted(data, key = lambda x: len(x))
>>> bylength
['ACTG', 'TTGGATC', 'ACCTGGCCA', 'TACGGCAGGAGACG']
```

In this example, the key is a very short function returning the length of its argument x. It is defined by the lambda keyword (Chapter 10, Box 10.6). The x variable takes the values of the elements of the list `data`. When x = 'ACTG', the lambda function returns 4 (the result of `len('ACTG')`) and so on for each string element of the list. Finally, the list is sorted in ascending order from the shortest sequence to the longest.

This example can be applied to any set of biological sequences. If you store a big number of sequences in a list (e.g., after reading them from a FASTA file) with this code, you can sort your sequences from the shortest one to the longest one (or vice versa).

If you have a table in the form of a nested list, you can use the key argument to specify the column by which you want to sort your table. The sorting instruction in Section 8.2.2, with a lambda function as argument for the `sorted()` function (replacing `itemgetter`), would read

```
table = sorted(table, key = lambda col: col[1])
```

8.4 EXAMPLES

Example 8.1 Sort a Table by the First Column, Then by the Second, Then by the Third, and So On

Here, the table in Section 8.2.2 is sorted by the first column to the last column:

```
from operator import itemgetter
# read table
in_file = open("random_distribution.tsv")
table = []
```

```
for line in in_file:
    columns = line.split()
    columns = [float(x) for x in columns]
    table.append(columns)
table_sorted = sorted(table, key=itemgetter(0, 1, 2, 3, 4,\
    5, 6))
print table_sorted
```

Source: Adapted from code published by A.Via/K.Rother under the Python License.

table (a nested list) has seven columns, and if you want to sort in reversed order, you have to add the reverse = True argument.

Example 8.2 Sort the Output of BLAST According to a Parameter of Your Choice (e.g., Sequence Identity Percentage)

This example uses the input file shown in Figure 8.2. The figure note explains how this file can be obtained by running BLAST locally. In this example, the BLAST output is sorted in descending order by the third column (col[2]), which contains the sequence identity percentage in the form of a floating-point number.

```
from operator import itemgetter
input_file = open("BlastOut.csv")
output_file = open("BlastOutSorted.csv","w")
# read BLAST output table
table = []
for line in input_file:
    col = line.split(',')
    col[2] = float(col[2])
    table.append(col)
table_sorted = sorted(table, key = itemgetter(2), \
    reverse = True)
# write sorted table to an output file
for row in table_sorted:
    row = [str(x) for x in row]
    output_file.write("\t".join(row) + '\n')
output_file.close()
```

Source: Adapted from code published by A.Via/K.Rother under the Python License.

```
sp|060218|AK1BA_HUMAN,gi|223468663|ref|NP_064695.3|,100.00,316,0,0,1,316,1,316,0.0,654
sp|060218|AK1BA_HUMAN,gi|119388973|pdb|1ZUA|X,100.00,316,0,0,1,316,2,317,0.0,654
sp|060218|AK1BA_HUMAN,gi|3150035|gb|AAC17469.1|,99.68,316,1,0,1,316,1,316,0.0,653
sp|060218|AK1BA_HUMAN,gi|30584339|gb|AAP36418.1|,99.68,316,1,0,1,316,1,316,0.0,652
sp|060218|AK1BA_HUMAN,gi|60832697|gb|AAX37021.1|,99.68,316,1,0,1,316,1,316,0.0,652
sp|060218|AK1BA_HUMAN,gi|114616054|ref|XP_001140450.1|,99.05,316,3,0,1,316,1,316,0.0,649
sp|060218|AK1BA_HUMAN,gi|297681560|ref|XP_002818524.1|,99.05,316,3,0,1,316,1,316,0.0,649
sp|060218|AK1BA_HUMAN,gi|27436418|gb|AAO13380.1|,98.73,316,4,0,1,316,1,316,0.0,645
sp|060218|AK1BA_HUMAN,gi|383413321|gb|AFH29874.1|,97.47,316,8,0,1,316,1,316,0.0,640
sp|060218|AK1BA_HUMAN,gi|384943758|gb|AFI35484.1|,97.47,316,8,0,1,316,1,316,0.0,638
sp|060218|AK1BA_HUMAN,gi|109068267|ref|XP_001100959.1|,97.15,316,9,0,1,316,1,316,0.0,638
sp|060218|AK1BA_HUMAN,gi|109068279|ref|XP_001102064.1|,96.20,316,12,0,1,316,1,316,0.0,637
sp|060218|AK1BA_HUMAN,gi|332224512|ref|XP_003261411.1|,97.15,316,9,0,1,316,1,316,0.0,635
sp|060218|AK1BA_HUMAN,gi|402913955|ref|XP_003919409.1|,95.89,316,13,0,1,316,1,316,0.0,633
sp|060218|AK1BA_HUMAN,gi|402864885|ref|XP_003896672.1|,96.20,316,12,0,1,316,1,316,0.0,632
sp|060218|AK1BA_HUMAN,gi|380790225|gb|AFE66988.1|,95.89,316,13,0,1,316,1,316,0.0,632
sp|060218|AK1BA_HUMAN,gi|109068275|ref|XP_001101597.1|,95.25,316,15,0,1,316,1,316,0.0,629
sp|060218|AK1BA_HUMAN,gi|397484839|ref|XP_003813574.1|,90.46,346,3,2,1,316,1,346,0.0,629
sp|060218|AK1BA_HUMAN,gi|109068273|ref|XP_001101418.1|,94.62,316,17,0,1,316,1,316,0.0,625
sp|060218|AK1BA_HUMAN,gi|402913957|ref|XP_003919410.1|,94.94,316,16,0,1,316,1,316,0.0,622
sp|060218|AK1BA_HUMAN,gi|296210580|ref|XP_002752014.1|,93.35,316,21,0,1,316,1,316,0.0,619
sp|060218|AK1BA_HUMAN,gi|109068285|ref|XP_001102522.1|,93.67,316,20,0,1,316,1,316,0.0,612
```

FIGURE 8.2 Portion of a BLAST output (BlastOut.csv) obtained with the following command line (see Recipe 11):

```
%blastp -db nr -query AK1BA_HUMAN.fasta -outfmt 10 -out BlastOut.csv
```

where -outfmt 10 is the option to generate a comma-separated output. Values in the output separated by the comma are (where q = query, s = subject): qID, sID, seqID, alignLength, mismatches, gapopen, qStart, qEnd, sStart, sEnd, e-value, Bitscore.

```
PDB ID,Chain ID,Exp. Method,Resolution,Chain Length
"1A4F","A","X-RAY DIFFRACTION","2.00","141"
"1C7C","A","X-RAY DIFFRACTION","1.80","283"
"1CG5","A","X-RAY DIFFRACTION","1.60","141"
"1FAW","A","X-RAY DIFFRACTION","3.09","141"
"1HDA","A","X-RAY DIFFRACTION","2.20","141"
"1IRD","A","X-RAY DIFFRACTION","1.25","141"
"1KFR","A","X-RAY DIFFRACTION","1.85","147"
"1QPW","A","X-RAY DIFFRACTION","1.80","141"
"1SPG","A","X-RAY DIFFRACTION","1.95","144"
"1UX8","A","X-RAY DIFFRACTION","2.15","132"
"1WXR","A","X-RAY DIFFRACTION","2.20","1048"
"2AA1","A","X-RAY DIFFRACTION","1.80","143"
"2D5X","A","X-RAY DIFFRACTION","1.45","141"
"2DHB","A","X-RAY DIFFRACTION","2.80","141"
"2H8F","A","X-RAY DIFFRACTION","1.30","143"
"2IG3","A","X-RAY DIFFRACTION","2.15","127"
"2LHB","A","X-RAY DIFFRACTION","2.00","149"
"2QMB","A","X-RAY DIFFRACTION","2.80","142"
"2W72","A","X-RAY DIFFRACTION","1.07","141"
"2WY4","A","X-RAY DIFFRACTION","1.35","140"
"3AEH","A","X-RAY DIFFRACTION","2.00","308"
"3ARL","A","X-RAY DIFFRACTION","1.81","152"
"3CY5","A","X-RAY DIFFRACTION","2.00","141"
"3D4X","A","X-RAY DIFFRACTION","2.20","141"
"3EOK","A","X-RAY DIFFRACTION","2.10","141"
"3MJU","A","X-RAY DIFFRACTION","3.50","141"
"3VRG","A","X-RAY DIFFRACTION","1.50","141"
"4ESA","A","X-RAY DIFFRACTION","1.45","143"
"4HBI","A","X-RAY DIFFRACTION","1.60","146"
```

FIGURE 8.3   PDB tabular report for chains A of hemoglobin. *Note:* The report was obtained from the RCSB advanced search (www.rcsb.org/pdb/search/advSearch.do) with the *hemoglobin* keyword by removing entries with 100% sequence identity. The chosen options for the tabular report are chain, experimental method, resolution, chain length.

Example 8.3 Sort Hemoglobin PDB Entries on the Basis of Their RMSD (from a RCSB Report)

This example uses the input file shown in Figure 8.3. See the figure note to know how this input file can be obtained from the RCSB resource (http://www.rcsb.org/). Actually, in this example, sorting is carried out first by RMSD (fourth column of the input table), then by the sequence length of the protein (fifth column of the input table).

```
from operator import itemgetter
input_file = open("PDBhaemoglobinReport.csv")
output_file = open("PDBhaemoglobinSorted.csv","w")
```

```
table = []
header = input_file.readline()
for line in input_file:
    col = line.split(',')
    col[3] = float(col[3][1:-1])
    col[4] = int(col[4][1:-2])
    table.append(col)

table_sorted = sorted(table, key = itemgetter(3, 4))

output_file.write(header + '\n')
for row in table_sorted:
    row = [str(x) for x in row]
    output_file.write('\t'.join(row) + '\n')
output_file.close()
```

Source: Adapted from code published by A.Via/K.Rother under the Python License.

Notice that you only need to convert `col[3]` and `col[4]` into numbers, therefore the corresponding variables have been simply reassigned to their numerical values after removing the double quotes and the newline character for the last column.

What if you wanted to sort by column 3 in ascending order and by column 4 in descending order? In this case, you can help yourself by inverting the sign of column 4 before sorting:

```
col[4] = -col[4]
```

Alternatively, a customized `cmp()` function may turn out to be useful (see Box 8.1).

8.5 TESTING YOURSELF

Exercise 8.1 Sorting a Table by Its Second Column

Write a program that reads the table with Lowry data (see Table 7.2) from a text file, sorts it by the second column, and writes the first three rows of the sorted table to a new file.

Exercise 8.2 Sorting by Sequence Length

Sort a multiple sequence FASTA file by the sequence length (from the longest to the shortest).

Hint: You first have to parse the file as you learned in Chapter 4 and create a list of lists, each line containing three elements (header, sequence, sequence length). Then you can sort the list according to the third element of the sublists and finally write the sorted list to a file.

Exercise 8.3 Sorting from Excel Files

Choose one of your Excel files, save it as a comma-separated text file, and use Python to sort it in ascending order going from the last column to the first.

Hint: Pay attention to the presence of a header line.

Exercise 8.4 Sorting FASTA Sequence Records in Alphabetical Order

Read a multiple sequence FASTA file and store its content in a dictionary {ac_number:sequence}. Use the AC numbers from the headers as dictionary keys. Sort the dictionary keys in alphabetical order and print the key:value pairs.

Exercise 8.5 Sort a BLAST Output by E-value in Ascending Order

Use the .csv output of BLAST as described in Example 8.2.

## CHAPTER 9

Pattern Matching and Text Mining

L EARNING GOAL: You can find patterns in sequences and natural language texts.

9.1 IN THIS CHAPTER YOU WILL LEARN

- How to express a consensus sequence using a regular expression

- How to use Python regular expression tools to search substrings in strings

- How to search for the occurrence of a functional motif in a protein sequence

- How to search a given word or a set of words in a text (e.g., scientific abstracts)

- How to identify motifs in nucleotide sequences (e.g., transcription factor or miRNA binding sites)

9.2 STORY: SEARCH A PHOSPHORYLATION MOTIF IN A PROTEIN SEQUENCE

9.2.1 Problem Description

A sequence functional motif is defined as a short amino acid or nucleotide sequence containing one or more residues involved in a function.
Phosphorylation sites, mannosylation sites, recognition motifs,

glycosylation sites, transcription binding sites, etc. are examples of functional motifs. Sequence functional motifs can be represented using a special notation called *regular expressions*. A regular expression, sometimes called *regexp*, is a string syntax composed of characters and metacharacters that represent *sets* of strings. In other words, if you want to represent several strings with a single expression, it is necessary to introduce rules and symbols that allow "meta" meanings like wildcards, repeated characters, and logical groups. An example often used in biology is the character *N* in DNA sequences. The sequence AGNNT could be AGAAT, AGCTT, AGGGT, or one of many other alternatives. Regular expressions work in a similar way but use a more complex set of special characters.

Suppose you want to represent, with a single expression, the following peptide strings: "AFL," "GFI," "AYI," "GWI," "GFI," "AWI," "GWL," and "GYL." If you use a symbolic expression such as "[AG]" to indicate that in a position of a string you might find either "A" or "G," you can use the expression "[AG][FYW][IL]" to represent all of the peptides above.

Notice that we are using not the literal meaning of "[" and "]" but a "meta" meaning. In this case, "[" and "]" are called *metacharacters*, and the expression encoding a set of strings, through the use of characters and metacharacters, is called a *regular expression*.

Another example is represented by functional motifs, which are usually short and may contain invariable and variable positions. For instance, you can represent a Ser/Thr phophorylation motif as [ST]Q. This expression, when searched in a protein sequence, will have a hit in correspondence of two different subsequences: "SQ" and "TQ," i.e., either a serine or a threonine followed by a glutamine. The first position of the motif is variable, while the second position is conserved. There are several publicly available resources dedicated to functional motifs (e.g., ELM: http://elm. eu.org; PROSITE: http://prosite.expasy.org/; etc.). Searching occurrences of a functional motif in a protein sequence, or in a data set of protein sequences, can be used as a procedure to infer the function of proteins. This is exactly what tools such as ScanProsite (http://prosite.expasy.org/ scanprosite/) do.

The following program simulates one of the functions of ScanProsite; i.e., the program searches for a phosphorylation motif in a protein sequence, and returns the first occurrence of the motif.

9.2.2 Example Python Session

```
import re
seq = 'VSVLTMFRYAGWLDRLYMLVGTQLAAIIHGVALPLMMLI'
pattern = re.compile('[ST]Q')
match = pattern.search(seq)
if match:
    print '%10s' % (seq[match.start() - 4: match.end() + 4])
    print '%6s' % match.group()
else:
    print "no match"
```

Source: Adapted from code published by A.Via/K.Rother under the Python License.

9.3 WHAT DO THE COMMANDS MEAN?

In the first line of the program, a module called `re` is imported. The `re` module provides metacharacters, rules, and functionalities to write and interpret regular expressions and match them to string variables. You can find an exhaustive tutorial on Python regular expressions by A.M. Kuchling at http://docs.activestate.com/activepython/2.5/python/regex/regex.html.

In the line

```
pattern = re.compile('[ST]Q')
```

the phosphorylation motif to be searched, in the form of the string `'[ST]Q'`, is converted into a new object by the `compile()` method of the `re` module. The conversion is mandatory, otherwise characters such as `"["` would be interpreted as a simple square bracket and not as a metacharacter with a precise meaning. The `re.compile()` function returns a *regular expression object*. The `re` module provides methods to handle regular expression objects.

9.3.1 Compiling Regular Expressions

`compile()` is the method to compile a string and convert it into a regular expression object (the `RegexpObject`).

```
>>> import re
>>> regexp = re.compile('[ST]Q')
>>> regexp
<_sre.SRE_Pattern object at 0x22de0>
```

The string can also be recorded in a variable:

```
>>> motif = '[ST]Q'
>>> regexp = re.compile(motif)
```

It is possible to pass arguments (compilation flags) to the `compile()` method, thus modifying some aspects of how regular expressions work. For instance, you can ignore uppercase and lowercase by

```
>>> regexp = re.compile(motif, re.IGNORECASE)
```

See Appendix A, Section A.2.17, subsection "Regular Expression Compilation Flags."

In the Python session in Section 9.2.2, the compiled pattern is searched in the sequence stored in the variable `seq` and printed to the screen using a number of methods: `search()`, `group()`, `start()`, and `end()`. We will discuss now what these methods do and what they return.

9.3.2 Pattern Matching

Once your regular expression is compiled, and you have a `RegexpObject`, you can search for its matches in a string using `RegexpObject` methods.

`RegexpObject` methods return `Match objects`, the content of which can be extracted using `Match object` methods. See Appendix A, Section A.2.17, subsection "re Module Methods." This is not conceptually different from file reading: when you want to access the content of a file, you first have to open the file, thus creating a "file object," then you have to use methods of file objects (e.g., `read()`, `readline()`, etc.) to read the content of the file.

The `search()` function scans a string, looking for a location where the regular expression matches for the first time. This means that the `search()` method will return at most one single match object per sequence. We use the `group()` method of match objects to print the first match:

```
>>> motif = 'R.[ST][^P]'
>>> regexp = re.compile(motif)
>>> print regexp
<_sre.SRE_Pattern object at 0x57b00>
>>> seq = 'RQSAMGSNKSKPKDASQRRRSLEPAENVHGAGGGAFPASQRPSKP'
>>> match = regexp.search(seq)
>>> match
<_sre.SRE_Match object at 0x706e8>
>>> match.group()
'RQSA'
```

Source: Adapted from code published by A.Via/K.Rother under the Python License.

The regular expression 'R.[ST][^P]' will match a substring with an arginine (R) in the first position, any amino acid in the second position (.), either a serine or a threonine in the third position ([ST]), and any amino acid but a proline in the last position ([^P]). Notice that the search() method returns not the matching substring directly but a Match object, which encodes the matching substring and its start and end positions along the sequence. This information can be retrieved using the following methods of a match object:

- match.group() returns the matching substring.
- match.span()   returns a tuple containing the (start, end) positions of the match.
- match.start()   returns the start position of the match.
- match.end()   returns the end position of the match.

Should you only be interested in finding the regular expression match starting at the first position of a sequence, you can use the method match().

Here, we propose the previous example, using match() instead of search():

```
>>> match1 = regexp.match(seq)
>>> match1
<_sre.SRE_Match object at 0x70020>
>>> match1.group()
'RQSA'
```

Notice that the match and match1 variables have identical values in this specific case.

In summary, both the search() and the match() methods return a Match object, which can be assigned to a variable in order to use its content through the Match object methods group(), span(), start(), and end():

```
>>> match1.span()
(0, 4)
>>> match1.start()
0
>>> match1.end()
4
```

Notice that UNIX/Linux provides a command to search for regular expressions matches in files (see Box 9.2).

Q & A: WHAT IF I WANT TO FIND ALL MATCHES OF A REGULAR EXPRESSION IN A STRING AND NOT ONLY THE FIRST ONE?

The re module offers two methods for this purpose: findall(), which returns a list containing all the matching substrings, and finditer(), which finds all the Match objects corresponding to the regular expression matches and returns them in the form of an *iterator*. More generally, an iterator is a "container" of objects that can be traversed in Python using a for loop. In this specific case, the iterator contains a set of Match objects, which can be individually accessed using Match object methods, such as group(), span(), start(), and end():

```
>>> all = regexp.findall(seq)
>>> all
['RQSA', 'RRSL', 'RPSK']
>>> iter = regexp.finditer(seq)
>>> iter
<callable-iterator object at 0x786d0>
>>> for s in iter:
...     print s.group()
...     print s.span()
...     print s.start()
...     print s.end()
...
RQSA
(0, 4)
0
4
RRSL
(18, 22)
18
22
RPSK
(40, 44)
40
44
```

9.3.3 Grouping

It is possible to divide a regular expression in subgroups, each matching a different component of interest. Suppose that, in the previous example, you wanted to know what amino acid type is matched by the ".". We can create a group delimiting the "." with round brackets and then get the matching amino acid type using the group() method, as follows:

```
import re
seq = 'QSAMGSNKSKPKDASQRRRSLEPAENVHGAGGGAFPASQRPSKP'
pattern1 = re.compile('R(.)[ST][^P]')
match1 = pattern1.search(seq)
print match1.group()
print match1.group(1)
pattern2 = re.compile('R(.{0,3})[ST][^P]')
match2 = pattern2.search(seq)
print match2.group()
print match2.group(1)
```

Source: Adapted from code published by A.Via/K.Rother under the Python License.

The output of the program is:

```
RRSL
R
RRRSL
RR
```

The group() method with no argument or the argument equal to 0 always returns the complete matching substring, whereas subgroups are numbered from left to right in increasing order (starting with 1).

Notice that subgroups could be nested, and to know the corresponding number, you have to count the number of open round brackets from left to right.

```
>>> p = re.compile('(a(b)c)d')
>>> m = p.match('abcd')
>>> m.group(0)
'abcd'
>>> m.group(1)
'abc'
>>> m.group(2)
'b'
```

You can also pass multiple arguments to the group() method. In this case, it will return a tuple containing the values for the corresponding groups:

```
>>> m.group(2, 1, 2)
('b', 'abc', 'b')
```

Finally, the groups() method returns a tuple with the substrings corresponding to all subgroups:

```
>>> m.groups()
('abc', 'b')
```

It is also possible to assign a name to each subgroup in order to selectively retrieve its content. For example, you can label a first group of a regular expression with the name w1 and a second group with w2 and later retrieve the identity of the match of each group using the group() function with the group name (w1 or w2) as argument:

```
>>> pattern = 'R(?P<w1>.{0,3})[ST](?P<w2>[^P])'
>>> regexp = re.compile(pattern)
>>> m1 = regexp.search(seq)
>>> m1.group('w1')
'RR'
>>> m1.group('w2')
'L'
```

The group label must be put between < and > symbols (i.e., <name>) and inserted in the round brackets of the group preceded by ?P (i.e., (?P<name>...)). A group is selectively accessible passing the label to the group() function as an argument (group(<name>)).

9.3.4 Modifying Strings

The re module also provides three methods that allow modifying strings: split(s), sub(r, s, [c]), and subn(r, s, [c]).

The method split(s) splits the string s at the matches of a regular expression. In the following example, a string will be split at all "|" symbols. Notice that the character "|" is also a metacharacter in the regular expression syntax (see Box 9.1). To tell Python to interpret it as a normal character, you have to put a backslash ("\") before the metacharacter. This is a general rule to make Python distinguish metacharacters from normal characters.

```
import re
separator = re.compile('\|')
annotation = 'ATOM:CA|RES:ALA|CHAIN:B|NUMRES:166'
columns = separator.split(annotation)
print columns
```

Source: Adapted from code published by A.Via/K.Rother under the Python License.

BOX 9.1 CHARACTERS AND METACHARACTERS

Not every character in a regular expression is a metacharacter. The regular expression metacharacters are

[] ^ $ \ . | * + ? { } ( )

[]

Square brackets are used to indicate a class of characters. For example, if you search a match of [abc] in a string s, you will find it if s contains 'a', 'b', or 'c'.

In particular, [a-z] matches the class of alphabet characters from a to z, whereas [0-9] matches the integers between 0 and 9.

^

[^a] indicates the complement of a, i.e., every character different from a. ^a (not enclosed in square brackets) indicates that a match exists in s only if a is in the first position of s.

$

a $ indicates that a match exists in s only if a is in the last position of s.

\

The meaning of \ depends on whether \ is followed by a metacharacter or a character. In the first case, it "protects" the metacharacter by restoring its literal meaning; in the second case, its meaning depends on the character that follows.

- \d corresponds to [0-9];
- \D corresponds to [^0-9];
- \s corresponds to [\t\n\r\f\v], i.e., any whitespace character
- \S corresponds to [^\t\n\r\f\v], i.e., any character that is not a whitespace
- \w corresponds to [a-zA-Z0-9], i.e., any alphanumeric character
- \W corresponds to [^a-zA-Z0-9], i.e., any character that is not alphanumeric.

.

This corresponds to any character except the newline character.

|

This is the OR operator. If placed between two regular expressions, matches will be searched either with the regexp on its left or with the one on its right.

( )

Round brackets are used to create subgroups in a regular expression.

Repetitions: * + ? { }

These metacharacters are used to find a match with repeated things.

*	The preceding character can be matched zero or more times. a*bc will match "bc," "abc," "aabc," "aaabc," "aaaabc," etc.
+	The preceding character can be matched one or more times. a+bc will match "abc," "aabc," "aaabc," "aaaabc," etc. but not "bc."
?	The preceding character can be matched zero times or once. can-?can will match both "can-can" and "cancan."
{m,n}	This qualifier means that at least *m* and at most *n* repetitions of the preceding character will be matched.

BOX 9.2 GREP: THE UNIX/LINUX COMMAND TO SEARCH WORDS IN FILES

grep is the UNIX/Linux command for searching text files for lines matching a regular expression.

grep ArticleTitle PMID.html

will return all the lines of the PMID.html text file matching the word ArticleTitle.

You can use the asterisk to indicate any character:

grep Ar*le mytext.txt

will return all the mytext.txt lines having at least a word starting with Ar and ending with le.

You can use more complicated regular expressions. For example,

grep ^'>' 3G5U.fasta

will return all the lines of the file 3GU.fasta starting with the character '>'. In this case you have to use quotation marks because '>' is a metacharacter in UNIX/Linux (it is the redirection character).

This code will produce a list with the split elements from the `annotation` string:

```
['ATOM:CA', 'RES:ALA', 'CHAIN:B', 'NUMRES:166']
```

The `RegexpObject` method `sub(r, s, [c])` returns a new string where nonoverlapping occurrences of a given pattern in the s string are all replaced with the value of `r` (if the optional argument c is not specified). In the following example, the pattern is '\|' (encoded in the separator `RegexpObject`), and it will be replaced by '@' in the s string:

```
import re
separator = re.compile('\|')
annotation = 'ATOM:CA|RES:ALA|CHAIN:B|NUMRES:166'
new_annotation = separator.sub('@', annotation)
print new_annotation
```

Source: Adapted from code published by A.Via/K.Rother under the Python License.

This results in:

```
ATOM:CA@RES:ALA@CHAIN:B@NUMRES:166
```

If the pattern is not found in the s string, s is returned unchanged. The optional argument c is the maximum number of pattern occurrences to be replaced; c must be a nonnegative integer. For example, set c to 2 in the previous example:

```
new_annotation = separator.sub('@', annotation, 2)
print new_annotation
```

Only the first two separators are replaced:

```
ATOM:CA@RES:ALA@CHAIN:B|NUMRES:166
```

The method `subn(r, s, [c])` does the same but returns a tuple of two elements, where the first element is the new string and the second is the number of replacements that were performed:

```
new_annotation = separator.subn('@', annotation)
print new_annotation
```

which results in:

```
('ATOM:CA@RES:ALA@CHAIN:B@NUMRES:166', 3)
```

9.4 EXAMPLES

Example 9.1 How to Convert a PROSITE Regular Expression into a Python Regular Expression

PROSITE (http://prosite.expasy.org/) is a resource for protein domains, families, and functional sites, which are described by either signature patterns (i.e., regular expressions) or profiles (i.e., tables of position-specific amino acid weights and gap costs). The regular expression syntax used in PROSITE (see http://prosite.expasy.org/scanprosite/scanprosite-doc.html#pattern_syntax), however, is different from the one used in Python. It can be very useful to be able to automatically convert one into the other.

The following simple script performs this task:

```
pattern = '[DEQN]-x-[DEQN](2)-C-x(3,14)-C-x(3,7)\
    -C-x-[DN]-x(4)-[FY]-x-C'
pattern = pattern.replace('{', '[^')
pattern = pattern.replace('}', ']')
pattern = pattern.replace('(', '{')
pattern = pattern.replace(')', '}')
pattern = pattern.replace('-', '')
pattern = pattern.replace('x', '.')
pattern = pattern.replace('>', '$')
pattern = pattern.replace('<', '^')
print pattern
```

Source: Adapted from code published by A.Via/K.Rother under the Python License.

The PROSITE regular expression in the example corresponds to the calcium-binding EGF-like domain signature (PROSITE ID: EGF_CA; PROSITE AC: PS01187).

Example 9.2 How to Find Transcription Factor Binding Sites in a Genomic Sequence

Suppose you have a list of transcription factor binding sites (TFBSs) and want to find out if and where they occur in the genome of a given organism. This can be easily done using Python regular expression tools. You need the text file with the nucleotide sequence of the genome you want to search and the list of TFBSs in a format that can be read by a computer program. For example, the Transcription Factor Database (http://cmgm.stanford.edu/help/manual/databases/tfd.html) uses the following format:

```
UAS(G)-pMH100 CGGAGTACTGTCCTCCG ! J Mol Biol 209: 423-32 (1989)
TFIIIC-Xls-50 TGGATGGGAG ! EMBO J 6: 3057-63 (1987)
HSE_CS_inver0 CTNGAANNTTCNAG ! Cell 30: 517-28 (1982)
ZDNA_CS 0 GCGTGTGCA ! Nature 303: 674-9 (1983)
GCN4-his3-180 ATGACTCAT ! Science 234: 451-7 (1986)
```

In this example, the 'TFBS.txt' file contains this list of TFBSs, and the 'genome.txt' file contains the sequence in FASTA format of, e.g., a whole chromosome of a eukaryotic organism of your choice.

```python
import re
genome_seq = open('genome.txt').read()
# read transcription factor binding site patterns
sites = []
for line in open('TFBS.txt'):
    fields = line.split()
    tf = fields[0]
    site = fields[1]
    sites.append((tf, site))

# match all TF's to the genome and print matches
for tf, site in sites:
    tfbs_regexp = re.compile(site)
    all_matches = tfbs_regexp.findall(genome_seq)
    matches = tfbs_regexp.finditer(genome_seq)
    if all_matches:
        print tf, ':'
        for tfbs in matches:
            print '\t', tfbs.group(), tfbs.start(), tfbs.end()
```

Source: Adapted from code published by A.Via/K.Rother under the Python License.

Example 9.3 Extract the Title and the Abstract Text from a PubMed HTML Page

If you go to a PubMed abstract web page (e.g., http://www.ncbi.nlm. nih.gov/pubmed/18235848), you can easily access the corresponding HTML source code. For example, this can be done through the "Develop" → "Show Page Source" link in the Safari menu or "Tools" → "Web Developer" → "Page Source" in the Firefox menu. Spend a few minutes exploring this page. You will see that the title of the paper is enclosed between the tags <h1> and </h1>, whereas the text of the abstract is enclosed between <h3>Abstract</h3><div class = ""><p> and </p>. These details are relevant for the selective extraction of the title and the abstract from a PubMed HTML abstract page.

The example script opens the HTML web page from a Python script and parses it in order to selectively fetch some parts of it (the title and the abstract in this case).

```python
import urllib2

import re
pmid = '18235848'
url = 'http://www.ncbi.nlm.nih.gov/pubmed?term=%s' % pmid
handler = urllib2.urlopen(url)
html = handler.read()
title_regexp = re.compile('<h1>.{5,400}</h1>')
title_text = title_regexp.search(html)
abstract_regexp = re.compile('<h3>Abstract</h3><div class\
    = ""><p>.{20,3000}</p></div>')
abstract_text = abstract_regexp.search(html)
print 'TITLE:', title_text.group()
print 'ABSTRACT:', abstract_text.group()
```

Source: Adapted from code published by A.Via/K.Rother under the Python License.

The `urllib2` module (see Recipe 13) provides tools to connect to a URL and retrieve its content. `urlopen()` is the method for URL opening (i.e., the method that establishes the connection). Its argument must be a URL, and, similarly to the `open()` built-in function for file opening, it returns a file-type Python object (a `handler`), which, as such, owns a number of methods that can be used to read its content (`read()`, `readline()`, `readlines()`, `close()`). The `read()` method reads the `handler` content as a single string of text.

Once the HTML source code is stored in a variable (`html`) in the form of a single string, we can use the tools provided by the `re` module to parse it.

In this example, by having a look at the HTML at hand (you should save the HTML text to a file and manually examine it), you can identify `<h1>` and `</h1>` as unique delimiting tags of the title. Therefore, a regular expression is defined as follows:

```
<h1>.{5,400}</h1>
```

This regular expression will match any text between 5 and 400 characters delimited by `<h1>` and `</h1>`. This choice univocally identifies the title. It must be noticed that the number of characters allowed between the two tags must be adapted to the maximum number of characters that can occur in titles of scientific papers. A similar procedure is applied for the selection of the abstract.

If you want to recursively extract the title and abstract from a list of PMIDs, you have to put the code in a for loop:

```
pmids = ['18235848', '22607149', '22405002', '21630672']
for pmid in pmids:
    url = 'http://www.ncbi.nlm.nih.gov/pubmed?term=%s'+%pmid
    ...
```

The list of PMIDs can be read from a text file and stored in the Python list.

Example 9.4 Detect a Specific Word or a Set of Words in a Scientific Abstract

You can use what you learned in Example 9.3 to detect a word or a set of words in a scientific abstract. More generally, this example can be applied to perform very simple text mining and can be compared to the "find" tool available in Microsoft Word.

```
import urllib2
import re
# word to be searched
keyword = re.compile('schistosoma')
# list of PMIDs where we want to search the word
pmids = ['18235848', '22607149', '22405002', '21630672']
for pmid in pmids:
    url = 'http://www.ncbi.nlm.nih.gov/pubmed?term=%s' +%pmid
    handler = urllib2.urlopen(url)
    html = handler.read()
    title_regexp = re.compile('<h1>.{5,400}</h1>')
    title = title_regexp.search(html)
    title = title.group()
    abstract_regexp = re.compile('<h3>Abstract</h3><p>.\
        {20,3000}</p></div>')
    abstract = abstract_regexp.search(html)
    abstract = abstract.group()
    word = keyword.search(abstract, re.IGNORECASE)
    if word:
        # display title and where the keyword was found
        print title
        print word.group(), word.start(), word.end()
```

Source: Adapted from code published by A.Via/K.Rother under the Python License.

If you want to identify all the occurrences of a word in a text, you can use the `finditer()` method:

```python
import urllib2
import re
# word to be searched
word_regexp = re.compile('schistosoma')
# list of PMIDs where we want to search the word
pmids = ['18235848', '22607149', '22405002', '21630672']
for pmid in pmids:
    url = 'http://www.ncbi.nlm.nih.gov/pubmed?term=%s' +%pmid
    handler = urllib2.urlopen(url)
    html = handler.read()
    title_regexp = re.compile('<h1>.{5,400}</h1>')
    title = title_regexp.search(html)
    title = title.group()
    abstract_regexp = re.compile('<h3>Abstract</h3><p>.\
        {20,3000}</p></div>')
    abstract = abstract_regexp.search(html)
    abstract = abstract.group()
    words = keyword.finditer(abstract)
    if words:
    # display title and where the keyword was found
    print title
    for word in words:
        print word.group(), word.start(), word.end()
```

Source: Adapted from code published by A.Via/K.Rother under the Python License.

9.5 TESTING YOURSELF

Exercise 9.1 Detecting Disulphide Bridge Patterns

Find all the Uniprot (SwissProt) sequences with pairs of cysteines separated by at most four residues.

Hint: Download Uniprot (SwissProt) sequences in FASTA format from http://www.uniprot.org/.

Hint: Search the following regular expression: `C.{1,4}C`.

Exercise 9.2 Parsing Moby Dick

Copy and paste to a text file the text from Herman Melville's *Moby Dick* (available from www.gutenberg.org). Is the word *captain* or *whale* used more frequently in the book? Write a program to search the occurrences of words in both lowercase and uppercase.

Hint: Remember that to search a character in both lowercase and uppercase, you can use the regular expression [Aa] or the re.IGNORECASE flag.

Exercise 9.3 Searching Phosphorylation Sites in Human Kinases

Search for threonine and serine phosphorylation sites in Uniprot (SwissProt) human kinases.

Hint: For the sake of simplicity, you can use the regular expression shown in this chapter ('R.[ST][^P]') for phosphorylation sites.

Hint: You have to parse the file downloaded in Exercise 9.1 and filter out all records that do not have the keywords *kinase* and *Homo sapiens* in the header.

Exercise 9.4 Manually Identify Suitable HTML Tags

Print (or save to a file) a PubMed HTML page for a given publication, examine the source code of the page carefully, and try to identify unique tags delimiting the part of the HTML text containing the authors of the paper.

Hint: You could just print the content of the HTML variable of Example 9.3 or go to a PubMed abstract page and fetch the source code.

Hint: Each author is associated with a web link, so the HTML text containing authors' names will look a bit messed up.

Exercise 9.5

Write a regular expression to extract the authors from the HTML page of Exercise 9.4.

Hint: A good start for such a regular expression might be <div class = "auths">.

II SUMMARY

We have played our cards. In Part II you encountered all parts of the Python language. You can perform calculations. You have a repertoire of data types that you can store in *variables:* integers, floating-point numbers, strings, lists, tuples, dictionaries, and sets. You can manipulate them using *functions.* You know that some functions are built-in parts of Python, others are parts of *objects* (e.g., strings) and have to be specified with the dot syntax, and others need to be imported from extra *modules* (e.g., the `math` module). You know *control flow* structures: `for` loops that repeat instructions, the `if` statement that makes decisions, and `while` loops that combine the two. You know how to *read* and *write* files. You know how to *parse* information from text. You know how to search, sort, and filter your data. You know how to manipulate *tables* that come as nested lists, nested dictionaries, or a mixture of both. Finally, you know how to extract information by *pattern matching.*

The list of the Python language components is close to complete. This means that at this stage you may have acquired all you need to manage your biological data. In fact, with them, you can practically write most of the software needed for data analysis.

III

Modular Programming

INTRODUCTION

As mentioned in the summary of Part II, we have played all our cards. Or have we? Still, this book contains four more parts. What should we write?

Now imagine you have a complete genome of a biological species and the number of chromosomes and base pairs. Does that tell you how the organism works? In programming, as in biology, having a complete list of components is not enough for a thorough picture of the language structure. Part III of this book is all about structure (see the figure). You will learn how to put together the existing parts in a reasonable way. Python offers many tools to modularize your code. Writing a good program is not much different from undertaking a research project. For your project to be successful, you must have clear goals in mind and a general, though flexible, plan. At least you must plan enough to start. Then you should improve in small steps by dissecting the main aims in specific programming tasks. Each single task needs testing and feedback, if possible. Then you have to connect and harmonize all parts. A way to success is to be open to ask for help and constantly get feedback from more experienced colleagues. Working in a small team (even with a single colleague or a friend) usually improves the pleasure of working and the likelihood of achieving good results.

All these principles also apply to writing programs. Before starting a new program, think what exactly should be your input and output. What

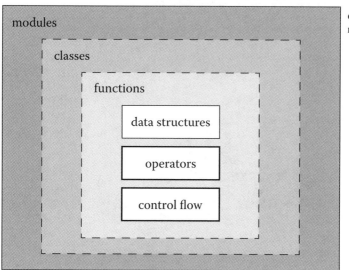

FIGURE The Python structure. *Note:* Modules are the files where you write your code. They may or may not contain classes (this is the meaning of the dashed line), which may or may not contain functions (again the dashed line). Functions can be made up of data structures, operators, and control flow structures. Data and control flow structures, operators, and functions can be written to a module without necessarily belonging to a function or a class.

precisely do you want your program to do? Roughly plan the shape of the program, at least enough to start writing it, and structure it in small chunks, using ideas and objects you will learn in this part of the book. Avoid writing a sequence of instructions one after the other up to the end of the program; this approach may work for programs with few lines, but it can become a mess for longer programs. When a subtask (e.g., a function) is ready, carefully analyze it for the presence of errors (the so-called debugging) and test it on a single well-known piece of data before applying it to a whole genome or database. In particular, separately check that each single part (e.g., each function) of the program does exactly what you expect it to do. Do not write the whole program before debugging and testing it! When all the subtasks are achieved and the corresponding pieces of code are debugged and work properly, you can connect all the parts, for example, by calling the various functions and using the result of a function as input to the following one. When everything is connected and your program is ready, test it again on a single data sample (e.g., a file with few sequences) before running it on the whole Uniprot. During the whole process, do not hesitate to ask for help and feedback from your colleagues, friends, and even Internet forums. Last but not least, consider that working in small groups (ideally in pairs) might be very effective, especially in the learning phase. The authors' experience in training courses showed that learning in pairs is more fruitful than learning alone.

In Chapter 10, you will learn how to write your own functions. Functions are subprograms that take a number of input parameters, work on them, and return a result. Functions help you to divide a bigger program into smaller parts. In Chapter 11, you will learn how to use classes. Classes are containers for both data and functions. They let you connect a well-defined data structure with operations on it. By using classes, you can create components that you can reuse in other programs easily. You will also learn how to create your own modules. In Chapter 12, you will learn about debugging. Programs rarely work perfectly in the first attempt. This is why it is important to know how to eliminate program errors and how to trace down bugs that are not visible at first sight. You can also empower your program to handle exceptions by itself. Chapter 13 focuses on the R package for Python. You not only will learn how to handle a big package that consists of many modules but also will apply it to run statistical tests and generate plots. In Chapter 14, program pipelines are introduced. A pipeline is an even bigger organizational unit composed of several programs connected to each other. You will learn how to run other programs from Python and manipulate files and directories using the os module. Finally, Chapter 15 takes a perspective on programming considering the human factor. When you are facing a task or question, what steps can you take to make a program out of it? The chapter gives you a small tool kit to divide problems into smaller units, maintain quality programs, and develop a program by gradual, constant improvement over time.

By the end of Part III, you will be able to build programs that are bigger, better organized, more efficient, more maintainable, and tidier than your first programs.

## CHAPTER 10

Divide a Program into Functions

LEARNING GOAL: You can use functions to organize your programs better.

10.1 IN THIS CHAPTER YOU WILL LEARN

- How to write your own functions

- How to extract the sequence from the coordinates of a protein structure

- How to selectively extract information from a PDB file

- How to calculate the distance between atoms in a protein or DNA three-dimensional structure

10.2 STORY: WORKING WITH THREE-DIMENSIONAL COORDINATE FILES

10.2.1 Problem Description

Protein or nucleotide three-dimensional (3D) structures are stored in PDB files. PDB files are text files that contain both annotation and atomic coordinates (x, y, z) of biological molecules. Crystallographers or NMR spectroscopists collect this information from structure determination experiments and submit it to the Protein Data Bank (http://www.rcsb.org), which contains about 88,000 structures at the time of writing. The format of PDB files is described in Box 10.1, and a sample is shown in Appendix C, Section C.6, "An Example of a PDB File Header (Partial)," and Section C.7, "An Example of PDB File Atomic Coordinate Lines (Partial)."

In the previous chapters, you learned how to parse protein or nucleotide sequence files. Here, you will learn how to parse protein structure files. The

---

BOX 10.1   PDB RECORDS

An example of a partial PDB record is reported in Appendix C, Section C.6, "An Example of a PDB File Header (Partial)," and Section C.7, "An Example of PDB File Atomic Coordinate Lines (Partial)." The first part of the record, called the *header*, includes several annotation lines, including source organism, experimental technique (X-ray, NMR, etc.), cross-references, experimental details, the sequence of the original molecule, mutated residues if any, etc. The second part of the record reports the atomic coordinate lines for standard groups. Each of the atomic coordinate lines describes exactly one atom. They start with the "ATOM" keyword and are separated into columns containing the following details:

```
COLUMNS     DEFINITION
_ _ _ _ _ _ _ _ _ _ _ _ _ _ _ _ _ _ _ _ _ _ _ _ _ _ _ _ _ _
1 - 6       Record name "ATOM".
7 - 11      Atom serial number.
13 - 16     Atom name.
17          Alternate location indicator.
18 - 20     Residue name.
22          Chain identifier.
23 - 26     Residue sequence number.
27          Code for insertion of residues.
31 - 38     x Orthogonal coordinates for X in Angstroms.
39 - 46     y Orthogonal coordinates for Y in Angstroms.
47 - 54     z Orthogonal coordinates for Z in Angstroms.
55 - 60     Occupancy.
61 - 66     Temperature factor.
77 - 78     Element symbol.
79 - 81     Charge on the atom.
```

This format can be translated into a Python string as follows:

```
pdb_format = '6s5s1s4s1s3s1s1s4s1s3s8s8s8s8s6s6s6s4s2s3s'
```

In fact, 6s (s stands for string) corresponds to columns 1–6, 5s to columns 7–11, and so on.

The complete description of the PDB file format is available from the wwPDB at www.wwpdb.org/docs.html.

following program extracts the amino acid sequence from the 3D coordinates of a protein structure. More precisely, it reads the protein sequence as three-letter codes from the ATOM records of a PDB file, converts them into one-letter codes, and writes this sequence to a FASTA file separately for each chain. Because this is a complex task, it is easier to manage by dividing it into subtasks. In Python, such subtasks can be formulated using functions.

10.2.2 Example Python Session

```python
import struct
pdb_format = '6s5s1s4s1s3s1s1s4s1s3s8s8s8s6s6s10s2s3s'
amino_acids = {
    'ALA':'A', 'CYS':'C', 'ASP':'D', 'GLU':'E',
    'PHE':'F', 'GLY':'G', 'HIS':'H', 'LYS':'K',
    'ILE':'I', 'LEU':'L', 'MET':'M', 'ASN':'N',
    'PRO':'P', 'GLN':'Q', 'ARG':'R', 'SER':'S',
    'THR':'T', 'VAL':'V', 'TYR':'Y', 'TRP':'W'
    }

def threeletter2oneletter(residues):
    '''
    Converts the three-letter amino acid,
    which is the first element of each
    list in the residues list,
    to a one-letter amino acid symbol
    '''
    for i, threeletter in enumerate(residues):
        residues[i][0] = amino_acids[threeletter[0]]

def get_residues(pdb_file):
    '''
    Reads the PDB input file, extracts the
    residue type and chain from the CA lines
    and appends both to the residues list
    '''
    residues = []
    for line in pdb_file:
        if line[0:4] == "ATOM":
            tmp = struct.unpack(pdb_format, line)
            ca = tmp[3].strip()
            if ca == 'CA':
                res_type = tmp[5].strip()
                chain = tmp[7]
                residues.append([res_type, chain])
    return residues

def write_fasta_records(residues, pdb_id, fasta_file):
    '''
    Write a FASTA record for each PDB chain
```

```
    '''
    seq = ''
    chain = residues[0][1]
    for aa, new_chain in residues:
        if new_chain == chain:
            seq = seq + aa
        else:
            # write sequence in FASTA format
            fasta_file.write(">%s_%s\n%s\n" % (pdb_id, chain,\
                seq))
            seq = aa
            chain = new_chain
    # write the last PDB chain
    fasta_file.write(">%s_%s\n%s\n" % (pdb_id, chain, seq))
def extract_sequence(pdb_id):
    '''
    Main function: Opens files, writes files
    and calls other functions.
    '''
    pdb_file = open(pdb_id + ".pdb")
    fasta_file = open(pdb_id + ".fasta", "w")
    residues = get_residues(pdb_file)
    threeletter2oneletter(residues)
    write_fasta_records(residues, pdb_id, fasta_file)
    pdb_file.close()
    fasta_file.close()
# call the main function
extract_sequence("3G5U")
```

Source: Adapted from code published by A.Via/K.Rother under the Python License.

10.3 WHAT DO THE COMMANDS MEAN?

The program shown in Section 10.2.2 does one main thing: it extracts the amino acid sequence from the atom coordinate lines of a protein structure file. The input file must be in the directory where you run the script. For each PDB chain, a different record in FASTA format is written to the output file. To achieve this result, the program accomplishes three subtasks. Each of these subtasks is carried out by a separate, reusable function:

get_residues(pdb_file): This reads the PDB input file, extracts the amino acid three-letter code and chain type, and stores them in a Python list called residues as pairs [res_type, chain] for each amino acid residue. An if condition (if ca == 'CA') selecting atoms with the name 'CA' makes sure each residue is used only once (you can see in Section C.7, "An Example of PDB File Atomic Coordinate Lines (Partial)," that the residue name is repeated for each atom belonging to the same residue). For parsing the lines, the `struct` module is used, which will be discussed in Section 10.3.3.

`threeletter2oneletter(residues)`: This converts the amino acid three-letter codes into one-letter codes. The function reads a list of [`res_type`, `chain`] pairs and replaces the first element, which is an amino acid three-letter code, with the corresponding one-letter code. A dictionary (see Chapter 5) is used to convert amino acid three-letter codes into one-letter codes. The built-in function `enumerate(residues)` generates tuples of the form (`n`, `residues[n]`), where `n` is an integer number counting up from zero:

```
>>> data = [['ALA', 'A'], ['CYS', 'A']]
>>> for i, j in enumerate(data):
...        print i, j
...
0 ['ALA', 'A']
1 ['CYS', 'A']
>>>
```

`write_fasta_records(residues, pdb_id, fasta_file)`: This formats the list of residues to strings in FASTA format (adding headers and newline characters (\n) to the sequence) and writes one sequence entry per PDB chain to an output file.

In the function `extract_sequence(pdb_id)`, the "main function" of the program, the previous three functions are called and a number of other operations are done: opening the input file, generating an amino acid sequence for each chain of the PDB file, and writing the output file.

But what are functions useful for in Python? How can you write and use them effectively?

A function is useful every time you need to write reusable code. For example, the code for extracting the residues from a PDB file (the `get_residues()` function) could be reused to count the amino acid frequencies instead of writing a FASTA file. This can be done if you write the code into a function and pass an opened PDB file as argument when you call the function.

You have already met functions several times. For example, `math.log()` and `math.sqrt()` are mathematical functions defined in the `math` module (see Chapter 1). They act on their argument (the value in parentheses) by calculating its logarithm and its square root, respectively. You have also met some built-in functions, i.e., functions built into the Python interpreter that are always available (see Box 10.2). If you want to use the `len()` function, you do not need to import a module first or link it to an object via a dot. You just have to pass a sequence (string, list, tuple, dictionary) to the function as an argument, and it will return the length of the sequence. You can always recognize a function in Python code by the presence of round brackets delimiting its arguments, i.e., objects written in the brackets (and separated by a comma) that may be necessary for the function to perform a given task.

BOX 10.2   BUILT-INS

There are many objects (modules, functions, classes) that are predefined in Python; some of them are automatically imported when you invoke the Python interpreter. These are called *built-ins*. For example, you never need to import the functions `len()`, `sum()`, and `range()` (see Box 10.3). When you use strings, you are implicitly using the built-in object `"str"` (i.e., you do not need to define what a string is); it is the same with lists, dictionaries, and numbers. In general, whenever you use something without defining it, it means that you are using a built-in object. You can find a table listing Python built-in functions in Appendix A, "Command Overview."

The Python interpreter provides you with hundreds of functions, either built-in or stored in specific modules (see the Python Standard Library at http://docs.python.org/library/). Box 10.5 describes two very useful built-in functions, `range()` and `xrange()`. However, you can also define your own functions.

10.3.1 How to Define and Call Functions

Defining a Function

The instructions to define a new function are as follows:

```
def my_function(arg1, arg2,…):
        '''documentation'''
        <instructions>
        return value1, value2, …
```

`my_function` is the function name. It can be any name of your choice, except for reserved words (see Chapter 1, Box 1.3). It is good practice to use a name suggesting what the function does. For example, a function for adding two numbers could be called `add(num1, num2)`.

`(arg1, arg2, ...)` are called the arguments of the function and are optional (i.e., a function can perform tasks without taking arguments, for example, printing a predefined text). Arguments are passed to the function upon function call.

`documentation` is an (optional) description as a triple-quoted comment.

`<instructions>` are the instructions that are executed when the function is called. They define what the function does.

`return` is the instruction that makes the interpreter stop executing further instructions in the function and go back to the program line from which the function was called.

`value1, value2, ...` are the values (the result) returned by the function. A function can just perform a number of tasks without returning any value (in this case, it will be called a "procedure," but this is a matter of terminology).

For instance, a function that calculates the sum of two numbers can be defined like this:

```
def addition(num1, num2):
        '''calculates the sum of two numbers'''
        result = num1 + num2
        return result
```

Source: Adapted from code published by A.Via/K.Rother under the Python License.

The `addition` function takes two arguments, adds them arithmetically, and returns the result.

Functions are useful to organize your scripts, in particular, if you need to repeat a task (e.g., a complex calculation) several times. Ten syntactical reminders have been collected in Box 10.5. To use a function, you need to call it at least once.

Calling a Function

If you want the interpreter to execute the `<instructions>` block, you have to "call" the function. To call a function, you have to write its name followed by round brackets. If the function requires any arguments, you need to pass exactly that many values or variables with the call. For

BOX 10.3 LOOPS WITH RANGE() AND XRANGE()

Two built-in functions very useful in for loops are range() and xrange(). range(n,m) creates a list of integers ranging from n to m–1, and xrange(n,m) creates an iterator. If n is omitted, 0 is used as a default value.

```
>> for i in range(5):
... print i**2
...
0
1
4
9
16
>>>
```

If you want to execute a loop a high number of times (e.g., 1,000), it is not comfortable to write a full sequence (tuple or list) of 1,000 elements ranging from 0 to 999. xrange() is very similar to range(), even though it does not return a list but yields the same values as the corresponding list only at the time they are needed. Therefore, xrange() consumes less memory and can be used with big numbers.

Moreover, both methods make it possible to specify a so-called step, if you want that your index only runs, for example, on the even numbers of a list of integers.

For example,

```
range([start], stop, [step])
```

returns the list:

```
[start, start + step, start + (2 * step),..., stop - 1]
```

If start is omitted, then the default value is 0. If step is omitted, the default value is 1. If step > 0, the *last element* of the list will be the highest value of

```
start + (i * step) < = stop
```

If step < 0, the *last list element* will be the smaller value of

```
stop < = start + (i * step)
```

In other words,

```
range(i,j) = [i,i+1,...,j-1]
range(k) = [0,1,...,k-1]
range(i,j,1) = [i,i+1,i+21,...,j-1]          (i<j)
range(i,j,-1) = [i,i-1,i-21,...,j+1]          (i>j)
```

instance, to call the `addition` function defined earlier, you need to give two numbers as arguments:

```
result = addition(12, 8)
print result
(writes 20)
```

10.3.2 Function Arguments

Function arguments are a way to pass data to a function (see Box 10.4). A Python function can have multiple arguments or none at all. Almost every Python object can be passed as an argument to a function. The result of a function call can be the argument of a function, too:

BOX 10.4   MULTIPLE FUNCTION ARGUMENTS AND RETURNED VALUES ARE TUPLES

The sequence of arguments passed to a function is a tuple, and functions return multiple values in the form of tuples as well:

```
>>> def f(a, b):
...     return a + b, a * b, a - b
...
>>> f(10, 15)
(25, 150, -5)
# You can also put the result in a variable
>>> result = f(10,15)
>>> result
(25, 150, -5)
>>> sum, prod, diff = f(20, 2)
>>> sum
22
>>> prod
40
>>> diff
18
```

**BOX 10.5   TEN THINGS TO REMEMBER ABOUT
PYTHON FUNCTIONS**

1. The statement to define a function is `def`.
2. A function must be defined and called using round brackets.
3. The body of a function is a block of code that is initiated by a colon character followed by indented instructions.
4. The last indented statement marks the end of a function definition.
5. You can pass arguments to a function (see Section 10.3.2). Multiple arguments have the form of a tuple (see Box 10.4).
6. You can define variables in the body of a function.
7. The statement `return` exits a function, optionally passing back a value to the caller. Multiple values have the form of a tuple (see Box 10.4).
8. A `return` statement with no values is possible, as well as a function with no `return` statement. In both cases the default returned value is `None`.
9. You can insert a documentation string in quotation marks in the body of a function. This string is ignored upon function call but can be retrieved using the `__doc__` attribute of the function object.
10. When a function is called, a local namespace is automatically created. The variables defined in the body of a function live in its *local* namespace and not in the *global* namespace of the script or module. When a function is called, names of the objects used in its body are first searched in the function namespace, and subsequently, if they are not found in the function body, they are searched in the global namespace of the script or module.

```
def increment(number):
    '''returns the given number plus one'''
    return number + 1
def print_arg(number):
    '''prints the argument'''
    print number
print_arg(increment(5))
```

Source: Adapted from code published by A.Via/K.Rother under the Python License.

This code prints the number 6. Even a function can be passed as an argument to another function (see Box 10.6).

BOX 10.6 LAMBDA FUNCTIONS (OR ANONYMOUS FUNCTIONS)

You can use the `lambda` statement to create small anonymous functions. These functions are called *anonymous* because they are not declared in the standard manner by using the `def` statement. Lambda functions are particularly useful when they are used as arguments of other functions.

```
>>># Traditional function (in one line of code):
>>> def f(x): return x**2
>>> print f(8)
64
>>># Same result obtained with a lambda function:
>>> g = lambda x: x**2
>>>
>>> print g(8)
64
>>> (lambda x: x**2)(3)
9
>>>
```

Lambda functions do not contain a `return` statement: they contain an expression, the value of which is always returned.

A lambda function can be defined everywhere, even in the argument of another function without assigning a name to it.

A lambda function can have an arbitrary number of arguments and returns the value of a single expression.

There are four kinds of arguments in Python: required arguments, keyword arguments, default arguments, and variable-length arguments.

Required arguments: One or multiple parameters must be passed to a function. The order of the arguments in the call must be exactly the same as the order in the function definition:

```
def print_funct(num, seq):
    print num, seq

print_funct(10, "ACCTGGCACAA")
```

The output of this program is:

```
10 ACCTGGCACAA
```

Keyword arguments: It is possible to assign a name to the arguments of a function. In this case, the order is not important:

```
def print_funct(num, seq):
    print num, seq

print_funct(seq = "ACCTGGCACAA", num = 10)
```

This program produces the same output as the one above.

Default arguments: It is also possible to use default (optional) arguments. These optional arguments must be placed in the last position(s) of the function definition:

```
def print_funct(num, seq = "A"):
    print num, seq

print_funct(10, "ACCTGGCACAA")
print_funct(10)
```

The output of these two function calls is:

```
10 ACCTGGCACAA
10 A
```

Default arguments should always be immutable types (numbers, strings). Never use lists or dictionaries as default arguments, because they can lead to very nasty bugs.

Variable-length arguments: The number of arguments can be variable (i.e., changing from one function call to the other); they are indicated by the symbol * (for a tuple) or a ** (for a dictionary):

```
def print_args(*args):
    print args

print_args(1,2,3,4,5)
print_args("Hello world!")
print_args(100, 200, "ACCTGGCACAA")
```

This program prints tuples with the arguments passed to the function:

```
(1, 2, 3, 4, 5)
('Hello world!',)
(100, 200, 'ACCTGGCACAA')
```

When you use the ** symbol, you have to provide both keys and values for the returned dictionary:

```
def print_args2(**args):
    print args

print_args2(num = 100, num2 = 200, seq = "ACCTGGCACAA")
```

Here, the function call prints a dictionary:

```
{'num': 100, 'seq': 'ACCTGGCACAA', 'num2': 200}
```

10.3.3 The struct Module

The struct module provides methods that make it possible to convert a string into a tuple on the basis of a customized format, or vice versa. It is a Python built-in module, which can be very useful to work with PDB files.

The struct method pack(format, v1, v2, ...) returns a single string made up of the v1, v2, ... values packed according to the format string that specifies which conversion characters will be used to pack the v1, v2, ... values into a string. For example,

```
format = '2s3s'
```

means a two-character string (in 2s, s stands for "string") followed by a three-character string (3s).

If the conversion characters in fmt have the form of strings (s), the arguments v1, v2, ... must be strings:

```
>>> import struct
>>> format = '2s1s1s1s1s'
>>> a = struct.pack(format,'10','2','3','4','5')
>>> a
'102345'
```

The method unpack(format, string) unpacks a string into a tuple, according to the format encoded by format. The string must contain the same number of characters present in the format string.

```
>>> import struct
>>> format = '1s2s1s1s'
>>> line = '12345'
>>> col = struct.unpack(format, line)
>>> col
('1', '23', '4', '5')
```

The `struct` method `calcsize(fmt)` returns the total number of characters of a given formatting string:

```
>>> import struct
>>> format = '30s30s20s1s'
>>> struct.calcsize(format)
81
```

Source: Adapted from code published by A.Via/K.Rother under the Python License.

Based on the `struct` module and on the table provided in Box 10.1, we can write the following formatting string for the PDB ATOM lines:

```
pdb_format = '6s5s1s4s1s3s1s1s4s1s3s8s8s8s6s6s6s4s2s3s'
```

By unpacking the ATOM lines of a PDB file using such format, we obtain a tuple where each element corresponds to one of the PDB columns described in Box 10.1. For example, here we consider only the first ATOM line of a PDB file:

```
>>> import struct
>>> line = 'ATOM 1 N ILE A 16 11.024 3.226 26.760 1.00 16.50 N '
>>> format = '6s5s1s4s1s3s1s1s4s1s3s8s8s8s6s6s10s2s3s'
>>> col = struct.unpack(format, line)
>>> col
('ATOM ', ' 1', ' ', ' N ', ' ', 'ILE', ' ', 'A', '
16', ' ', ' ', ' 11.024', ' 3.226', ' 26.760', '
1.00', ' 16.50', ' ', ' N', ' ', ' ')
```

You can observe that the 6th element of `col` corresponds to the amino acid three-letter code (`'ILE'`) and that the 8th element corresponds to the PDB chain (`'A'`). The 12th, 13th, and 14th elements correspond to the atom x-, y-, and z-coordinates, respectively.

In summary, the Python program in Section 10.2.2 opens a PDB file (in `extract_sequence()`), reads it and extracts the amino acid residue types and chain from its ATOM lines (in `get_residues()`), "translates" amino acid three-letter codes into one-letter codes (in `threeletter2oneletter()`), and uses the latter to write to a file the PDB sequence in FASTA format (in `write_fasta_records()`). Notice that, in order to make the program simpler, the amino acid sequences were not formatted to 64-character-long lines (according to the typical FASTA format). You could try to modify the program in this sense.

10.4 EXAMPLES

Example 10.1 How to Write a Function That Calculates the Distance between Two Points in Cartesian Space

Calculating the distance between two points was introduced in Example 1.1. Here, the same calculation is made more flexible using a function. The coordinates of two points, p1 and p2, can be passed to a Python function in the form of lists or tuples:

```python
from math import sqrt
def calc_dist(p1, p2):
    '''returns the distance between two 3D points'''
    dx = p1[0] - p2[0]
    dy = p1[1] - p2[1]
    dz = p1[2] - p2[2]
    distsq = pow(dx, 2) + pow(dy, 2) + pow(dz, 2)
    distance = sqrt(distsq)
    return distance

print calc_dist([3.0, 3.0, 3.0], [9.0, 9.0, 9.0])
```

Source: Adapted from code published by A.Via/K.Rother under the Python License.

The function call results in:

```
10.3923048454
```

If you place this function in a separate Python file distance.py, you can import it from other programs.

Example 10.2 How to Write a Function That Takes as Input Any Number of Arguments and Returns a String of Tab-Separated Arguments Ending with a Newline Character

When creating output files, you will often find it useful to join elements by tabulators. However, the number of elements to be written may vary. The following example function uses variable-length arguments to take care of that.

Notice that we have to convert each single argument into a string type using the built-in function str() on each argument passed to the function. As usual, '\t' is the Python metacharacter to insert a

TAB into a string, and '\n' is the Python metacharacter to insert a newline character into a string.

```python
def tuple2string(*args):
    '''returns all arguments as a
    single tab-separated string'''
    result = [str(a) for a in args]
    return '\t'.join(result) + '\n'
```

This function can be used to generate a file containing a nucleotide substitution matrix (the frequency values reported in the following example are approximate):

```python
outfile = open("nucleotideSubstitMatrix", "w")
outfile.write(tuple2string('', 'A', 'T', 'C', 'G'))
outfile.write(tuple2string('A', 1.0))
outfile.write(tuple2string('T', 0.5, 1.0))
outfile.write(tuple2string('C', 0.1, 0.1, 1.0))
outfile.write(tuple2string('G', 0.1, 0.1, 0.5, 1.0))
outfile.close()
```

Source: Adapted from code published by A.Via/K.Rother under the Python License.

The nucleotideSubstitMatrix output file will read

```
     A     T     C     G
A   1.0
T   0.5   1.0
C   0.1   0.1   1.0
G   0.1   0.1   0.5   1.0
```

Example 10.3 How to Identify Specific Residues in a PDB File

Here you will learn how to extract and write the atomic coordinates of the trypsin active site to a file. The trypsin active site is constituted by the well-known triad: Asp 102, His 57, and Ser 195. To this aim, you have to go to the PDB archive and download and save the atom coordinates of trypsin to a file. In this example, we will use the PDB file 1TLD.pdb, which reports the crystal structure of the bovine beta-trypsin at 1.5 Å resolution.

```python
import struct
pdb_format = '6s5s1s4s1s3s1s1s4s1s3s8s8s8s6s6s10s2s3s'

def parse_atom_line(line):
    '''returns an ATOM line parsed to a tuple '''
    tmp = struct.unpack(pdb_format, line)
```

```
    atom = tmp[3].strip()
    res_type = tmp[5].strip()
    res_num = tmp[8].strip()
    chain = tmp[7].strip()
    x = float(tmp[11].strip())
    y = float(tmp[12].strip())
    z = float(tmp[13].strip())
    return chain, res_type, res_num, atom, x, y, z

def main(pdb_file, residues, outfile):
    '''writes residues from a PDB file to an output file.'''
    pdb = open(pdb_file)
    outfile = open(outfile, "w")
    for line in pdb:
        if line.startswith('ATOM'):
            res_data = parse_atom_line(line)
            for aa, num in residues:
                if res_data[1] == aa and res_data[2] == num:
                    outfile.write(line)
    outfile.close()

residues = [('ASP', '102'), ('HIS', '57'), ('SER', '195')]
main("1TLD.pdb", residues, "trypsin_triad.pdb")
```

Source: Adapted from code published by A.Via/K.Rother under the Python License.

The code uses two functions to do the work: one that parses a single line from the PDB file, and another for processing the entire file. Notice that we used the string method `strip()` to remove the blank spaces before and/or after the actual characters of residue type and number. Furthermore, some variables returned by `parse_atom_line()` are not used in this example: they have been defined in order to make the function more general and for reuse in Examples 10.4 and 10.5. For more compact code, the values returned by `parse_atom_line()` are collected in a single variable, `res_data`, which is a tuple.

Example 10.4 How to Calculate the Distance between Two Atoms in a PDB Chain

Now we will combine functions. We will import and use the `calc_dist()` function defined in Example 10.1 (assuming that it was stored in the `distance.py` module) and the `parse_atom_line()` function defined in Example 10.3 (assuming that it was saved in the `parse_pdb.py` module) to calculate the distance between the CA atoms of residue 123 and 209 of chain A. Even though the program does a lot in order to extract the information, the code is short because most of the work is done in the imported functions:

```
from math import sqrt
from distance import calc_dist
from parse_pdb import parse_atom_line

pdb = open('3G5U.pdb')
points = []
while len(points) < 2:
    line = pdb.readline()
    if line.startswith("ATOM"):
        chain, res_type, res_num, atom, x, y, z = \
            parse_atom_line(line)
        if res_num == '123' and chain == 'A' and atom == 'CA':
            points.append((x, y, z))
        if res_num == '209' and chain == 'A' and atom == 'CA':
            points.append((x, y, z))
print calc_dist(points[0], points[1])
```

Source: Adapted from code published by A.Via/K.Rother under the Python License.

Example 10.5 How to Calculate the Distance between All CA Atoms in a PDB Chain

If you now want to calculate the distance between all the CA atoms in a PDB chain (e.g., chain A), you can write a function that collects all the CA coordinates and then calculates the distance between all the possible pairs of CA atoms:

```
from math import sqrt
from distance import calc_dist
from parse_pdb import parse_atom_line

def get_ca_atoms(pdb_file):
    '''returns a list of all C-alpha atoms in chain A'''
    pdb = open(pdb_file)
    ca_list = []
    for line in pdb:
        if line.startswith('ATOM'):
            data = parse_atom_line(line)
            chain, res_type, res_num, atom, x, y, z = data
            if atom == 'CA' and chain == 'A':
                ca_list.append(data)
    pdb_file.close()
    return ca_list
```

```
ca_atoms = get_ca_atoms("1TLD.pdb")
for i, atom1 in enumerate(ca_atoms):
    # save coordinates in a variable
    name1 = atom1[1] + atom1[2]
    coord1 = atom1[4:]
    # compare atom1 with all other atoms
    for j in range(i+1, len(ca_atoms)):
        atom2 = ca_atoms[j]
        name2 = atom2[1] + atom2[2]
        coord2 = atom2[4:]
    # calculate the distance between atoms
        dist = calc_dist(coord1, coord2)
        print name1, name2, dist
```

Source: Adapted from code published by A.Via/K.Rother under the Python License.

10.5 TESTING YOURSELF

Exercise 10.1 Counting FASTA Records

Write a function that reads a file containing several protein records in FASTA format and returns the total number of records in the file. Call the function passing the input filename as argument and print the result on the screen.

Hint: You can count the number of '>' occurring in the input FASTA file.

Exercise 10.2 Saving Sequence Records in Separate FASTA Files

Write a program that reads a file containing several protein records in FASTA format. Write a function saving each record to a separate file and name each file <AC>.fasta, where AC is the accession number of the sequence record. The function argument must be a single sequence record from the input FASTA file.

Hint: Read the file (e.g., the one shown in Appendix C, Section C.4, "A Multiple Sequence File in FASTA Format") line by line. If a line starts with '>', use the split() method of strings to select the AC number.

Hint: Use tricks you learned in Chapter 4 to extract the sequence from the record.

Hint: Build a list of two elements [AC, sequence] and pass this list to a function.

Hint: The function will open a new file and use the AC for its name and then will write the sequence to it.

Hint: If you want, you can use the fastAformat() function to suitably format the sequence before writing it to the input file.

Exercise 10.3 Identifying the Two Closest Residues in a Protein Structure

Print the name and PDB residue number of the two closest residues belonging to the same chain of a PDB structure. Take as distance between two amino acids the distance between their CA atoms.

Hint: You have to add a few instructions to the script in Example 10.5 to get your result. You can initialize a variable (maxval) to a very high number (e.g., 10,000) and reset maxval to the value of a tmp variable (recording the distance between pairs of CA atoms) when tmp < maxval. This ensures that, at the end of all comparisons, the value recorded in maxval will be the lowest possible.

Hint: There is a recipe (Recipe 15) that does this for you.

Exercise 10.4 PDB Complex Interfaces

Identify the residues at the interface between two PDB chains and write them to a file.

Hint: You can modify the script in Example 10.5 to select all residues i (belonging to chain A, i ∈ A) and all residues j (belonging to chain B, j ∈ B) the distance of which is < 6 Å.

Hint: There is a recipe (Recipe 16) that does this for you.

Exercise 10.5 Disorder, Secondary Structure, and Solvent Accessibility Predictor

Write a single script including Example 5.2 and Exercises 5.4 and 5.5. Put the task of each exercise in a different function and use raw_input() to allow the user to choose which function to call for a given protein sequence.

Hint: You can use as input a sequence filename and a number (e.g., 1, 2, or 3) specifying which predictor will be called. In the script, you may use something like the following:

```
sequence = raw_input("Type the sequence filename: ")
predictor = raw_input("1 (disorder), 2 (sse), or 3
    (accessibility): ")
F = open(sequence)
if predictor == '1': prediction = disorder(F)
elif predictor == '2': prediction = sse(F)
else: prediction = accessibility(F)
```

## CHAPTER 11

Managing Complexity with Classes

L EARNING GOAL: You can group data and functions into classes.

11.1 IN THIS CHAPTER YOU WILL LEARN

- How to represent complex things by a Python class

- How to create objects from a class

- How to use a class as a data container

- How to define methods for classes

- How to print objects using the `__repr__` class method

- How to build complex structures involving more than one class

11.2 STORY: MENDELIAN INHERITANCE

11.2.1 Problem Description

In 1856, the monk Gregor Mendel carried out experiments on peas that laid the very foundation for genetics. When crossing different strains of peas, he observed characteristic proportions of the phenotypes. His work contributed to the knowledge that in addition to the visible phenotype, a hidden genotype exists in each individual pea (see Figure 11.1).

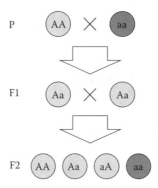

FIGURE 11.1   Visible and hidden genotypes in crossing two strains of peas.

Writing a program that calculates pea phenotypes over a few genera-tions is not easy. You need to manage many small things: the genotype of each pea, how the phenotype relates to a genotype, and how a new genera-tion of peas is created. Moreover, you need to make decisions: How many phenotypic traits should be considered? What data types are to be used? What should the output look like? Even though the underlying science is trivial, its representation as program code is complex.

As your programming skills increase, you may discover that writing large programs that contain just a series of commands like loops and vari-able assignments are difficult to handle. Debugging and modifying a long program consisting of a single big block of code is ineffective and unsat-isfactory. You will feel likely frustrated because the connection between the code lines and the biological problem at hand is not clear anymore.

Wouldn't it be better if you could describe the things you are program-ming more explicitly? For instance, if you know that a pea is green if it has two alleles for green color in its genotype, could you tell it to your program independently of the rest of the program? *Classes* are program-ming structures that help describe how things work in the real world, and they keep the complexity in check.

11.2.2 Example Python Session

Classes represent real or abstract objects. This chapter explains how to structure data and functions into classes. The example program uses a class to simulate hybridization experiments on peas like Gregor Mendel did. The class manages the genetic component determining pea color. The dominant allele (carrying yellow color) is represented by "G", and the recessive allele (carrying green color) is represented by "g".

The class Pea contains statements that define how each individual pea behaves. Each pea has its own genotype ("GG," "Gg," "gG," or "gg"), which tells what the phenotype is ("yellow" or "green"). Finally, each pea can create offspring together with a second pea.

```
class Pea:

    def __init__(self, genotype):
        self.genotype = genotype

    def get_phenotype(self):
        if "G" in self.genotype:
            return "yellow"
        else:
            return "green"

    def create_offspring(self, other):
        offspring = []
        new_genotype = ""
        for haplo1 in self.genotype:
            for haplo2 in other.genotype:
                new_genotype = haplo1 + haplo2
                offspring.append(Pea(new_genotype))
        return offspring

    def __repr__(self):
        return self.get_phenotype() + ' [%s]' % self.genotype

yellow = Pea("GG")
green = Pea("gg")
f1 = yellow.create_offspring(green)
f2 = f1[0].create_offspring(f1[1])
print f1
print f2
```

Source: Adapted from code published by A.Via/K.Rother under the Python License.

The output of the program is:

```
[yellow [Gg], yellow [Gg], yellow [Gg], yellow [Gg]]
[yellow [GG], yellow [Gg], yellow [gG], green [gg]]
```

11.3 WHAT DO THE COMMANDS MEAN?

The program creates a yellow pea (yellow = Pea("GG")) and a green pea (green = Pea("gg")), hybridizes them (yellow.create_offspring(green)), and prints the phenotypes of all their children (f1). Then two of the children (f1[0] and f1[1]) are hybridized again, and the second child generation (f2) is also printed. All that happens in the final paragraph of the program. One of the purposes of the `Pea` class is to make that last paragraph easy to understand. The `Pea` class has a lot in common with the real peas examined by Gregor Mendel: it contains a genotype (defined in `__init__()`), it has a phenotype that depends on the genotype ("yellow" or "green," calculated in `get_phenotype()`), and it requires a second pea to create the next generation of peas (in `create_offspring()`). A difference to real peas is that the pea class is deterministic: hybridizing two identical peas will always result in the same four children, and the children cover all possible combinations of genotypes. Taken together, the `Pea` class defines how a genotype translates to a phenotype in a structure independent of the rest of the program.

11.3.1 Classes Are Used to Create Instances

A class is an abstract representation of real or imaginary things. The class defines how the things it represents will behave, but it does not contain any specific data. Concrete data are in objects generated from a class that are called *instances*. For instance, the `Pea` class defines "A pea has a genotype." The class does not have any particular genotype in itself. Each instance has an exactly defined genotype ("GG", "Gg", "gG", or "gg"). Therefore, the `Pea` class is the "idea" of all peas in the platonic sense (all peas have a genotype, require other peas to create next generations of peas, etc.), whereas a concrete pea in Python is an instance (e.g., the `"yellow = Pea("GG")"` pea characterized by the `"GG"` genotype). To use a class, you first need to define it, then write a constructor (i.e., an `__init__()` function), and finally create instances from that class.

Defining a Class
Classes are always defined by the `class` keyword, followed by the name of the class. The line starting with `class` ends with a colon:

```
class Pea:
```

The entire following indented code block belongs to that class. The next instruction that is on the same indentation level as the `class` statement does not belong to the class anymore. A class block can be followed by any regular Python command: instructions, a function, or another class.

The Constructor `__init__()`
The *constructor* is a special function that defines what kinds of data your class should contain.

In Python, the constructor is called __init__ () (with two underscores before and after init). In the Pea class, the constructor defines a single data item:

```
def __init__(self, genotype):
        self.genotype = genotype
```

That means that a pea has the genotype specified by the second argument of the constructor. The constructor __init__ is invisibly called whenever you create a new instance. Then a specific genotype is stored in an internal variable via the self.genotype = genotype assignment. Such internal variables are called *attributes* (see Section 11.3.2). The attributes work much like normal variables, only that each variable is preceded by self. You can use default parameters in the constructor, as you would in a normal function. When you have written the constructor, your class is ready to be used.

Q & A: I FIND THE TWO UNDERSCORES AT THE BEGINNING OF __
init__ ARE A LITTLE UGLY. CAN I GIVE THE FUNCTION A DIFFERENT
NAME?

No, the constructor has to be named __init__ (). Python calls the constructor automatically when you call a class. The name needs to be __init__ (), so that Python can recognize it.

How to Create Instances
When creating instances, you feed the class with concrete data (see Figure 11.1). To create instances, you call the class similarly as you do a function, passing as arguments all the parameters required by the constructor but one. In fact, you do not have to explicitly provide a value for the self argument. self is used to tell the class which instance has called it. For example, in

```
yellow = Pea("GG")
```

yellow is the (variable of the) instance that is calling the Pea class. This name is implicitly passed as first argument to the __init__ () constructor so that by typing

```
print yellow.genotype
```

you will get "GG" printed. In other words, all the variables defined in a class become attributes of the instances calling that class. The concrete values of the attributes depend on the specific instances, not on the class. "GG" is the concrete value of the `genotype` attribute for the `yellow` instance. More information on class and instance attributes is provided in Section 11.3.2. The instances resulting from calling the same class with different parameters share the same structure but contain different data. In the example, two pea instances are created at the beginning:

```
yellow = Pea("GG")
green = Pea("gg")
```

These commands create two pea instances (stored in the `yellow` and `green` variables), each having a different genotype ("GG" and "gg", respectively). Each instance is stored in its own variable: they are really different peas, but both can be used in the same way; they have the same attributes and methods of the `Pea` class (see following).

A class works like a function that creates instances.* The created objects have the same basic structure but different content. Whenever you create a new instance, already existing instances do not change. You can have as many instances in parallel as you wish.

11.3.2 Classes Contain Data in the Form of Attributes

Data inside instances are stored in *attributes*. You can access them by using the dot syntax, e.g., `yellow.genotype`. You can use all attribute names that have been defined in the constructor. In the case of the `Pea` class defined earlier, you can access the genotype attribute for any given `Pea` instance:

```
yellow = Pea('GG')
print yellow.genotype
```

Figure 11.2 illustrates the attributes defined in the `Pea` class. Attributes can be changed dynamically like variables. For example, you could update the genotype of a `Pea` instance by reassigning its value:

```
yellow.genotype = 'Gg'
```

Attributes can have any type: integer and floating-point numbers, strings, lists, dictionaries, and even other objects.

---

* This is not exactly how computer scientists define a class, because other programming languages treat classes more strictly, but in Python this is the bottom line of what a class does.

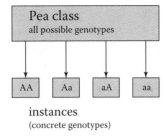

instances
(concrete genotypes)

FIGURE 11.2 Classes versus instances: In the pea example, each Pea instance has its own, individual genotype. The Pea class defines a placeholder for the genotype attribute, but it has no value by itself; it represents general properties of all peas.

Q & A: How Can I Decide Which Attributes Can Go in a Class and Which Not?

Classes are a way to represent tabular data. The attributes of a class correspond to the columns, and each row corresponds to an instance. If you consider creating a class, think of it as a table first. For example, you could group genotypes for different peas in a table:

Genotype	Phenotype
GG	yellow
Gg	yellow
gG	yellow
gg	green

The Pea class is designed in such a way that the genotype is an attribute. The phenotype is calculated from the genotype, but it could also be an attribute (e.g., if the calculation would take very long). If compiling a table with such information makes sense to you, representing each row as an instance of a class will also make sense.

11.3.3 Classes Contain Methods

The functions within a class are called *methods*. Methods are used to work with the information within a class. For instance, the Pea class contains a method to calculate the phenotype from the genotype:

```
def get_phenotype(self):
    if "G" in self.genotype:
        return "yellow"
    else:
        return "green"
```

In this case, the get_phenotype() method uses the data from the genotype attribute. self refers to the instance that calls the method, so that the method will produce a different output for each instance. This is why you need an instance before calling a method (otherwise get_pheno-type() wouldn't know which genotype to use to calculate the phenotype):

```
yellow = Pea('Gg')
print yellow.get_phenotype()
```

Figure 11.3 illustrates that the Pea class has three methods: get_phe-notype(), create_offspring(), and __repr__(). Methods can have default values and optional parameters in the same way that functions do. One class can have methods that analyze, edit, or format the data. Grouping functions as methods in a class is a way of dividing a big program into smaller logical units.

The Parameter self

The main difference between a normal function and a method is that methods contain the self parameter. As mentioned earlier, the self parameter contains the instance for which the method was called. With self you can access all attributes of a class (e.g., self.genotype) and methods (e.g., self.get_phenotype()). The self parameter is auto-matically passed to the method. Therefore, you always call a method with one fewer parameter than there is in the method definition.

11.3.4 The __repr__ Method Makes Classes and Instances Printable

When you print an object created from a class, Python normally prints something like

```
<Pea object a FFFFx234234ou>
```

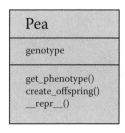

FIGURE 11.3 Methods of the Pea class.

This is not very informative. You can print objects in a more meaningful way by adding a special method called __repr__ () to a class:

```
def __repr__(self):
    return self.get_phenotype() + ' [%s]' %self.genotype
```

This method of the Pea class returns a string containing the genotype and phenotype of an instance. The __repr__ () method takes no parameters except self. No print statement is needed inside the method definition.

Whenever you print a Pea instance, __repr__ () will be called automatically. You do not need to call it explicitly. The __repr__ () method is also called when you print a list containing Pea instances or convert an instance to a string using str(). When writing your own __repr__ () method, you don't have to include all information in the returned string, only what you think is necessary for a concise report. Generally __repr__ () helps to nicely format your data. Therefore, it should be one of the first methods you implement in a new class.

Q & A: What if I Want to Create Different Kinds of Output from My Data?

You can create several methods, one for each output format. For instance, you could add an extra method to the Pea class for tab-separated output:

```
def get_tab_separated_text(self):
    return '%s\t%s' % (self.genotype, self.get_phenotype())
```

Now you can use the __repr__ () format to print a concise summary of a Pea instance and get_tab_separated_text() for generating a tabular report for a generation of peas:

```
for pea in f2:
    print pea.get_tab_separated_text()
```

11.3.5 Using Classes Helps to Master Complex Programs

Using classes helps to get structured, extensible code even for small tasks. But designing classes well is difficult. When you define a class, you only need to decide which attributes and methods the class should contain. As soon as you have two classes, you also need to think whether class A should know about B, B about A, or both. With even more classes, the number of available options quickly explodes.

As a nonexpert programmer, you don't need to worry about these relationships too much. You can build classes as isolated data containers with a few helpful methods and let your main program do the rest. If this makes your data management easier, it is already a big achievement. Good methods to begin with are elementary operations that don't require any additional data, like the `get_phenotype()` method. Typical examples are methods that write data from an instance to a file, calculate simple statistics, or compare two instances of the same class.

When you gather more experience, you can experiment with classes that reference other classes or inherit from them (see Example 11.3). In Python it is also possible to customize how operators like +, -, <, > work for your classes to facilitate, e.g., sorting. Professional design principles for defining classes efficiently are as follows. Each class should be responsible for exactly one thing, known as the *Single Responsibility Principle*. Code that repeats should be avoided, known as the *Don't Repeat Yourself Principle*. To help you create a well-structured architecture using classes, *Design Patterns* offer proven ways to make classes work together (http://sourcemaking.com/). The purpose of these sophisticated structures is to make your program easier to read and understand, not to make it more complicated.

Classes are good for controlling complexity in your programs. The most important thing is to group your data into separate objects and let them talk to each other. Using classes allows you to see how things in your program behave independently of what your program does as a whole. This makes it possible to separate responsibilities more clearly than by using functions, because in the instance of a class, data and methods are strongly connected to each other. The `Pea` class in the previous example is responsible for translating a genotype into a phenotype, regardless of the program it is used in. A good class helps your code become more self-explanatory and reusable.

Q & A: I Have Heard That I Should Write My Entire Program with Classes

Using classes has many advantages when writing a big program, especially when a team of programmers is working together. But you are probably writing small programs most of the time, and you don't have to coordinate five or more people. If your program is just one or two screen pages long, you may not even need classes, but lists and dictionaries will be sufficient. Only if your program keeps growing will it pay off to introduce more structure, for instance, by using classes. Some languages like Java or Smalltalk require everything written as classes. Python is not that strict. You can mix classes with functions and unstructured code. Use this to your advantage: use classes to subdivide the complex parts of your program, and use plain code for the simple parts.

11.4 EXAMPLES

Example 11.1 Importing a Class from a Module

When you want to reuse your classes, it is good to place them in separate modules. Then, you can import them and create objects in different Python programs. For instance, to reuse the Pea class, you need to do the following:

1. Create a new text file pea.py.
2. Paste the class definition there (the entire block from the class keyword until the end of the __repr__ () method).
3. Import the class from another Python program in the same directory. You can also import classes from a different directory by appending the *path* of that directory to the sys.path variable, which is a list (see Chapter 14, Section 14.3.3).

Creating instances of the Pea class in a separate program is very short:

```
from pea import Pea
green = Pea('gg')
print green
```

Source: Adapted from code published by A.Via/K.Rother under the Python License.

This way, you need to maintain only one copy of your class definition.

Example 11.2 Combining Two Classes

The attributes of a class can store anything normal Python variables can. This allows you to combine two or more classes to create more

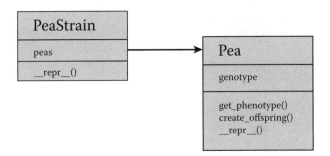

FIGURE 11.4   The PeaStrain class managing instances of the Pea class.

sophisticated structures. For instance, you could have a PeaStrain class that manages a group of peas (see Figure 11.4):

```
from pea import Pea

class PeaStrain:
    def __init__(self, peas):
        self.peas = peas

    def __repr__(self):
        return 'strain with %i peas'%(len(self.peas))
yellow = Pea('GG')
green = Pea('gg')
strain = PeaStrain([yellow, green])
print strain
```

Source: Adapted from code published by A.Via/K.Rother under the Python License.

The strain instance contains a list with two Pea instances. You can access that list directly as an attribute as well:

```
print strain.peas
```

Example 11.3 Creating Subclasses

You can extend the functionality of a class by inheriting other classes. The attributes and methods of a class can be inherited from other classes. The class inherited from is called *base class* or *parent class*, and the inheriting class is called *derived class* or *subclass*. For instance, to allow peas to contain comments, you could define a class `CommentedPea` inheriting from the `Pea` class:

```
from pea import Pea

class CommentedPea(Pea):

    def __init__(self, genotype, comment):
        Pea.__init__(self, genotype)
        self.comment = comment

    def __repr__(self):
        return '%s [%s] (%s)' % (self.get_phenotype(),
        self.genotype, self.comment)
yellow1 = CommentedPea('GG', 'homozygote')
yellow2 = CommentedPea('Gg', 'heterozygote')
print yellow1
```

Source: Adapted from code published by A.Via/K.Rother under the Python License.

The program writes:

```
yellow [GG] (homozygote)
```

The `CommentedPea` subclass has the same methods and attributes as the `Pea` class, plus the ones you add. For instance, the `get_phenotype()` method of the `Pea` class still works in the same way as before. Note that the `__init__()` method calls the method of its parent class. You can redefine methods in a subclass to replace or adapt functionality. For instance, the `__repr__()` method added to `CommentedPea` returns the comment as well. Taken together, subclassing extends the functionality of existing classes. You can think of subclassing in the sense that a `CommentedPea` is a special kind of `Pea`, not that the `Pea` is the biological parent of a `CommentedPea`.

11.5 TESTING YOURSELF

Exercise 11.1 Create a Class

You have a table with information about ion channel protein structures.[*] The table contains a protein name, an identifier of the structure in the

---

[*] Hildebrand, P.W., R. Preissner, and C. Froemmel. *Structural features of transmembrane helices.* FEBS Letters 559, 2004, pp. 145–151.

PDB database, and the average backbone torsion angles φ and ψ of the transmembrane helices.

Ion Channel Name	PDB Code	Mean φ Angle	Mean ψ Angle
Potassium channel	1jvm	354.2	351.7
Mechanosensitivity channel	1msl	359.2	345.7
Chloride channel	1kpl	361.3	344.6

Create a class definition to represent ion channel proteins and implement the __init__ () method. The attributes of the class should correspond to the column names of the table.

Exercise 11.2 Create Objects from a Class

Create three ion channel objects that contain the data from the table in Exercise 11.1. Print each of the objects. Also print its attributes from the main program using the dot syntax. Do not write additional methods to the class.

Exercise 11.3 Make a Class Printable

Add a method to the class that returns a well-formatted string representation of an ion channel. Print the three ion channels again.

Exercise 11.4 Create a Separate Module for a Class

Create a second Python file that imports the ion channel class. Move the commands to the new file and import the class there. Make sure that the program produces the same output as before.

Exercise 11.5 Implement a Class with Methods

Create a class DendriticLengths that manages a list of dendritic lengths like the ones in Chapter 3. The constructor should take a filename as a parameter and create a list of dendritic lengths as an attribute. The class should have three methods: one for calculating the average length, one for calculating the standard deviation, and a __repr__ () method that returns the number of items as a string. Write the class in such a way that the following code works without modification:

```
>>> from neurons import DendriticLengths
>>> n = DendriticLengths('neuron_lengths.txt')
>>> print n
Data set with 9 dendritic lengths
>>> print n.get_average()
184.233666667
>>> print n.get_sttddev()
151.070213316
```

## CHAPTER 12

Debugging

L EARNING GOAL: You can detect and eliminate program errors.

12.1 IN THIS CHAPTER YOU WILL LEARN

- What to do when your program does not work

- How to find and fix typical Python errors

- How to write programs where errors are easy to find

- Who to ask for help

12.2 STORY: WHEN YOUR PROGRAM DOES NOT WORK

12.2.1 Problem Description

Human programmers are not perfect. We make mistakes, because we over-look a small detail, because we are tired, or simply because the problem we are trying to solve is complicated. Even very experienced programmers make mistakes all the time. It is a normal part of writing programs. The bigger the program, the more errors it will contain. Accepting that your programs will contain errors, you can think about the next logical questions: How do you fix the program? What can you do when your program does not work? How can you find and eliminate errors? How can you know that you fixed all errors? The latter is especially relevant for scientific studies, because if they are based on erroneous calculations, the results are not worth much.

In this chapter, you will encounter different types of errors in Python and what you can do to fix or avoid them. You will learn how to fix three kinds of errors: *syntax errors*, *runtime errors*, and *logical errors*. Syntax errors are wrong symbols in the program code that Python does not recognize to the point that your program won't even start. Runtime errors are errors in the code that cause your program to stop abruptly while it is being executed. Logical errors mean that your program finishes normally, but the results are wrong because the program does something different from what you had in mind. You will learn strategies that help you write programs that contain fewer errors and where errors are easier to find. When you get stuck despite these efforts, it is time to ask for help (see Box 12.1).

To learn to know all aspects of debugging, you will analyze a broken program. The program is supposed to sort dendritic lengths from a text

---

BOX 12.1 ASK FOR HELP

Debugging is difficult. Many programmers say it is much more difficult than programming itself. When you are fully immersed into your code, it can be hard to spot the problem. When you are stuck, you get lost in details and overlook things obvious to other people. In that situation a fresh pair of eyes often helps. That fresh pair of eyes could belong to the following:

- *An experienced programmer.* They like to solve difficult problems.
- *A similarly experienced colleague to whom you can explain the program.* The explanation often helps you find where your thoughts might have taken a wrong turn. Or your colleague might come up with a good question that leads you to the wrong piece of code.
- *A nonprogrammer.* Explaining what you are trying to do to a nonprogrammer requires you to simplify conceptually, and this might help you as well.
- *Yourself.* You may simply overlook an easy programming bug because you are tired. If you are fighting with the same error for more than 20 minutes, it is time for a break. Not a break on your web browser. A real break. Switch off the screen, and come back after a while. Then consider again what your program should do, and then take a look to see whether it really does that.

Admitting that you are stuck at the moment and talking to another person are often the fastest ways to get a hint that solves the problem right away, or they allow you to take a deep breath and approach your code with fresh energy.

file into three categories. The input text file contains two columns with primary and secondary dendritic lengths:

```
Primary       16.385
Primary       139.907
Primary       441.462
Secondary     29.031
Secondary     40.932
Secondary     202.075
Secondary     142.301
Secondary     346.009
Secondary     300.001
```

The program should count how many neurons are shorter than 100 μm, how many are longer than 300 μm, and how many are in between.

12.2.2 Example Python Session

```python
def evaluate_data(data, lower=100, upper=300):
    """Counts data points in three bins."""
    smaller = 0
    between = 0
    bigger = 0

    for length in data:
        if length < lower:
            smaller = smaller + 1
        elif lower < length < upper:
            between = between + 1
        elif length > upper:
            bigger = 1
    return smaller, between, bigger

def read_data(filename):
    """Reads neuron lengths from a text file."""
    primary, secondry = [], []

    for line in open(filename):
        category, length = line.split("\t")
        length = float(length)
        if category == "Primary"
            primary.append(length)
        elif category == "Secondary":
            secondary.append(length)
    return primary, secondary

def write_output(filename, count_pri, count_sec):
    """Writes counted values to a file."""
    output = open(filename,"w")
```

```
    output.write("category <100 100-300 >300\n")
    output.write("Primary : %5i %5i %5i\n" % count_pri)
    output.write("Secondary: %5i %5i %5i\n" % count_sec)
    output.close()
primary, secondary = read_data('neuron_data.xls')
count_pri = evaluate_data(primary)
count_sec = evaluate_data(secondary)
write_output_file('results.txt',count_pri,count_sec)
```

Source: Adapted from code published by A.Via/K.Rother under the Python License.

12.3 WHAT DO THE COMMANDS MEAN?

The program is supposed to analyze a two-column table of dendritic lengths. The input table is in the neuron_data.txt file. When you try to run the program, you will notice that Python terminates abruptly with an error message:

```
File "neuron_sort.py", line 23
    if category == "Primary"
       ^
SyntaxError: invalid syntax
```

On first sight, the code looks good. What is wrong, and how can the program be fixed? At this point you should start working your way through the errors one by one.

12.3.1 Syntax Errors

You have probably seen messages like

```
SyntaxError: invalid syntax
```

before. SyntaxError means that the Python interpreter did not understand a particular line of code and stopped immediately. The reason for the error is usually that a keyword or special character is misspelled or in the wrong position. For example, instead of print the code contains prin, or instead of defining a list with commas [1, 2, 3] the code is written with semicolons [1; 2; 3].

A syntax error is the programming mistake that is easiest to find. You can find most syntax errors by looking at the code. Python helps you by giving not only the line number (line 23 in this case) but also a ^ symbol indicating where in the line the problem occurred:

```
File "neuron_sort.py", line 23
    if category == "Primary"
      ^
```

If you don't spot the error right away, check the corresponding line in your text editor. Make sure you are using a text editor that displays the line numbers (practically all editors except Windows Notepad do that). Also, Python syntax-highlighting like in the IDLE editor helps you spot mistakes before you try to run your program. Some advanced editors like Eric immediately mark syntax errors with big red dots. Some syntax errors may be hard to find. In Box 12.2 is a checklist of things you can try to spot. If you have tried everything listed there and still haven't found the error, it is time to ask other programmers. They all started programming because they like to solve difficult problems.

BOX 12.2   HOW TO TACKLE A SYNTAXERROR

- Check the line before the one highlighted in the syntax error message.
- If there is an `if`, `for`, or `def` statement, is there a colon (:) at the end of the line?
- If there is a string starting earlier, is it closed properly (check this especially if you are using multiline strings with """ and ''')?
- If there is a list, dictionary, or tuple stretching over multiple lines, is there a closing bracket?
- Check whether spaces and tabs for indentation are mixed in the code. They look the same; therefore, it is much better to consequently use spaces from the very beginning.
- Comment the line or the entire section where the error occurs. Does the syntax error disappear? If you see a different error message, you have localized the problem.
- Remove the line or section where the error occurs. Does the syntax error disappear now?
- Are you trying to run Python 2.7 code with Python 3.x? The `print` command in Python 2.x is a `print()` function in Python 3.x. In this and other cases, code is incompatible between Python 2 and 3.

In the program, the colon at the end of line 23 is missing. The correct line should be

```
if category == "Primary":
```

12.3.2 Runtime Errors

When you add the missing colon and run the program again, you should see a different message:

```
  File "neuron_sort.py", line 37, in <module>
      primary, secondary = read_data('neuron_data.xls')
  File "neuron_sort.py", line 20, in read_data
      for line in open(filename):
IOError: [Errno 2] No such file or directory:
'neuron_data.xls'
```

If there are no syntax errors in your code, Python tries to execute your program line by line. All error messages you see from that point on are called runtime errors. Generally they mean that Python tried to execute a particular line of code, but something went wrong during execution. On the good side, at least you know now that the program is syntactically correct.

There are many possible reasons for this category of error. Maybe the program tried to use a variable that doesn't exist, or a file was not found. In any case, you need to analyze what exactly has happened in order to fix the error. A common strategy to do this is to read the error message *from the bottom*. You need to identify two things there:

- The type of error in the last line (an IOError in this case)

- The line where it occurred in the innermost function (line 20 in this case; line 37 just calls the function that leads to line 20)

Python knows many kinds of runtime errors. We discuss the most important next.

IOError

Your program tried to communicate with an input or output device (a file, directory, or website), but something went wrong. With files the most common reason is that the file or directory name is misspelled. It is also possible that the program could not read or write a given file because you as a user have no permission or the file is already open. With web pages, the reason can be a wrong URL or a problem with the Internet connection.

Before you consider any of the other reasons, check the name of the file, directory, or web page. Print the name from the program to the screen by adding a `print` instruction right before the line in which the error occurs:

```
print filename
for line in open(filename):
```

Check the filename twice, and pay attention to whitespaces and uppercase and lowercase letters. With files, also double-check in your file browser or a terminal whether the file is really in the directory where you think it is. Check whether your program really started in the directory you expect it to be in (having multiple copies of your program makes this error very common).

If you are reading a website, copy and paste the address to your browser. When you double-check filenames carefully, IOErrors are usually easy to fix. Chances are that you won't see the same problem again. In the example program in Section 12.2.2 there is such a spelling mistake. It is enough to change the filename from neuron_data.xls to neuron_data.txt to get rid of the IOError.

NameError

After fixing the previous bug, there is a new error message:

```
Traceback (most recent call last):
    File "neuron_sort.py", line 37, in <module>
        primary, secondary = read_data('neuron_data.txt')
    File "neuron_sort.py", line 26, in read_data
        secondary.append(length)
NameError: global name 'secondary' is not defined
```

A NameError indicates that the name of a variable, function, or another object is unknown to Python at the moment it is encountered. The line where the error occurs is important, because the namespace, the set of all known names and variables currently available in your program, changes all the time, e.g., when new variables are defined, when the program jumps into a function and back, or when an external module is imported. Frequent reasons for a `NameError` are the following:

- *A name was not imported.* You forgot to import a variable or function from a different module. Unless you use `import *`, it is easy to find out whether you imported a particular name (therefore prefer `import name` over `import *`).

- *A variable has not been initialized.* For instance, if you want to use a variable for counting and have a line

```
counter = counter + 1
```

then you will need to initialize that variable somewhere above:

```
counter = 0
```

This can be tricky if your variable initialization depends on a condition, e.g.,

```
if a > 5:
    counter = 0
counter = counter + 1
```

This piece of code will work if a is initially 6 or bigger; otherwise it will terminate with an error because `counter` has not been defined. Therefore, variable initialization should always be unconditional.

- *A variable or function name is misspelled.* This is a very frequent bug. What can make it difficult to track is that the misspelling is not always in the line given in the error message. For instance, if you check line 26 in the previous script, you will find that `secondary` is spelled correctly. So you need to check the code above it to see whether the variable *definition* is also spelled correctly.

A good diagnostic tool for `NameErrors` is to add the line

```
print dir()
```

to your program in the line before the error occurs. Then you will see the list of variables that are known to Python (except for the built-in functions). In this case:

```
['category', 'filename', 'length', 'line', 'primary',
'secondry']
```

Now it is easier to see that secondry is misspelled. When you search for the misspelled version in your code, you will find it in line 18. Thus, the NameError can be fixed by replacing secondry with secondary in line 18. Another NameError occurs in line 40 and can be fixed by changing write_output_file to write_output in the same line. Now the neuron_sort.py program runs without errors and produces an output file. Other types of errors that occur frequently in Python include ImportError, ValueError, and IndexError. They are presented in Examples 12.1–12.3.

12.3.3 Handling Exceptions

Errors that occur because of misspelled filenames like the IOError in the example program in Section 12.2.2 are very common. Another situation where you can expect problems are when your program expects a certain data format that is not always there. If you know that such problems can arise, it would be good if your program could anticipate them and react accordingly instead of you having to start debugging each time a problem occurs. *Exception handling* in Python does exactly that.

The try...except Statement
The try...except statement allows you to catch program errors before they stop the program. The set of statements where you anticipate a runtime error is inserted into an indented block starting with try, whereas statements that react to the error are inserted in an indented block starting with except. try...except statements allow handling exceptions by specifying in the except statement what type of exception you want to take care of. For instance, you can use the except statement to react on specific error types:

```
try:
    a = float(raw_input("Insert a number:"))
    print a
except ValueError:
    print "You haven't inserted a number. Please retry."
    raise SystemExit
```

Source: Adapted from code published by A.Via/K.Rother under the Python License.

In the previous example, the *expected* exception (ValueError) is given as argument to the except statement: the corresponding indented block will be entered only in case of ValueError exceptions; any other type of exception will not be handled. The raise SystemExit creates an exception that terminates the program in a controlled way. If you want to handle more than one exception, you can add as many except blocks as the number of exceptions you want to handle.

If you do not specify any arguments in the except statement, the first exception in the try block—no matter what type—will cause the program to execute the except block. This ensures a controlled termination of your program in all cases but gives you less control over what is happening (because you cannot determine what kind of runtime error in the try block caused the execution of the except block).

The else *Statement*

An else statement can be optionally added after a try...except block. The set of statements controlled by else are executed if no exception has been generated in the try block, i.e., if none of the except statements have been executed.

```
try:
    <statements>
except:
    <statements>
else:
    <statements>
```

A practical use for the else statement is to read a filename from the keyboard and process a file only if the file can be opened:

```
try:
    filename = raw_input("Insert a filename:")
    in_file = open(filename)
except IOError:
    print "The filename %s has not been found." % filename
    raise SystemExit
else:
    for line in in_file:
        print line
        in_file.close()
```

Source: Adapted from code published by A.Via/K.Rother under the Python License.

Using `else` allows you to fully exploit exception handling and better organize your code by putting in the `try` block only those statements the exceptions of which you want to control.

Summarizing, the `try...except` statement allows your program to react to errors. First, the `try` block is executed. If the interpreter does not meet any exception, the `except` statement is ignored and the `else` block (if present) is executed. If an exception occurs during the execution of the `try` block, the rest of the block is ignored and the interpreter jumps to an `except` statement with the right type of exception. If no `except` statement can handle the error, execution terminates with an error message.

12.3.4 When There Is No Error Message

Some program errors do not produce an error message. These are silent bugs. Your program runs, but it does not do what it should. This is the most challenging situation in debugging, because you have to figure out that there is something wrong in the first place and then locate and eliminate the problem. Now, you could start reading the code, try to analyze it step by step, check variable and function names, etc. But this is very difficult, because you don't know what to look for. You need more information.

Where to Start?

> I have no data yet. It is a capital mistake to theorise before one has data. Insensibly one begins to twist facts to suit theories, instead of theories to suit facts.[*]

To track a bug, you need to gather as much information as possible. Use the deductive approach found in Sherlock Holmes's investigations. When facing a mysterious case, the sleuth usually starts collecting facts and then excludes possibilities. Finally, he draws logical conclusions. In programming, if function A works fine and function C as well, the problem must be in B. Thus, you need to watch the parts of your program at work. There are three ways to do so: (1) comparing input and output of your program, (2) adding `print` statements, and (3) using the Python debugger.

---

[*] From *The Adventures of Sherlock Holmes* by Sir Arthur Conan Doyle.

Comparing Input and Output

A first starting point is to compare the input and output of your program. Using a small example file for testing helps at this point. When you run the neuron_sort.py program with the input file given previously, you obtain a file results.txt, containing three lines:

```
category <100 100-300 >300
Primary: 1 1 1
Secondary: 2 2 1
```

The input file neuron_data.txt contains nine lines. Thus, the table should contain a sum of nine counts as well. You can see that there are only eight. More precisely, the last column should have a "2" for secondary neurons. Concluding, there is a bug causing one neuron length >300 not being counted. The rest of the table seems unaffected.

Adding print *Statements*

You can add print statements almost anywhere in your code to print variables and results of functions or simply to indicate that a given line has been reached. If you choose this strategy to track a silent error, add a print statement before and after a possibly erroneous piece of code. Then run the program and check the output manually. By moving the print statements up and down, you can narrow down the location of an error.

In the neuron_sort.py program, one out of six secondary neurons was not counted properly. The first print statement could be added to the last paragraph of the script to check whether the individual functions are working:

```
primary, secondary = read_data('neuron_data.txt')
print secondary
count_pri = evaluate_data(primary)
count_sec = evaluate_data(secondary)
print count_sec
write_output('results.txt',count_pri,count_sec)
```

The first print statement writes a list of all secondary neuron lengths to the screen:

```
[29.031, 40.932, 202.075, 142.301, 346.009, 300.001]
```

You can see that all six numbers from the input file are in the list. From that you can conclude that the read_data() function works correctly.

The second print statement writes a tuple with the counts for the three length bins to the screen:

```
(2, 2, 1)
```

From the input data, it can be seen that this tuple should be (2, 2, 2). Now you can conclude that the problem must be in the evaluate_data() function. You need to examine the code in there more closely. You can do that by adding more print statements there or by using the Python debugger.

Using the Python Debugger

The Python debugger is a tool that allows you to execute your program step by step and see what it does in each single line. It gives you a shell from which you can check and modify variables and execute lines one by one. To use the Python debugger, you need to insert two lines into your program:

```
import pdb
pdb.set_trace()
```

When these lines are reached, Python holds the execution and gives you control of the program in a shell window with a few extra commands available:

- "n" executes the next line.
- "s" executes the next line but does not descend into functions.
- "l" shows where in the code the program currently is.
- "c" continues execution normally.

Apart from that, everything you can do in the regular Python shell works as well. To analyze the bug in the evaluate_data() function of neuron_sort.py, you can start the debugger in that function.

```
def evaluate_data(data, lower = 100, upper = 300):
    """Counts data points in three bins."""
    import pdb
    pdb.set_trace()
```

When you start the program, the debugger is started:

```
> neuron_sort.py(6)evaluate_data()
-> smaller = 0
(Pdb)
```

(Pdb) is the prompt of the debugger. The line displayed is the next one to be executed. Now you can check what parameters the function got by typing data at the (Pdb) prompt:

```
(Pdb) data
[16.385, 139.907, 441.462]
```

This is the list of data for the primary neurons. You already know the problem is with the secondary neurons, so you can type "c" to continue the program. The program finishes running for the primary neurons and starts running for the secondary neurons. After a moment, when the evaluate_data() function is called for the second time, the debugger stops at the same point again. This time, we have the secondary neurons:

```
(Pdb) data
[29.031, 40.932, 202.075, 142.301, 346.009, 300.001]
```

Now you can track the execution line by line by typing "n" a couple of times. You will see that the lines starting from

```
-> for length in data:
```

are repeating in the debugger output, once for each of the six items in data. You can check the value of length anytime after executing the for statement with "n" by typing length at the debugger prompt:

```
(Pdb) n
> neuron_sort.py(11)evaluate_data()
-> between = 0
(Pdb) n
> neuron_sort.py(12)evaluate_data()
-> bigger = 0
(Pdb) n
> neuron_sort.py(14)evaluate_data()
-> for length in data:
(Pdb) n
> neuron_sort.py(15)evaluate_data()
-> if length < lower:
```

```
(Pdb) length
29.030999999999999
(Pdb)
```

The screen output of the debugger shows which lines are executed for each value from the data list. For the first two values, these are as follows:

```
-> if length < lower:
-> smaller += 1
```

For the second two,

```
-> if length < lower:
-> elif lower < length < upper:
-> between += 1
```

Here, you can see that the if length < lower: condition, for that length value, returns False. So, after this line, the second if condition is executed instead. For the last two values,

```
-> if length < lower:
-> elif lower < length < upper:
-> elif length > upper:
-> bigger = 1
```

At that point, you may notice that the bigger = 1 line probably should look like the other two additions. When you change the line to

```
-> bigger += 1
```

the program will run without mistakes.

12.4 EXAMPLES

Example 12.1 ImportError

When you see an ImportError, it means that Python tried to import a module but failed. There can be two reasons for this. Either the module was not found, or the module did not contain what you tried to import. There are a few things you can check:

- You can check the spelling of the module name.
- You can verify the directory you started the program from. Is the module you want to import where you expect it to be?

If you import from a Python library you have installed manually, you can try

```
import sys
print sys.path
```

- The directory you want to import from should be listed there. Is it really? If not, you need to add it to your PYTHONPATH variable (on the UNIX or Windows level) or to append it to the sys.path list.
- Check whether you can import the module itself in the first place: import X.
- Then try whether you can import variables and functions from within from X import Y.
- If you import from a subdirectory (e.g., import tools. parser), does the directory contain the mandatory Python file __init__ .py?
- You can check for duplicate names. If the name of your module or function is the same as the name of a function imported from the Python Standard Library, this means trouble. Change the name of your module, and see whether the problem persists. This kind of problem is very hard to detect if you use import *, so it's better that you don't use it.

Example 12.2 `ValueError`

A `ValueError` occurs when two variables for an operation are incompatible. For instance, when you try to add an integer to a string. The reason could be that you forgot to convert your data using the `int()` or `float()` function while reading a file. Another possibility is that you forgot to write the [i] index to access the elements of a list. Python then tries to work with the entire list instead of its contents. Adding a list to a number also gives you a `ValueError`. A good diagnostic tool for `ValueError`s is to add a `print` statement before the line of the error. For instance, if adding a and b gives you an error, you can see whether their types are compatible by

```
print a, b
result = a + b
```

or

```
print type(a), type(b)
result = a + b
```

If in the example a is [1, 2, 3] (or type(a) is "list") and b is 4 (or type(b) is "int"), which does not fit, you will see that in the output.

Example 12.3 **IndexError**

An IndexError occurs when Python fails to find an element in a list or dictionary. For instance, if you have a list with three items, accessing the fourth item from zero by [3] leads to an error:

```
>>> data = [1, 2, 3]
>>> print data[3]
Traceback (most recent call last):
    File "<pyshell#3>", line 1, in <module>
        print data[3]
IndexError: list index out of range
```

When you have large lists and dictionaries, variables for the indices, or more complex structures, IndexErrors are more complicated to analyze. You can add print statements to display the entire data, other variables you use, or the keys of a dictionary using the keys() function. If the problem is more complex, you will have to do a more thorough analysis of the errors. Then, the situation is similar to errors that give you no error message at all.

Example 12.4 Writing Readable Code

Your job as a programmer is not to write error-free programs. Nobody can. Not even the best programmers can write a program in the first attempt that is error free. Your job is to write programs in such a way that your own mistakes are easier to find. Generally, code is made more readable by good code modularization (as described in Chapters 10 and 11), by good organization of a programming project (see Chapter 15), and by well-formatted code. Good formatting is characterized by the following:

- Use descriptive names for variables and functions. A line like

    ```
    for line in sequence_file:
    ```

    tells much more about what your program does than

    ```
    for l in f:
    ```

- Variable names are better if they explicitly describe the kinds of data, not just the types: `sequence` is better than `text`, `seq_length` is better than `number`.
- Function names should start with a verb (expressing the function action) and contain one to three words: `read_sequence_file` is easier to read than `read` or `seq_file`. You cannot write programs as you would write English text, but occasionally you can get pretty close to that.
- *Write comments.* Comments help anyone reading your code to understand what the program does. The most important thing is a short description at the very beginning of a program or function. However, you don't need to comment everything. A lot can be made understandable by naming variables and functions well. When your program evolves, the comments will go out of date quickly. As a rule of thumb, use comments as headings for paragraphs in your program, and document lines you find difficult.
- *Avoid the import * statement.* Whenever you import something, add the explicit names of all objects you import. For example, instead of

```
from math import *
```

it is better to write

```
from math import pi, sin, cos
```

Your code will be much easier to analyze.
- Use a uniform code formatting style like PEP8 (see Chapter 15).

12.5 TESTING YOURSELF

Exercise 12.1 Debugging Python Session in Section 12.2.2

Copy the example in Section 12.2.2 to a text file and retrace all the actions described in this chapter to debug it. In particular, see what happens when you use the `pdb` Python debugger.

Exercise 12.2 Use of `try...except` Statements

Once you have debugged the program, let it react to a missing input file. Add a suitable `try...except` statement to Section 12.2.2 that catches `IOErrors` and reacts with a neatly printed error message.

Exercise 12.3 Exception Handling in Dealing with Files and Numbers

Write a script that reads a set of numbers from a column of a text file, converts the numbers to floating-point numbers, and calculates their mean (see Chapter 3, Example 3.1) and standard deviation (see Chapter 3, Example 3.2). Add some lines to the input file containing "–" instead of a number. Which error would this create? Use exception handling to skip those lines but print a warning message.

Exercise 12.4 Nested `try...except` Statements

To obtain a greater control of errors, you can use nested `try...except` statements by inserting an additional `try...except` block into an existing `except` or `else` block. Modify the script from Exercise 12.3 to use a nested `try...except` block to handle both wrong filenames and nonnumber symbols in the input data.

Exercise 12.5 Exception Handling in Dealing with Standard Input and Numbers

Do the same exercise as in Exercise 12.3 but read the numbers from the keyboard instead of a file. Do you have to use the same `try...except` blocks? Why?

Hint: Use `raw_input()` to read the numbers from the standard input.

When reading the standard input, insert a condition to stop inserting numbers. For example,

```
input_numbers = []
number = None
while number != 'q':
        number = raw_input("Insert a number: ")
        input_numbers.append(number)
```

## CHAPTER 13

Using External Modules: The Python Interface to R

L EARNING GOAL: You can use Python to do statistical analyses with R.

13.1 IN THIS CHAPTER YOU WILL LEARN

- How to run R commands from a Python script

- How to save R output into a Python variable

- How to generate an R object (e.g., a vector) from a Python object (e.g., a tuple)

- How to automatically generate R plots from Python

13.2 STORY: READING NUMBERS FROM A FILE AND CALCULATING THEIR MEAN VALUE USING R WITH PYTHON

13.2.1 Problem Description

Biologists have a constant need to statistically analyze their data and plot them. R (www.r-project.org/) is one of the most frequently used pieces of software for statistical computing and graphical analyses. In many situations, you may find it very useful to be able to call R from a Python script.

For example, if you have to calculate the mean value and the standard deviation of several distributions of numbers, each recorded in a different file, and then you want to automatically create one or more plots, you can delegate many tasks of your calculation to R and use Python to connect them. In this chapter, we assume that you already know how R works. If not, we strongly suggest that you familiarize yourself with the basics of R before reading this chapter.

Python has two modules, RPy and RPy2, to connect with R. RPy2 is a redesigned version of RPy. All examples in this chapter use RPy2, which we recommend to use. The module must be downloaded, installed, and imported into a script or a Python session (see Box 13.1 for RPy2 installation).

BOX 13.1   INSTALLING THE PYTHON INTERFACE TO R

Installing RPy or RPy2 may be the most difficult thing in this entire chapter. In fact, you have to choose a release of RPy or RPy2 that is consistent with the R and Python versions that are installed on your computer.

If `easy_install` is available on your computer, you can just type in a UNIX/Linux shell

```
sudo easy_install rpy2
```

`easy_install` is a Python module that lets you automatically download, build, install, and manage Python packages. To check if the package is available on your computer, go to the command line terminal and type

```
easy_install
```

If you get the warning (or a similar one)

```
error: No urls, filenames, or requirements specified (see- help)
```

instead of

```
easy_install: Command not found.
```

it means that you already have `easy_install`. Otherwise go to https://pypi.python.org/pypi/setuptools.

In the following session, simple R actions are performed, such as creating vectors, creating matrices, reading data from a file, and calculating the mean of a set of numbers. Once you get the philosophy behind the use of R via Python, you will find it easy to access any R function from Python.

13.2.2 Example Python Session

Python commands:
```
import rpy2.robjects as robjects
r = robjects.r
pi = r.pi
x = r.c(1, 2, 3, 4, 5, 6)
y = r.seq(1,10)
m = r.matrix(y, nrow = 5)
n = r.matrix(y, ncol = 5)
f = r("read.table('RandomDistribution.tsv', sep = '\t')")
f_matrix = r.matrix(f, ncol = 7)
mean_first_col = r.mean(f_matrix[0])
```

Source: Adapted from code published by A.Via/K.Rother under the Python License.

Equivalent R commands:
```
> p = pi
> x = c(1,2,3,4,5)
> y = seq(1,10)
> m = matrix(y, nrow = 5)
> n = matrix(y, ncol = 5)
> f = read.table('RandomDistribution.tsv',sep = '\t')
> f_matrix = matrix(f, ncol = 7)
> mean_first_col = mean(f[,1])
```

Figure 13.1 shows how the file `RandomDistribution.tsv` looks.

6071	103	0.0169659034755	40	0.00658870037885	276	0.0454620326141
6106	109	0.0178512938094	38	0.00622338683262	265	0.0433999344907
6148	93	0.015126870527	65	0.0105725439167	261	0.0424528301887
6119	114	0.018630495179	32	0.00522961268181	239	0.0390586697173
6118	87	0.0142203334423	47	0.00768224910101	287	0.0469107551487
6154	104	0.0168995775106	52	0.00844978875528	277	0.0450113747156
6154	118	0.019174520637	31	0.00503737406565	258	0.0419239519012
6143	94	0.0153019697216	23	0.00374409897444	281	0.0457431222530
6120	120	0.0196078431373	26	0.00424836601307	261	0.0426470588235
6142	108	0.0175838489092	45	0.00732660371215	290	0.0472158905894
6129	107	0.017457986621	36	0.00587371512482	262	0.0427475934084
6117	126	0.0205983325159	37	0.00604871669119	285	0.0465914664051
6171	138	0.0223626640739	40	0.00648193161562	255	0.0413223140496
6121	140	0.0228720797255	25	0.00408429995099	257	0.0419866034962
6090	107	0.0175697865353	39	0.00640394088670	270	0.0443349753695
6123	106	0.0173117752736	45	0.00734933855953	260	0.0424628450106
6139	141	0.0229679100831	53	0.00863332790357	225	0.0366509203453
6122	118	0.0192747468148	38	0.00620712185560	265	0.0432865076772
6084	99	0.0162721893491	33	0.00542406311637	260	0.0427350427350
6094	113	0.0185428290121	21	0.00344601247128	259	0.0425008204792
6139	102	0.0166150838899	27	0.00439811044144	289	0.0470760710213

FIGURE 13.1   Portion of the RandomDistribution.tsv file.

13.3 WHAT DO THE COMMANDS MEAN?

13.3.1 The `robjects` Object of `rpy2` and the `r` Instance

We assume that the module `rpy2.py` is installed on your computer (see Box 13.1) and that you already know how R works. The module to be imported to use the `rpy2` package is `robjects`:

```
import rpy2.robjects as robjects
```

The `r` object of the `rpy2.robjects` module (`robjects.r`) represents the "bridge" from Python to R. In the example, `robjects.r` has been assigned to the variable `r` to avoid writing `robjects.r` every time an R function is used:

```
r = robjects.r
```

13.3.2 Accessing an R Object from Python

At this point, you can start using R in Python. You can access R objects from Python in three ways: (1) accessing an R object as an attribute of the `r` object, using the dot syntax; (2) using the `[]` operator on `r` like you would use a dictionary; and (3) calling `r` like you would do with a function, passing the R objects as arguments. In all cases, the result is an R vector.

Accessing an R Object as an Attribute of the `r` Object,
Using the Dot Syntax
In R you can access, for instance, the `pi` object (which in R is a vector of length 1, the value of which is 3.141593) as follows:

```
> pi
[1] 3.141593
```

In Python, you can get `pi` by typing

```
>>> import rpy2.robjects as robjects
>>> r = robjects.r
>>> r.pi
<FloatVector - Python:0x10c096950/R:0x7fd1da546e18>
[3.141593]
```

This makes perfect sense, with r being the Python interface to R: R objects are basically attributes of the r object, and you can access them by simply using the dot syntax. Notice that if you use the print statement, the result will look a bit different:

```
>>> print r.pi
[1] 3.141593
```

And since r.pi is a vector of length 1, if you want to get its numerical value, you have to use indexing:

```
>>> r.pi[0]
3.141592653589793
```

Accessing an R Object Using the [] Operator on r
Like You Would Use a Dictionary
You can think of R object names and their values as key:value pairs of a dictionary and retrieve the value of 'pi' as follows:

```
>>> pi = r['pi']
>>> pi
<FloatVector - Python:0x10f4343b0/R:0x7f8824e47f58>
[3.141593]
>>> pi[0]
3.141592653589793
```

Calling r Like You Would Do with a Function,
Passing the R Object as an Argument
Another way to access the value of an R object is by calling the r object as you would do with a function, passing as argument the R object name:

```
>>> pi = r('pi')
>>> pi[0]
3.141592653589793
```

In summary, the r object works like an object with attributes through the dot syntax, like a dictionary, and like a function to achieve the same result. Notice that in all these cases, the result is a vector, the value(s) of which can be accessed using the [] operator as you do for Python lists or tuples.

Nearly everything in R is a vector or a matrix (a vector of vectors). Therefore, it is important to learn how to manipulate such objects, how to extract their elements, and how to convert R objects into Python objects and vice versa.

13.3.3 Creating Vectors

Similarly to the R pi object, R functions for vector building can be called as attributes of robjects.r (remember that robjects.r has been stored in the r variable) using the dot syntax:

```
>>> print r.c(1, 2, 3, 4, 5, 6)
[1] 1 2 3 4 5 6
```

Remember that R vectors can be generated using the c() function. As in the case of the pi object, you can use two additional ways to get R vectors from the r object: interpreting r as a dictionary or as a function.

To use the dictionary-like approach, you can translate the R function c() into a Python function using the [] operator:

```
>>> print r['c'](1, 2, 3, 4, 5, 6)
[1] 1 2 3 4 5 6
```

This means that you can call the arguments of c() after you have converted it to the Python function r['c']. You can also do it in two steps by assigning the r['c'] function to a variable first and then calling it as you usually call functions in Python:

```
>>> c = r['c']
>>> print c(1, 2, 3, 4, 5, 6)
[1] 1 2 3 4 5 6
```

If you want to use the r object as a function instead, you can do it as follows:

```
>>> print r('c(1,2,3,4,5,6)')
[1] 1 2 3 4 5 6
```

Notice that the argument in the r call is converted to a string type (using single quotation marks).

These three approaches work for any R functions, e.g., if you want to generate a vector in Python using the R function seq():

```
>>> y = r.seq(1, 10)            #using the dot syntax
>>> print y
```

```
[1]  1 2 3 4 5 6 7 8 9 10
>>> s = r['seq']                    #dictionary-like
>>> print s(1,10)
[1]  1 2 3 4 5 6 7 8 9 10
>>> print r('seq(1,10)')            #function-like
[1]  1 2 3 4 5 6 7 8 9 10
```

Q & A: WHICH OF THE THREE WAYS TO ACCESS R OBJECTS SHOULD I USE?

Our suggestion is this: the simpler the better, but much depends on your preferences. You might even decide to mix the different ways to get R objects in a Python program. For example, r.pi looks slightly simpler than r('pi'), but you may prefer the latter. In all cases, you have to remember that the result of retrieving an R object in Python is always an R vector; therefore, you have to use indexing to specifically access its elements.

13.3.4 Creating Matrices

You can create a matrix in R as follows:

```
> y = seq(1,10)
> matrix(y, ncol = 5)
         [,1]    [,2]    [,3]    [,4]    [,5]
[1,]       1       3       5       7       9
[2,]       2       4       6       8      10
>
```

In Python, you have to convert both seq() and matrix() R functions into Python objects using robjects.r. You can do it in the same three ways as in Section 13.3.3.

Accessing R Functions as Attributes of robjects.r Using the Dot Syntax

```
>>> import rpy2.robjects as robjects
>>> r = robjects.r
>>> y = r.seq(1,10)
>>> print r.matrix(y, ncol = 5)
         [,1]    [,2]    [,3]    [,4]    [,5]
[1,]       1       3       5       7       9
[2,]       2       4       6       8      10
```

```
>>> print r.matrix(y, nrow = 5)
       [,1]    [,2]
[1,]      1       6
[2,]      2       7
[3,]      3       8
[4,]      4       9
[5,]      5      10
```

Notice that you can also reassign the function to a variable and then use it:

```
>>> import rpy2.robjects as robjects
>>> r = robjects.r
>>> y = r.seq(1,10)
>>> m = r.matrix
>>> print m(y, nrow = 5)
       [,1]    [,2]
[1,]      1       6
[2,]      2       7
[3,]      3       8
[4,]      4       9
[5,]      5      10
```

Accessing R Functions Using the [] Operator on robjects.r Like You Would Use a Dictionary

```
>>> import rpy2.robjects as robjects
>>> r = robjects.r
>>> y = r['seq'](1,10)
>>> print r['matrix'](y, ncol = 5)
       [,1]    [,2]    [,3]    [,4]    [,5]
[1,]      1       3       5       7       9
[2,]      2       4       6       8      10
```

Also in this case you can reassign the function to a variable and then use it:

```
>>> import rpy2.robjects as robjects
>>> r = robjects.r
>>> y = r['seq'](1,10)
>>> m = r['matrix']
>>> print m(y, ncol = 5)
       [,1]    [,2]    [,3]    [,4]    [,5]
[1,]      1       3       5       7       9
[2,]      2       4       6       8      10
```

Calling `robjects.r` *Like You Would Do with a Function,*
Passing the R Functions as Arguments

```
>>> import rpy2.robjects as robjects
>>> r = robjects.r
>>> y = r('seq(1,10)')
>>> print r('matrix('+y.r_repr()+', ncol = 5)')
       [,1]    [,2]    [,3]    [,4]    [,5]
[1,]     1       3       5       7       9
[2,]     2       4       6       8      10
```

Notice that the arguments are passed to the `r` object in the form of strings. This implies that the following commands

```
>>> y = r('seq(1, 10)')
>>> print r('matrix(y, ncol = 5)')
```

will return an error message because y is not a string (but a vector of numbers), and in Python you cannot mix different data types. Thus, you first have to convert y into a string and then concatenate it to `'matrix()'`. The conversion into a string can be nicely done with the `r_repr()` method, which works on all R objects and returns a string representation that can be directly evaluated as R code:

```
>>> y = r.seq(1, 10)
>>> y.r_repr()
'1:10'
```

This applies in general to any R commands: you can write them as strings and then pass them as arguments to `robjects.r` when you call it. For example, the following R command

```
> f = read.table("RandomDistribution.tsv", sep = "\t")
```

can be written in Python as follows:

```
>>> import rpy2.robjects as robjects
>>> r = robjects.r
>>> f = r("read.table('RandomDistribution.tsv', sep = '\t')")
```

13.3.5 Converting Python Objects into R Objects

The previous examples show how to create R objects in Python using R functions in the form of `robjects.r` attributes, dictionary keys, or function arguments. The content of the resulting objects can be either accessed as you would access Python arrays (through the `[]` operator, e.g., `y[0]`) or reused in R functions (as `y` in `matrix()`). However, in many cases, it will turn out to be very useful to convert a Python object (e.g., a list or a tuple) into an R object, which can be used in R functions. In fact, suppose you read the table in Figure 13.1 from a file and save its content to a Python list of lists (e.g., using `readlines()`). What if you want to calculate with R the mean of the values of the table's first column? To this purpose, you can use the `FloatVector()` method of `robjects` that converts lists or tuples of floating numbers (or of strings containing floating numbers) into an R array of floating numbers.

```
>>> F = open('RandomDistribution.tsv')
>>> lines = F.readlines()
>>> l = []
>>> for line in lines:
...     l.append(float(line.split()[0]))
>>> R_vector = robjects.FloatVector(l)
>>> print r.mean(R_vector)
[1] 6127.931
```

The `StrVector()` and `IntVector()` methods of `robjects` convert Python lists (or tuples) into R arrays (i.e., readable by R functions) of strings and integers, respectively:

```
>>> float_vector = robjects.FloatVector([3.66, 2.16, 7.34])
>>> print float_vector
[1] 3.66 2.16 7.34
>>> float_vector = robjects.FloatVector(['3.66', '2.16', '7.34'])
>>> print float_vector.r_repr()
c(3.66, 2.16, 7.34)
>>> string_vector = robjects.StrVector(['atg', 'aat'])
>>> print string_vector
[1] "atg" "aat"
>>> print string_vector.r_repr()
c("atg", "aat")
>>> int_vector = robjects.IntVector(['1', '2', '3'])
>>> print int_vector
```

```
[1] 1 2 3
>>> int_vector = robjects.IntVector([1, 2, 3])
>>> print int_vector.r_repr()
1:3
```

Finally, it must be pointed out that R vector-like objects can be accessed with the delegator rx, which represents the R operator "[":

```
>>> print float_vector.rx()
[1] 3.66 2.16 7.34
>>> print string_vector.rx()
[1] "atg" "aat"
>>> print string_vector.rx(1)
[1] "atg"
>>> print int_vector.rx()
[1] 1 2 3
```

13.3.6 How to Deal with Function Arguments That Contain a Dot

If you want to calculate with R the mean of the values listed in the first column of the table in Figure 13.1, you can write the following:

```
> f = read.table("RandomDistribution.tsv", sep = "\t")
> f_matrix = matrix(f, ncol = 7)
> mean_first_col = mean(f[,1])
> mean_first_col
[1] 6127.931
```

This translates into Python as follows:

```
>>> import rpy2.robjects as robjects
>>> r = robjects.r
>>> f = r("read.table('RandomDistribution.tsv', sep = '\t')")
>>> f_matrix = r.matrix(f, ncol = 7)
>>> mean_first_col = r.mean(f_matrix[0])
[1] 6127.931
```

But what if you want to deal with, e.g., missing values in the input table? In R you would simply set the na.rm argument of the R mean() function to FALSE. However, in Python, the dot has a precise function, and its use for a different purpose would cause the program to behave wrongly or break. In other words, everything with R function arguments in Python works fine unless one of the argument names contains a dot (e.g., na.rm).

In this case, the standard choice consists of translating the dot into a "_" in the argument name:

```
> f = read.table('RandomDistribution.tsv', sep = '\t')
> m = mean(f[,7], trim = 0, na.rm = FALSE)
```

would become the following in Python:

```
>>> f = r("read.table('RandomDistribution.tsv', sep = '\t')")
>>> r.mean(f[3], trim = 0, na_rm = 'FALSE')
<FloatVector - Python:0x106c82cb0/R:0x7fb41f887c08>
[38.252747]
```

See Example 13.3 for more on this.

13.4 EXAMPLES

Example 13.1 Running a Chi² Test

The following script tests if the expression of two genes is correlated or independent. An input file (Chi-square_input.txt) is shown in Figure 13.2. The first column contains the sample number, and the second column contains the expression level (H = High, N = Normal) of two genes (GENE1, GENE2) in the samples.

SAMPLE	GENE1	GENE2
1	H	H
2	H	H
3	N	N
4	H	N
5	N	N
6	N	N
7	N	N
8	H	H
9	N	N
10	H	N
11	H	H
12	N	N
13	N	N
14	N	N
15	N	N
16	H	H
17	H	N
18	H	H
19	N	H
20	H	H
21	N	N

FIGURE 13.2 Content of the Chi-square_input.txt file used in Example 13.1. *Note:* The first column contains the sample number, and the second and third columns contain the expression level (H = High, N = Normal) of two genes (GENE1, GENE2) in each sample.

Notice that you can choose a shorter name for the imported module; e.g., you can write

```
import rpy2.robjects as ro
```

In this example, we use a short name for rpy2.robjects.

R session:
```
> h = read.table("Chi-square_input.txt",header = TRUE,sep \
    = "\t")
> names(h)
[1] "SAMPLE" "GENE1" "GENE2"
> chisq.test(table(h$GENE1,h$GENE2))

    Pearson's Chi-squared test with Yates' continuity \
        correction
data: table(h$GENE1, h$GENE2)
X-squared = 5.8599, df = 1, p-value = 0.01549

Warning message:
In chisq.test(table(h$GENE1, h$GENE2))  :
Chi-squared approximation may be incorrect
```

Corresponding Python session:
```
import rpy2.robjects as ro
r = ro.r
table = r("read.table('Chi-square_input.txt', header=TRUE,\
    sep='\t')")
print r.names(table)
cont_table = r.table(table[1], table[2])
chitest = r['chisq.test']
print chitest(table[1], table[2])
```

Source: Adapted from code published by A.Via/K.Rother under the Python License.

The result from the Chi-squared test looks like this:

```
    Pearson's Chi-squared test with Yates' continuity \
        correction
...
X-squared = 5.8599, df = 1, p-value = 0.01549
```

Notice that the following code does basically the same:

```
import rpy2.robjects as ro

r = ro.r
table = r("read.table('Chi-square_input.txt', header=TRUE,\
    sep='\t')")
```

```
contingency_table = r.table(table[1], table[2])
chitest = r['chisq.test']
print chitest(contingency_table)
```

Source: Adapted from code published by A.Via/K.Rother under the
Python License.

Example 13.2 Calculating Mean, Standard Deviation, *z*-score, and *p*-value of a Set of Numbers

```
R session:
> f = read.table("RandomDistribution.tsv", sep = "\t")
> m = mean(f[,3], trim = 0, na.rm = FALSE)
> sdev = sd(f[,3], na.rm = FALSE)
> value = 0.01844
> zscore = (m -value)/sdev
> pvalue = pnorm(-abs(zscore))
> pvalue
[1] 0.3841792
```

```
Corresponding Python session:
import rpy2.robjects as ro
r = ro.r
table = r("read.table('RandomDistribution.tsv',sep = '\t')")
m = r.mean(table[2], trim = 0, na_rm = 'FALSE')
sdev = r.sd(table[2], na_rm = 'FALSE')
value = 0.01844
zscore = (m[0] - value) / sdev[0]
print zscore
x = r.abs(zscore)
pvalue = r.pnorm(-x[0])
print pvalue[0]
```

Source: Adapted from code published by A.Via/K.Rother under the
Python License.

Notice that to extract a column from the input file, in Python
you have to count from 0. This means that column f[3] in R corresponds to column f[2] in Python. Moreover, the R objects
returned by robjects.r are vectors. Therefore, if you want to
utilize their value, you have to extract it using the [] operator. For
example, in this example, the *z*-score cannot be calculated directly
using

```
zscore = (m - value) / sdev
```

as you would do in R, because m and sdev are vectors.

Example 13.3 Creating Plots Interactively

Plots with R may or may not be made interactively. Here, we show how to create R plots interactively using functions such as `plot()` or `hist()`.

```
R session:
plot(rnorm(100), xlab = "x", ylab = "y")
```

```
Corresponding Python session:
import rpy2.robjects as ro
r = ro.r
r.plot(r.pnorm(100), xlab = "y", ylab = "y")
```

Another example:

```
R session:
f = read.table("RandomDistribution.tsv", sep = "\t")
plot(f[,2], f[,3], xlab = "x", ylab = "y")
hist(f[,4], xlab = 'x', main = 'Distribution of values')
```

```
Corresponding Python session:
import rpy2.robjects as robjects
r = robjects.r
table = r("read.table('RandomDistribution.tsv',sep = '\t')")
r.plot(table[1], table[2], xlab = "x", ylab = "y")
r.hist(table[4], xlab = 'x', main = 'Distribution of values')
```

Source: Adapted from code published by A.Via/K.Rother under the Python License.

Running this example could be frustrating because the plots will appear and immediately disappear from your screen due to the program execution completion. One trick to keep them on the screen for, say, five seconds each, is to use the `sleep()` method from the time module to suspend the program run for five seconds after the execution of each plot command:

```
import rpy2.robjects as ro
import time

r = ro.r
r.plot(r.rnorm(100), xlab = "y", ylab = "y")
time.sleep(5)

table = r("read.table('RandomDistribution.tsv',sep = '\t')")
```

```
r.plot(table[1], table[2], xlab = "x", ylab = "y")
time.sleep(5)

r.hist(table[4], xlab = 'x', main = 'Distribution of values')
time.sleep(5)
```

Source: Adapted from code published by A.Via/K.Rother under
the Python License.

Example 13.4 Saving Plots to Files

To plot to a file with R, you have to set a graphical device like png
or pdf. In Python you need to import importr, a method from
the rpy2.robjects.packages module. importr makes it possible to retrieve the grDevices object, the attributes of which are
grDevices.png and other devices you may need. After finishing
the plot, the graphical device must be closed using the dev.off() R
command. Here, we show the same examples as in Example 13.3, but
plots are saved to .png files:

```
import rpy2.robjects as ro
from rpy2.robjects.packages import importr
r = ro.r
grdevices = importr('grDevices')
grdevices.png(file = "RandomPlot.png", width = 512, \
    height = 512)
r.plot(r.rnorm(100), ylab = "random")
grdevices.dev_off()
```

RandomPlot.png is shown in Figure 13.3. Here is a second example:

```
import rpy2.robjects as ro
from rpy2.robjects.packages import importr
r = ro.r
table = r("read.table('RandomDistribution.tsv',sep = '\t')")
grdevices = importr('grDevices')
grdevices.png(file = "Plot.png", width = 512, height = 512)
r.plot(table[1], table[2], xlab = "x", ylab = "y")
grdevices.dev_off()
grdevices.png(file = "Histogram.png", width = 512, height = 512)
r.hist(table[4], xlab = 'x', main = 'Distribution of values')
grdevices.dev_off()
```

Source: Adapted from code published by A.Via/K.Rother under the
Python License.

Plot.png and Histogram.png are shown in Figure 13.4.

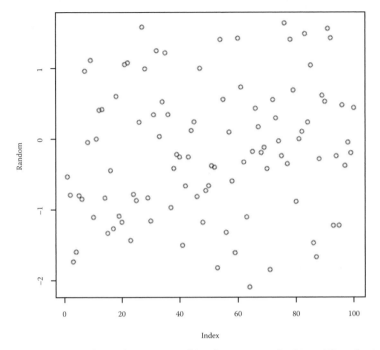

FIGURE 13.3   Random plot generated with RPy2 tools. *Note:* Plot obtained in the first part of Example 13.4 (RandomPlot.png).

13.5 TESTING YOURSELF

Exercise 13.1 Statistical Calculations

Calculate mean, standard deviation, z-score, and p-value of sets of values obtained from your experiments.

Exercise 13.2 Chi-square Test for Smokers and Nonsmokers and Lung Cancer

Carry out a chi-square test to check whether two variables x and y are independent, where x = yes/no (yes if the sample patient was a smoker) and y = yes/no (yes if the sample patient died of lung cancer). You can either retrieve patient samples from the Internet or invent them.

Exercise 13.3 Plot a Histogram and Save It to a .pdf File

Read a list of numbers from a file, use them to plot a histogram using R with Python, and save the plot to a .pdf file.

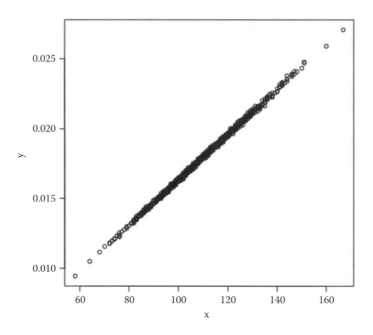

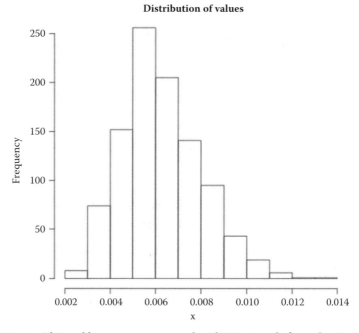

FIGURE 13.4   Plot and histogram generated with RPy2 tools from data in Figure 13.1. *Note:* Plots obtained in the second part of Example 13.4 (Plot.png and Histogram.png).

Exercise 13.4 Plot a Boxplot

Plot the boxplot of the second, fourth, and fifth columns of a table of your choice. Color it in orange, set *x* and *y* labels, and the plot title. Do it both interactively and saving the plot to a file.

Hint: `r.boxplot(f[1], f[3], f[5], col = "orange", xlab = "x", main = "Boxplot", ylab = "y")`

Exercise 13.5

Plot a heatmap of two sets of data.

## CHAPTER 14

Building Program Pipelines

Learning goal: You can pass parameters to your programs and run other programs from Python.

14.1 IN THIS CHAPTER YOU WILL LEARN

- How to make programs work together

- How to run other programs from Python

- How to pass parameters to a Python program

- How to navigate directories from Python

14.2 STORY: BUILDING AN NGS PIPELINE

14.2.1 Problem Description

For many research questions, one program is not enough. Often, you need to make two or more programs work together. Next-generation sequencing (NGS) data analysis is a representative example. NGS technologies are today widely used to study differential expression of genes, noncoding RNAs, mutations, and more. All NGS platforms perform massive parallel sequencing of cDNA or DNA molecules independently of the machine configuration and the chemistry used. In particular, they can be used to perform the "whole transcriptome shotgun sequencing" or RNA-seq, which consists of high-throughput cDNA sequencing aimed at collecting information about the RNA content of a sample.

The output of an RNA-seq experiment is a file containing the sequence of small fragments of cDNA (called "reads"). The reads have to be mapped on reference genomes and reassembled so you can obtain the overall sample sequence. For typical NGS data analysis pipelines, several computational tools have been made available in recent years. An example pipeline is shown in Figure 14.1, where the idea is that reads coming from the Illumina platform (or any other NGS platform) and stored in a text

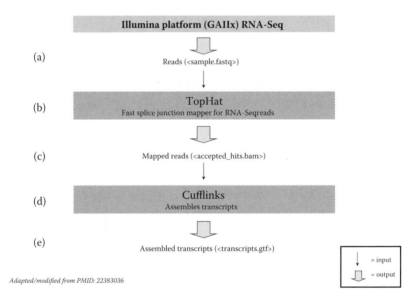

FIGURE 14.1 An example of an NGS data analysis pipeline. *Note:* (a) RNA-seq reads, generated by the Illumina platform (`GAIIx`), are stored in a text file that we called `sample.fastq`. (b) `sample.fastq` is the input of the TopHat program (http://tophat.cbcb.umd.edu/). TopHat is a splice junction mapper for RNA-seq reads. It aligns RNA-seq reads to mammalian-sized genomes (the reference genomes must be stored in a local directory so TopHat can access them), and then analyzes the mapping results to identify splice junctions between exons. (c) The TopHat output is a list of read alignments stored in a .bam file (`accepted_hits.bam`). (d) `accepted_hits.bam` is the input of the program Cufflinks from the Cufflinks package (http://cufflinks.cbcb.umd.edu/), which is made up of three programs for transcript assembling (`Cufflinks`), transcriptome comparison (`Cuffcompare`), and testing for differential expression and regulation in RNA-seq (`Cuffdiff`). (e) The output of the Cufflinks program is an assembled transcriptome (file `transcripts.gtf`), which can be compared to other assembled transcriptomes (e.g., from different sample cells) using the Cuffcompare program (see Chapter 6 and, in particular, Figure 6.1).

file (e.g., `sample.fastq`) can be read by a program (TopHat) that maps them onto a reference genome (which must be downloaded from the Internet and saved in a specific directory; see following text). The output of TopHat (`accepted_hits.bam`) can then be passed to a program (e.g., Cufflinks, from the Cufflinks package) that assembles the transcripts and generates a transcriptome in the form of a `transcripts.gtf` file for further use.

The transcriptomes (`transcripts.gtf` files) obtained from different samples (e.g., wild-type and cancer cells or replicas of the same cell type; see Figure 6.1) can be first filtered as described in Chapter 6, Section 6.2, and then compared to each other in order to reassemble them into a unique reference transcriptome. To this aim, the Cuffcompare program from the Cufflink package can be used (see Chapter 6).

14.2.2 Example Python Session

```
import os

tophat_output_dir = '/home/RNA-seq/tophat'
tophat_output_file = 'accepted_hits.bam'
bowtie_index_dir = '/home/RNA-seq/index'
cufflinks_output_dir = '/home/RNA-seq/cufflinks'
cufflinks_output_file = 'transcripts.gtf'
illumina_output_file = 'sample.fastq'

tophat_command = 'tophat -o %s %s %s' %\
    (tophat_output_dir, bowtie_index_dir,\
    illumina_output_file)
os.system(tophat_command)

cufflinks_command = 'cufflinks -o %s %s%s%s' %\
    (cufflinks_output_dir, tophat_output_dir, os.sep,\
    tophat_output_file)
os.system(cufflinks_command)
```

Source: Adapted from code published by A.Via/K.Rother under the Python License.

14.3 WHAT DO THE COMMANDS MEAN?

In the first line, the `os` module is imported. `os` is a module for using the operating system, i.e., running UNIX or Windows commands from a Python program. In Section 14.2.2, an `os` method (`os.system`) is used to run the `tophat` and the `cufflinks` programs. To run these programs from a UNIX command line, you would type (see Appendix D) the following:

```
tophat -o/home/RNA-seq/tophat/home/RNA-seq/index \
    sample.fastq cufflinks -o/home/RNA-seq/cufflinks \
    /home/RNAseq/tophat/accepted_hits.bam
```

The Python program needs to build these two commands as strings and execute them. Knowing that, you will see the rest of the program is quite simple: the six lines after the import are just variable assignments. In lines 8 and 10, two UNIX command strings are built and stored in variable names (`tophat_command` and `cufflinks_command`, respectively). In lines 9 and 11, the `system()` method of the `os` module is used to run the `tophat` and `cufflinks` programs, respectively. Notice that (1) the argument of `os.system()` is a single string consisting of the UNIX shell command needed to run the program; (2) for the sake of clarity, the parts of these strings are concatenated from the variables previously defined (in lines 2–17); and (3) `cufflinks` uses the `tophat` output as input. The value of the `os.sep` variable in line 11 is the character to connect directories and filenames ("/" on UNIX and Mac; "\" on Windows).

You can easily test programs using `os.system()`. If you copy and paste the `os.system()` string argument to a prompt in a UNIX shell, you will get the same result as running `os.system()` from a program with that argument. The `os` module provides a Python interface to the operating system, and the `os.system()` method executes a "system call"; i.e., it makes the operating system execute the command line specified in the method argument.

Q & A: I AM NEW TO UNIX, AND LEARNING TWO DIFFERENT COMMAND LINE TERMINALS CONFUSES ME. HOW DO I KNOW WHETHER I SHOULD USE THE UNIX OR PYTHON COMMAND LINE?

First, make sure you know where you are. If you open a terminal window and get a prompt that looks similar to "home/yourname>," you are in a UNIX shell. If you see a ">>>" prompt, you are in a Python shell. To learn Python, you do not need much knowledge of UNIX. For most examples in this book, you need to change directories (using the `cd` command) and be running Python (using `python` or `python <program name>`). Everything else can be done from Python itself.

What probably causes you the most trouble is writing directory names correctly, because the concept is new if you have used folders mostly by clicking on a desktop. The first thing to do when (1) you can't start a program, (2) a program does not find a file, or (3) something goes wrong but you don't know what, is to check the directory names. On the UNIX shell, type pwd and ls to see which directory you are in and which files are there. To do the same from a Python program, use

```
import os
print os.getcwd()
print os.listdir('.')
```

Also check Appendix D, Section D.3.4, to learn how to write absolute and relative *paths* correctly. Comfortably navigating files and directories in UNIX will potentially speed up your work. A one-hour online tutorial is a good investment, because what you learn there not only is useful for Python programming but helps you with all kinds of bioinformatics applications.

14.3.1 How to Use TopHat and Cufflinks

The usage of TopHat from the command line is

```
tophat -o <tophat dir> <index dir> <input file>
```

where tophat is the program name, -o is the option flag needed to specify the output directory (<tophat dir>), <index dir> is the name of the directory where you have stored the indexed genome you want to use as a reference to map your reads, and <input file> is the name of the Illumina output file, which must be in the directory where you run tophat.

The usage of Cufflinks is as follows (the program name, the option, and the argument should be self-explanatory):

```
cufflinks -o <cufflinks dir> <tophat dir>/accepted_hits.bam
```

TopHat and Cufflinks can be used in a much more sophisticated way of course, setting the several available options depending on the user's needs (see http://tophat.cbcb.umd.edu/manual.html and http://cufflinks. cbcb.umd.edu/manual.html). Here, a simplified version (though perfectly working) has been preferred. The code illustrated earlier is a simple but realistic example of what is called a *pipeline*.

14.3.2 What Is a Pipeline?

Pipelines are scripts used to connect programs to each other. Imagine you have two programs (e.g., tophat and cufflinks), and you want to run the second using as input the output (e.g., accepted_hits.bam) of the first (see Figure 14.1 b, c, d). Of course, you can do this manually (see

Appendix D). But, if you want to automate the process and do the same operation many times with different input files, you need a program that runs the two external programs for you. In other words, you need to build a pipeline.

Here is the previous example, iteratively used on several samples, followed by the call of os.system() to run cuffcompare (see Chapter 6):

```
import os
input_string = ''
for s in ('WT1', 'WT2', 'WT3', 'T1', 'T2', 'T3'):
    os.system('tophat -o ' + tophat_dir + s + ' ' +\
        index_dir + ' sample' + s + '.fastq')
    os.system('cufflinks -o ' + cufflinks_dir + s + ' ' +\
        tophat_dir + s + '/accepted_hits.bam')
    input_string = input_string + cufflinks_dir + s\
        + '/transcripts.gtf '
os.system('cuffcompare ' + input_string)
```

Source: Adapted from code published by A.Via/K.Rother under the Python License.

In this example, the tophat output corresponding to each different input file (sampleWT1.fastq, ..., sampleT1.fastq, ...) is saved to a different directory (/home/RNA-seq/tophatW1, etc.) with the same filename (accepted_hits.bam). This is because the user cannot modify the tophat output filename but can set different output directories.

Similarly, the cufflinks output for each different input file (/home/RNA-seq/tophatW1/accepted_hits.bam, etc.) is saved to a different directory (/home/RNA-seq/cufflinksWT1, etc.) with the same filename (transcripts.gtf).

A program that runs another program is also called a *wrapper*.

In Python, you have two possibilities to use programs written by other people. First, if it is a Python program, you can import its modules and use the functions within. Second, you can start programs from a Python program like you would in UNIX. The program must be installed and running on your computer, and the Python function you need to call in the wrapper is os.system(). In the previous examples, the TopHat, Cufflinks, and Cuffcompare programs must all be installed and running on your computer; the reference genome(s) must be stored in the index directory; the sampleWT1.fastq, ... , sampleT3.fastq files must be available; and all the output directories must exist.

14.3.3 Exchanging Filenames and Data between Programs

When you connect several programs that call each other, they need to exchange data. The second program needs to know the filename of the first program output. There are four ways by which you can make your programs communicate:

1. *Enter a filename using raw_input().* The program asks you to enter a filename or parameter each time the program runs. This approach is the easiest to implement. You retain full control of what happens, and, especially if you are learning, it may help to use the manual input to go through your pipeline step by step. But when you use the program a few times, entering the same things over and over quickly gets annoying.

2. *Use Python variables.* You can store filenames, parameters, and even data in your code directly as strings. This approach has the advantage that you have everything in one place. The disadvantage is that every time you want to use your pipeline on a different file, you will have to modify the program.

3. *Store filenames in a separate file.* You can write the input filenames in a separate file and then open and read them later from your program. This implies that you have to open, write, close, and save a file every time you want to use different filenames. This approach makes it easy to modularize your code if the program has become big.

4. *Use command-line parameters.* Command-line parameters are the short options or filenames you write in some UNIX commands (e.g., in the ls -l data/ command, there are two parameters: the -l is the option flag for turning the long output on, and data/ is the name of a directory). You can use the same approach to pass string parameters to Python programs. Python program parameters (arguments) are automatically stored in a list in the sys.argv variable. The following two lines allow you to print the list of parameters made available inside your program from the UNIX command line used to run the program (also see Section 14.3.6):

```
import sys
print sys.argv
```

14.3.4 Writing a Program Wrapper

The Python session in Section 14.2.2 and the code shown in Section 14.3.1 are examples of simple NGS pipelines. They can be improved in several ways. For example, by inserting `if` conditions to verify that, e.g., a directory exists before passing it to the system call

```
tophat_dir = '/home/RNA-seq/tophat'
index_dir = '/home/RNA-seq/index'
if os.path.exists(tophat_dir) and os.path.exists(index_dir):
    os.system('tophat -o ' + tophat_dir + ' ' + index_dir + \
        sample.fastq')
else:
    print 'You have to create tophat and/or index\
        directories before running your wrapper'
```

The `os.path.exists()` method returns `True` if its argument is an existing path on your computer.

You may also want to assign all path variables in a separate module, say `pathvariables.py`. This is a very good practice, and we strongly recommend it. Indeed, this ensures that when you change location to your files or programs, you do not have to go through all your programs and wrappers to reassign the content of path variables. The most common execution errors arise when you miss reassignments of path variables upon directory name or file location changes. To use path variables, you have to import the module in the program where you need them. Here is the previous example reformulated according to this practice:

```
from pathvariables import tophat_dir, index_dir
if os.path.exists(tophat_dir) and os.path.exists(index_dir):
    os.system('tophat -o ' + tophat_dir + ' ' + index_dir +\
        sample.fastq')
else:
    print 'You have to create tophat and/or index\
        directories before running your wrapper'
```

where the content of `pathvariables.py` is

```
tophat_dir = '/home/RNA-seq/tophat'
index_dir = '/home/RNA-seq/index'
cufflinks_dir = '/home/RNA-seq/cufflinks'
```

You can simply import the `pathvariables.py` module if it is in the same directory as your wrapper, but what if you want to import the module in several programs located in different directories? The easiest thing to do is to store the `pathvariables.py` module in a given directory and then append the *path* of that directory to the `sys.path` variable (which is a list object) in each program you want to import it to:

```
import sys
import os
sys.path.append('/home/RNA-seq/')
from pathvariables import tophat_dir, index_dir
if os.path.exists(tophat_dir) and os.path.exists(index_dir):
    os.system('tophat -o ' + tophat_dir + ' ' + index_dir +\
        'sample.fastq')
else:
    print '''You have to create tophat and/or index\
        directories before running your wrapper'''
```

Source: Adapted from code published by A.Via/K.Rother under the Python License.

14.3.5 Lag When Closing Files

One problem with pipelines is that, in general, given a program calling two (or more) external programs, the system call of the second program should occur when the execution of the first program has finished, especially if the second one uses as input the output of the first one. This is what should happen in general. In fact, `os.system()`, and other methods available for invoking subprocesses such as `os.popen()` or `subprocess.call()`, waits for the command to complete and then return a value corresponding to the program exit status (the `os.system()` returned value is 0 in case of success, 256 in case of failure). However, it may happen sometimes that, for example, writing an output file involves a lag, during which the next system call starts the execution of the subsequent program even though its input file is not yet there or completed. This generally causes problems.

There is a trick to avoid such inconvenience: you can insert an action and verify its success after a system call and before the subsequent system call. For example, you can open a dummy file, write something into it, and close it. Then, you can use the `os.path.exists()` method to check if the dummy file has been created:

```
import sys
import os
sys.path.append('/home/RNA-seq/')
from pathvariables import tophat_dir, index_dir, cufflinks_dir
# the tophat program creates an output file
os.system('tophat -o ' + tophat_dir + ' ' + index_dir\
    + 'sample.fastq')
```

```
# here we don't know whether the tophat output file is
# completed and available
# we open and close a dummy file, so the operating
# system catches up
lag_file = open('dummy.txt', 'w')
lag_file.write('tophat completed')
lag_file.close()
# read the output file
if os.path.exists('/home/RNA-seq/dummy.txt'):
    os.system('cufflinks -o ' + cufflinks_dir + ' '\
        + tophat_dir + '/accepted_hits.bam')
```

Source: Adapted from code published by A.Via/K.Rother under the Python License.

14.3.6 Using Command-Line Parameters

By passing parameters from the UNIX command line to your Python programs, you can make your programs easier and more flexible to use. Instead of changing the program or editing a file, you can vary the command by which you start the program from the UNIX console:

```
python ngs_pipeline.py dataset_one.fastq
python ngs_pipeline.py dataset_two.fastq
```

How can you get the filenames `dataset_one.fastq` and `dataset_one.fastq` into your Python program? The parameters (also called *arguments*) are automatically stored in a list in the `sys.argv` variable. To see how it works, you can try the following Python code (create a file `arguments.py`; it won't work from the Python shell):

```
import sys
print sys.argv
```

Now try calling this program with different arguments from a UNIX terminal:

```
python arguments.py
python arguments.py Hello
python arguments.py 1 2 3
python arguments.py -o sample.fastq
```

You should see that anything you enter in the UNIX command line ends up as a string in a normal Python list on the elements of which you can use all regular Python functions. A frequent practice is to use one of the command-line arguments as a filename. Note that the first element of the list is always the name of the program that you have called, so the first argument is `sys.argv[1]`. Using the information from the `sys.argv` list, you can pass command-line arguments into your program. If your program needs to manage many different command-line options, consider using the `optparse` module (see http://docs.python.org/2/library/optparse.html).

14.3.7 Testing Modules: `if __name__ == '__main__'`

Python modules are objects, and, as such, they have attributes and methods, which can be visualized using the built-in function `dir()`. Some attributes are module specific (e.g., `localtime` is a specific attribute of the `time` module), whereas three in particular refer to modules in general: `__file__`, `__doc__`, and `__name__`.

The `__file__` attribute returns the path of the module. The `__doc__` attribute returns the module documentation, if present. The `__name__` attribute returns the name of the imported file without the `.py` suffix if the module is imported and the string `'__main__'` if the module is executed. This may turn out to be very useful when you want to do different things depending on whether you execute or import the module. In fact, if you insert the following condition in your programs

```
if __name__ == '__main__':
        <statements>
```

you obtain that `<statements>` will be executed only if the module is run from the command line and not imported by means of an `import` statement. This trick is particularly useful if you want to test a module before integrating it in a larger script or if you want a module to be both executable and importable. For example, if you define one or more functions in a module and put their call in the "`if __name__ == '__main__':`" block, the functions will be executed only when you run the module from the command line, because in this case `__name__` is equal to `'__main__'`. On the other hand, if you import the module from other programs, `__name__` will be equal to the module name, and the functions will be imported without being executed.

14.3.8 Working with Files and Directories

Working with files in a pipeline usually involves some management tasks: reading all files from a directory, creating temporary files, checking whether a file really exists, etc. The Python module os is very useful for interacting with the operating system and has many functions that come in handy for working with files. In fact, it is a combination of two modules: os and os.path. Before you can use any of the modules, they need to be activated by

```
import os
import os.path
```

Notice that importing os should automatically activate os.path. Therefore, os.path should work even though you only import os. A very useful feature of the os.path module is that it helps operate with directory names. Probably the most frequently used function is os.path.split(filename), which separates a filename from directory names. If a program uses a filename, but you need to verify whether the file really exists before using it, the os.path.exists(filename) function will return True or False.

Reading Files from Directories

Among the most frequently used operations is the function os.listdir(directory), which reads all files from a given directory into a list. This is particularly useful in a pipeline if you want to automatically run a program on all the files in a given folder:

```
import os
for filename in os.listdir('data/'):
        os.system('<my_program>%s'%(filename))
```

Changing Directories

Some third-party programs require to be started from a particular directory. In the UNIX shell, you would go to that directory using the cd command before starting the program. The os.chdir() function does the same in Python:

```
os.chdir('../data/')
```

moves one level up and then descends into the data/directory.

You can determine the current directory with the os.getcwd() function:

```
print os.getcwd()
```

Making and Removing Directories

os functions to make and remove directories are `os.mkdir()` and `os.rmdir()`, respectively.

Removing Files

If a file needs to be deleted, you can do this by using `os.remove(filename)`. When you connect many programs, you will create many files: prepared input files, intermediate data, and log files, and often files appear that you have no idea what they are good for. These files tend to accumulate quickly and litter your directories. From time to time, it is good if you can get rid of them. This is where the `os.remove()` function comes in:

```
os.remove('log.txt')
```

Of course, you can delete a file only if it exists, so you might want to check that first:

```
if os.path.exists('log.txt'):
        os.remove('log.txt')
```

Creating Temporary Files

When you use temporary files for intermediate data, they get deleted automatically after you use them. The `tempfile` module allows you to create temporary files.

Q & A: What if One Part in the Middle of the Pipeline Breaks?

Generally, be prepared. The more programs you connect to each other, the more fragile the overall structure becomes. Make sure that you have a possibility to find out what happened. Test how the wrapper works for the first program you want to call, and check that the output generated, if any, looks as expected. Only then, add the call to the second program and so forth. Ensure your wrappers report what they are doing by adding a few print statements by default or writing a log file. For diagnosis, the intermediate data files are useful as well. When your pipeline crashes, you can check the intermediate data that was created last for the presence of any unusual things. You can also write your pipeline in such a way that it checks whether some of the result files are already present and skip calculating them again. In a pipeline consisting of many programs, your code is not necessarily responsible for the problem. You can manually call the programs in the pipeline one by one to find out whether the bug is in one of the external programs.

14.4 EXAMPLES

Example 14.1 Running T-COFFEE

The following code is a wrapper for the multiple sequence alignment (MSA) T-COFFEE[*] algorithm. However, the procedure is in principle the same for any other MSA algorithm, such as ClustalW. Before starting to write the wrapper, you need to go to the program web page (www.tcoffee.org/Projects/tcoffee/) and learn how the command line to execute it looks, what options are available, which are the mandatory arguments, which output files you can expect, etc. Then, you have to download the program, install it, and check that it runs from the command line as you expect. In some cases, it will be necessary to add the program directory *path* to the shell variable PATH (see Appendix D, Section D.3.5). Usually this information is given in the program user manual. Once the command line for the program is ready, you can use it as argument of the os.system() function:

```
import sys
import os
sys.path.append('pathmodules/')
from tcoffeevariables import tcoffeeout
cmd = 't_coffee -in="file.fasta" -run_name="' + \
    tcoffeeout + 'tcoffe.aln" -output=clustalw')
os.system(cmd)
```

Source: Adapted from code published by A.Via/K.Rother under the Python License.

In this Python wrapper, variables are assigned in the tcoffee-variable.py module, which is stored in the pathmodules directory (which is a subdirectory of the wrapper directory). -in is the option for the input file, -run_name is the option to assign a name to the output file, and -output is the option to set the output format (clustalw, in this case).

---

[*] C. Notredame, D.G. Higgins, and J. Heringa. "T-Coffee: A Novel Method for Fast and Accurate Multiple Sequence Alignment," *Journal of Molecular Biology* 302 (2000): 205–217.

Example 14.2   Write a Wrapper That Uses `bl2seq` to Align Two Sequences, the Uniprot ACs of Which Are Command-Line Arguments

The UNIX command to run the wrapper with two Uniprot ACs as arguments is

```
python Bl2seqWrapper.py F1B2B3 E2CXB4
```

The BLAST+ package must be installed and running on your computer (see Recipe 11). The following program defines two functions: `run_blastp()` and `get_seq()`. The former contains the system call to `blastp` (with options for the program, the input files, and the output file). The latter is a function to download the FASTA file of the input sequences from the Uniprot website. The function is called only if the sequences are not already present in your computer (in the `input_seq` directory). The *path* variables are stored in the `blastvariables.py` module, which is saved in the path-modules/ directory. The `if __name__ == '__main__'` trick makes it possible to import the wrapper from another program and individually use the functions defined therein, without executing the code enclosed in the `if __name__ == '__main__'` condition.

```python
import sys
import os
import urllib2

sys.path.append('pathmodules/')
from blastvariables import *

def run_blastp(seq1, seq2):
    os.system("blastp -query " + input_seq + seq1 +
            ".fasta -subject " + input_seq + seq2 +
            ".fasta -out " + seq1 + "-" + seq2 + ".aln")

def get_seq(seq1, seq2):
    for seq in (seq1, seq2):
        url = 'http://www.uniprot.org/uniprot/' + seq + '.fasta'
        handler = urllib2.urlopen(url)
        fasta = handler.read()
        out = open(input_seq + seq + '.fasta', 'w')
        out.write(fasta)
        out.close()
```

```
if __name__ == '__main__':
    try:
        seq1 = sys.argv[1]
        seq2 = sys.argv[2]
    except:
        print 'usage: BlastpWrapper.py seq1-UniprotAC \
            seq2-UniprotAC'
        raise SystemExit
    else:
        if os.path.exists(input_seq + seq1 + '.fasta') \
and os.path.exists(input_seq + seq2 + '.fasta'):
            run_blastp(seq1, seq2)
        else:
            get_seq(seq1, seq2)
            run_blastp(seq1, seq2)
```

Source: Adapted from code published by A.Via/K.Rother under the Python License.

14.5 TESTING YOURSELF

Exercise 14.1 Execute Two Programs Separately

Write a Python program, first.py, that writes a single number to a text file. Write a second program, second.py, that reads the number from the file and prints its square. Run both programs manually from the command line.

Exercise 14.2 Connect Two Programs to a Pipeline

Write a pipeline.py Python program that uses os.system() to first execute first.py and then execute second.py.

Exercise 14.3 Use a Command-Line Parameter

Extend the pipeline so that both first.py and second.py accept the input filename as a command-line parameter. This requires changing all three programs and using sys.argv.

Exercise 14.4 Read a Directory

Write a program that counts the number of files in your home directory (/home/username on UNIX). Use the os.listdir() function.

Exercise 14.5 Run BLAST and Parse the Output to a FASTA File

Write a program that executes a local BLAST query and writes the output to an XML file. Write a second program that reads the XML output and writes the alignments from each HSP to a separate FASTA file. Build a pipeline that allows you to execute the BLAST query and parse its output for any given input sequence using a single command. For instructions on how to run BLAST and parse its XML output, see Recipes 11 and 9, respectively.

## CHAPTER 15

Writing Good Programs

L EARNING GOAL: You can apply some software engineering techniques that make your programs better.

15.1 IN THIS CHAPTER YOU WILL LEARN

- How to divide a project into smaller tasks

- How to divide a program into functions and classes

- How to use pylint to improve your program code

- How to use Mercurial to keep track of program versions

- How to share your program with other people

- How to improve a program iteratively

- How to build your own modules and packages

15.2 PROBLEM DESCRIPTION: UNCERTAINTY

15.2.1 There Is Uncertainty in Writing Programs

By now, you probably have already written your first programs. You also understand that programming is a very powerful tool for a scientist. As your skills develop and your programs get more complex, your goal is hidden in a cloud of uncertainty. When you start, you have a certain idea about what your program should do. But as you are writing code, you eventually find out that what you really need is something else, so you introduce changes to the program. And while you do so, your target keeps moving. As a result, writing a big program resembles a gradual optimization process rather than a straight line. Planning everything in advance is risky and rarely works. Uncertainty is a characteristic of almost all software projects. In contrast, change is a constant: changes will be demanded by your supervisor, by your reviewers, and, most of all, by your own ideas. How can you create a good program under conditions of uncertainty and change? This chapter gives you some engineering practices that can help you create solid, working software.

Many programming projects start with an email like the one in Section 15.2.2.

15.2.2 Example Programming Project

```
Dear coworker,

At the conference last week, I had a great idea I'll tell you
about later. I made some BLAST queries with a couple of
sequences of Phenylalanine-tRNA-synthetases (PheRS) from
different organisms. The results need some cleanup, however.

I saved the sequences in a set of files in FASTA format in the
same directory. Each file contains many sequences in the
following format:

> gi|sequence name|species name
AMINQACIDSEQUENCE

Because the sequence diversity of the aminoacyl-tRNA
synthetase family is very high, there are many sequences from
other aaRS proteins among the results. We therefore need to
filter out sequences that do not have the right tRNA
specificity (marked by Phenylalanine, Phe etc. in the sequence
name). Ah, and there might be duplicate hits in the files that
need to be removed, too.

I think the results fit well into your project. Could you
please find all the sequences for Phe and put them into a
single FASTA file? We could then pass it to an alignment
program. Your new programming skills might come in handy
for that.

Best regards,
    Your supervisor

P.S.: If this works, we will probably do the same thing for the
other 19 amino acids as well.
```

15.3 SOFTWARE ENGINEERING

There is a whole discipline called software engineering that focuses on how to write good programs. Software engineering usually relates to programs consisting of tens of thousands or even millions of lines of code. For smaller programs, as the ones you are probably going to write, it comes down to three basic steps, referred to as *The Dogma of Programming* by Donald Knuth:

- First, make it work.

- Second, make it nice.

- Third, and only if it is really necessary, make it fast.

In the previous chapters, you learned a set of tools to write working programs. In this chapter, you will learn techniques for the second step: writing clean, transparent programs. These include breaking up a programming task into smaller steps, formatting your source code, using comments, keeping track of changes electronically, sharing your program with fellow scientists, and improving a program iteratively.

For the third point, making programs faster, you need a deeper knowledge about *algorithmics*. When you reach the point that your programs work well and are well organized but run too slowly, it is a good moment to read more books or ask experienced programmers for advice.

15.3.1 Dividing a Programming Project into Smaller Tasks

Before you start writing a program, there is a question to answer: What should the program do? It is worth considering that question before you start to write code, because the question is not a technical one. More precisely, the question is, What does the program need to do to provide a benefit? You must identify in advance how you expect the program to make your work more efficient, advance your research, or make someone's life better in some way. Because a program is a precise machine, it is important to have a precise goal, and it's best to write it down. Start by formulating the goal in simple nontechnical words. Often, you do not start with empty hands. You have a textual description from a grant proposal, meeting notes, or an email and usually some data. From the supervisor's email in Section 15.2.2, the following program purpose can be extracted:

> Clean up FASTA files with aaRS protein sequences so that they can be used to build a sequence alignment.

Although the overall goal of the project is not clearly stated in the email ("I'll tell you about later"), you know what the data are to be used for afterward. Without that information, the goal would be insufficient and incomplete: "Clean up FASTA files containing aaRS protein sequences" can be interpreted in different ways. If the goal is still unclear at that point, each question you ask will save you many hours of programming time.

With the overall goal set, you can start thinking of the program tasks more formally. In particular, you must know the answers to the following three questions.

What Is the Input?

What kinds of data does the program need to read? What options are essential? Should the user provide any additional parameters? Should there be some kind of interface, or will filenames simply be written into the code? In the aaRS project, the input is clearly described:

The program reads a directory with many FASTA files containing protein sequences. The sequence entries in the files have the format:

```
> gi|sequence name|species name
AMINQACIDSEQUENCE
```

With that, you could start writing a function for reading the files. But don't start yet!

What Is the Output?

What kinds of files does the program produce? Is there any output to the screen? Does the program need to write a logfile with details? Does the program need to open windows or create graphics? What should happen if something is wrong with the input data or parameters?

In the aaRS project, the output is clearly expected:

The program writes a single FASTA file with aaRS sequences.

There are, however, assumptions that you can make at this point. First, the format of the FASTA file should be the same as it is in the input. Second, the order of the sequences in the output does not matter. Third, the format described for the FASTA output must be readable by the program used for creating a sequence alignment. Each of these assumptions can be written down, checked, and discussed with project stakeholders. Eventually, they will come up with clarifications or important details that you did not think of before.

What Should Happen between Input and Output?
Obviously, this is where the program does its work. At first, you can describe the tasks the program should perform in a language that users of the program will understand. They should point out a clear benefit. Usually, the work a program does can be divided into several parts. For instance, the aaRS program needs to do the following:

> Remove all sequences that do not have Phenylalanine or Phe in its name.

> Remove all duplicate sequence entries (identical sequences).

The second point contains an assumption again: What exactly does "duplicate entries" mean? Does it mean that the sequence is identical or that the description is identical? Or both? While biological scientists may have an implicit consensus about many terms they use, they may disagree when the term needs to be cast into precise language. If in doubt, ask first, program second.

Writing down things the program should do in everyday language is called *collecting requirements*. A single requirement from the users' point of view is also called a *user story*. A single requirement or story should be short enough to fit on a paper card (two to three sentences). The four sentences we highlighted are requirements or user stories. Having clearly written requirements helps you to communicate about the program with the people who need it (including yourself) and with people you ask for advice and to track what you have already completed. Such documentation should be concise; a set of five to ten requirements on paper cards is enough to organize your work for a week of programming. Clear requirements will save you plenty of programming time.

15.3.2 Split a Program into Functions and Classes
When writing a small program, you can start coding right away. With a bigger program, it avoids trouble later to create scaffolding first: design

functions and/or classes. A good strategy is to define separate functions for the input, the output, and the work in between. The requirements from the previous sections provide the raw material for that. In a program of the size of the aaRS project, there may be even one function per requirement. A scaffolding for the program could look like this:

```python
def read_fasta_files(directory):
    '''
    Reads a directory with many FASTA files containing
    protein sequences.
    '''
    pass

def filter_phe(sequences):
    '''
    Removes all sequences that do not have
    Phenylalanine or Phe in their name.
    '''
    pass

def remove_duplicates(sequences):
    '''
    Removes all sequence records, having an identical
    sequence.
    '''
    pass

def write_fasta(sequences, filename):
    '''
    Writes a single FASTA file with aaRS sequences.
    '''
    pass

if __name__ == '__main__':
    INPUT_DIR = 'aars/'
    OUTPUT_FILE = 'phe_filtered.fasta'
    seq = read_fasta_files(INPUT_DIR)
    filter_phe(seq)
    remove_duplicates(seq)
    write_fasta(seq, OUTPUT_FILE)
```

Note that the text from the requirements found its way into the comments for each function. The pass statement does nothing. In this case, it just makes sure the function body is not empty. The if __name__ == '__main__': section is executed only when you run the program, not when you import it. It is a convention on how to write the main part of a program in Python (see Chapter 14, Section 14.3.7).

This program can already be executed. It does nothing, of course, but seeing that it works allows you to decide better about how exactly the program should work. For instance, you need to think how exactly the functions should exchange data. If you figure out you had misconceptions about the task, it is still easy to apply changes without losing your work.

One big advantage of having separate functions is that you can analyze and test each function independently. If a program is big and complicated, errors are difficult to find. On the other hand, if your program consists of functions not longer than 10 lines, you can easily go through the code of each function line by line and analyze it. You can add `print` statements or run a function separately with sample data to test it. The smaller your functions, the easier it gets. Therefore, breaking your program into functions makes debugging a lot easier. (See also Box 15.1.)

BOX 15.1   WRITING SMALLER PROGRAMS

A saying goes, "Today I had a successful day at programming: I deleted 300 lines of code" (Lorenzo Catucci). One virtue of programming is to write programs in such a way that errors are easy to find. In that sense, what is the best program possible? Is it the one you write with the most sophisticated code that you are capable of? Or is it the one where you paste a few simple lines together and let libraries do the rest?

The answer given by Python philosophy is that simpler is better. When you can achieve your goals without using functions, do it. When you can do something without complex data structures, do it. When you can cut the size of your program by half and it still does the job, do it. Ultimately, the best program ever is one that gets the job done with a single line of well-readable code. (Also see www.catb.org/esr/writings/unix-koans/ten-thousand.html.)

If you can get the job done by writing three small programs instead of a single big one and run them with

```
> python program1.py
> python program2.py
> python program3.py
```

there is nothing wrong with that as long as it works for you. If you later figure out that you need a bigger program to automatize, you can do that later. If your programs 1–3 do not take any arguments, you can execute them from another program with

```
import program1
import program2
import program3
```

If you have small, well-defined component programs, these three lines are all you need.

Dividing your program into smaller functions is an art. This is even more the case with classes. Don't expect to get it right in the first attempt. Often, functions and classes need to be broken down further, merged, and reassembled. Creating one function from each of your requirements is a starting point to create a first working version of your program.

15.3.3 Writing Well-Formatted Code

Messy code is hard to read. But what should cleaned-up code look like? In Python, there is a standardized convention to format code called PEP8. The PEP8 convention claims, for instance, the following:

- You should add a space after commas and arithmetic operators.

- Variables in functions should have lowercase names.

- Constants in modules should have uppercase names.

- After each function there should be two empty lines.

- Each function should have a documentation string as a comment.

While obeying every single one of these guidelines may seem overly strict, the coding standard makes sense as a whole. A standardized program code is much easier to read and understand by other programmers (e.g., those you ask for help or yourself when you return to your program after a long break).

The program *pylint* allows you to check how well the Python code complies with the PEP8 convention. After downloading (http://www.pylint.org/) and installing pylint, you can run it on any Python file from the UNIX command line by

```
pylint aars_filter.py
```

pylint analyzes the code and prints an overall score and a list of suggestions. You can then edit your program code and see how your score improves. It is not necessary to strive for the maximum score of 10.0 in all your Python files. Usually, with reasonable effort you can achieve well-readable code scoring at 7.0–9.0.

Docstrings

Documentation strings or *docstrings* are important components of formatting conventions. In Python, each function and class should have a triple-quoted comment in the line right after the function or class definition. This comment should briefly explain what the function or class does.

```python
class AARSFilter(object):
    '''
    Reads a set of FASTA files and removes duplicate
    sequence entries.
    '''
    pass
```

The description should be short and not too detailed, because the comments might go out of date quickly.

Comments

Using comments (text preceded by a # that is ignored by the interpreter) to describe a line or a block of code makes your programs much more readable. Comments should be very short and very informative at the same time and should be kept updated when you modify a program; otherwise they generate more confusion than help.

15.3.4 Using a Repository to Control Program Versions

When you write a program, it develops gradually. You introduce many changes over time. Most of them are good; some turn out to be wrong. Sometimes you need to change the program, but you are not sure whether the changes will be final or whether you will return to an original version. The intuitive solution is to create copies of your program code. In practice, such copies tend to accumulate over time:

```
my_program.py
my_program2.py
my_program3.py
my_program3_optimized.py
my_program_new_version.py
```

Which of these programs is the most recent one, and how do they differ? A frequent reason for strange import problems and other nasty bugs is that you are working with multiple copies of your program in different places. If you do, it is quite easy to mess something up. It's better that you don't do this in the first place. This kind of problem is completely solved by using a *code repository*. Repositories are the equivalent of a lab journal for programmers. They record changes you apply to your program over time together with short remarks. At any point you can return to an older version stored in the repository or switch back and forth between versions. You can put both program code and data in the repository. There are many programs that allow you to control versions of program code: Mercurial (http://mercurial.selenic.com/wiki/Mercurial), GIT (http://git-scm.com), Bazaar (http://bazaar.canonical.com/en/), and SVN (http://subversion. apache.org) are the most widely known ones. Here, we explain the basic usage of Mercurial. After you download and install Mercurial, go to the directory with your program code and type (in the command-line shell)

```
hg init
```

This tells Mercurial that you want to keep track of files in that directory. Now you can add each file to the version control system:

```
hg add aars_filter.py
hg add phe_sequences.py
```

Now Mercurial knows that you want to keep track of these two files and will remember changes on them. All other files will be ignored by Mercurial. Now you can continue to work normally on your files. When you finish editing them after some period of work (typically at the end of the day), you can record the changes using the commit command:

```
hg ci
```

Mercurial will prompt you to describe briefly what you have changed. Usually a single descriptive line is enough. After the hg ci command, Mercurial has memorized the current contents of all files. Even if you change, delete, or replace them, you will be able to restore their contents as if you would have a backup copy.

To return to an earlier version, use

```
hg log aars_filter.py
```

This command prints the log messages of all versions you committed earlier. You can look in the list for the version number to which you would like to return. Then update all files to that version number (e.g., 23):

```
hg up -r 23
```

After this command, all files will have the content they had when you committed version 23. If you want to share your source code with other people or want to back it up on a server, you can create a repository on the bitbucket (https://bitbucket.org) or sourceforge (http://sourceforge.net) websites. First, create a repository via the website instead of using hg init. You will need to enter a username and a password. The website will display you a command that you can copy and paste to a command shell:

```
hg clone https://bitbucket.org/myproject/
```

When you run this command in a command shell, an empty repository will be created on your computer. You can then add files and commit changes as described previously.

If you want to have a copy of the repository on the server, you can send the files to the server with

```
hg push
```

Later you can update changes from the server to your computer:

```
hg pull
hg up
```

After these two commands, your files will contain the latest version from the server or your local copy, whichever is more recent. Note that code on free accounts is visible to everyone. If you are storing unpublished research results there, make sure you have the consent of your collaborators and/or supervisor. A natural solution is to store the program code online but not the data. A detailed introduction to Mercurial can be found at http://mercurial.selenic.com/wiki/Tutorial.

15.3.5 How to Release Your Program to Other People

Once you decide that your program is good enough to give it to other people, how do you deliver it? To distribute a program, you create a stable version called a *release*. To create a release, you need to do three things. First, set a version number. When you communicate with your collaborators, this number will help you to find out whether they used the latest version without checking email or even comparing the code. It does not matter whether you start your version number at 0.1 or 1.0 or use the date as a version number, as long as you use increasing numbers consistently. Second, write a short README.TXT file, i.e., a short text file covering the following points:

- What is the name of the program?

- What is your name and how can you be contacted?

- What is the version number?

- What is the program good for? Describe it in 50–100 words.

- How can the program be used? A good idea is to include a simple command-line example.

- Under what conditions can the program be distributed? For example, if you write, "All rights reserved," everybody will have to explicitly ask you for permission. If you write, "Distributed under the conditions of the Python license," both free distribution and commercial uses are allowed. There are many more software licenses readily available. You can find and use one that fits your scope without hiring a lawyer.

Finally, create a zip file out of the directory with the program, including the README.TXT file. Share that file in whatever way is convenient for you and your collaborators. You don't need to set up a big web page in the first place.

More sophisticated approaches to creating releases include the following:

- Use the Python distutils library to autogenerate an install script.

- Use the py2exe program to create executable files for Windows that don't require your users to install Python.

- Manage releases on a sourceforge (http://sourceforge.net) or github (https://github.com) account.

If you want to know more about this topic, you can have a look at http://docs.python.org/2/distutils/introduction.html.

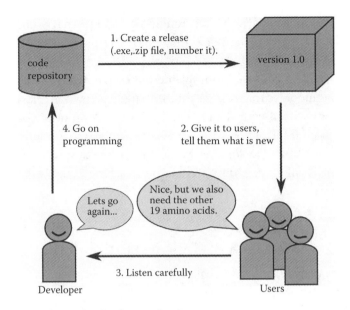

FIGURE 15.1   The cycle of software development.

Most of these measures won't be necessary for a program that is used by a few people. But if you have written a small program that you think is useful for a wider audience (but is not sufficient for publication on its own), you can consider making it available to advertise your project or engage in communication with your peers.

15.3.6 The Cycle of Software Development

At some point your stakeholders will ask when your program will be finished. However, from an engineering point of view, a good program is never finished. There are always more ideas to implement, more details to add. The key question is not whether the program is finished but whether it has an effect that can be scientifically proven. The methodology to develop software useful for scientific purposes requires not only epistemology (e.g., Thomas Kuhn)[*] but also entrepreneurial theory.[†] In the latter, development can be described by a circular model with three stages: plan, program, and prove (see Figure 15.1).

---

[*] T.S. Kuhn, *The Structure of Scientific Revolutions* (Chicago: University of Chicago Press, 1962).

[†] Eric Ries, *The Lean Startup: How Today's Entrepreneurs Use Continuous Innovation to Create Radically Successful Businesses* (New York: Crown Publishing, 2011), 103.

Plan. You first need to have a rough idea about what your program should do. Dissect the main goal into smaller tasks only that much that you can start programming (see Section 15.3.1).

Program. Next you need to implement the plan. Create the functions, classes, and modules that realize your initial idea. You don't have to cover all situations within your data in the first place. A proof of concept will give you more information in the long run and save you many hours of programming time. Do some initial testing to make sure your program contains no `SyntaxErrors` and does not crash on a simple data set (see Box 15.2).

BOX 15.2   SMALL DATA FILES FOR TESTING

When you are using long data files to perform calculations, and the output is equally long, you will have a hard time proving whether your program runs correctly. In contrast, if both input and output data files fit on half a screen page, you can quickly check whether the output meets your expectations.

You might say now, "But in my project, there are big data files." Of course there are. This is why you want to use a computer in the first place. The point is to use your program with *artificial input data for testing* before you use it for your real project data. For instance, in the program for sorting dendritic lengths (see Section 12.2.2), you could open a text editor and create the following input file (see Section 12.2.1):

```
Primary         16.385
Primary         139.907
Primary         441.462
Secondary       29.031
Secondary       40.932
Secondary       202.075
Secondary       142.301
Secondary       346.009
Secondary       300.001
```

This file contains 3 primary and 6 secondary neurons, and each of the three size categories (<100, 100–300, and bigger) contains three items. These numbers should cover many possible situations in the program. They produce the following output:

```
category     <100    100-300    >300
Primary:      1         1         1
Secondary:    2         2         2
```

In the output, you can see immediately whether the total number of counts is 9 or not. This helps you discover problems much sooner than by looking at the code. When you are sure your program works correctly for a small input file, you can try the big one.

Prove. Expose your program to a real-world situation. Does it work in the way you intended? Does the output provide an added value for your research? Do other people state that it is useful for them, or are they just being polite? What would the program have to do in order to contribute to your research objectives? Record your own observations and ask your collaborators for feedback. If possible, collect hard evidence (numbers, statistics) that prove the usefulness of your program.

Then go back to the planning phase and start over again. Your job as a programmer is to accelerate this circle. Start with a minimal program. Do not bother with optional features. Create a release early on. Try it in practice. Collect opinions and feedback. Then make the next small improvement. The best way to make better software is to make your program evolve faster. A fast development cycle is measured in days, hours, sometimes even minutes. To develop good programs, scientific peer review is far too slow. Only by creating many bad programs first will you create a good program.

This plan–program–prove circle is the essence of Agile, a modern software development philosophy (http://agilemanifesto.org/; see Figure 15.2). The Agile approach has been put to life in a number

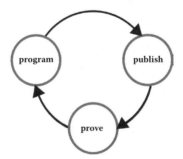

FIGURE 15.2 Iterative development with Agile. *Note:* The "Manifesto for Agile Software Development" is summarized in four points: (1) individuals and interactions over processes and tools, (2) working software over comprehensive documentation, (3) customer collaboration over contract negotiation, and (4) responding to change over following a plan.

of software development methodologies for professional development teams such as Scrum (www.scrum.org), eXtreme Programming (www. extremeprogramming.org), and Crystal (www.agilekiwi.com/other/ agile/crystal-clear-methodology/). Most Agile methodologies are too heavy for a scientific research group. However, many best practices can be adopted.[*] A toolbox of software engineering techniques used in bioinformatics research, such as automated testing, code reviews, and user stories has been described.[†]

15.4 EXAMPLES

Example 15.1 Creating Your Own Module

Long Python files are hard to read. A better strategy for bigger programs is to split the code into several files and then import them. These modules are easy to reuse. In practice, a module is a normal Python file, but it's one that mostly defines variables and functions. To clean up your program with modules you could, for instance, do the following:

- Collect all functions for file parsing in one module.
- Collect all functions for creating output in one module.
- Create a module for constants like filenames and parameters. When you give your constant names in uppercase, you will recognize everywhere that you are dealing with constants.
- Assign the paths of your data files to variable names in a separate module, say my_paths, and save it in a given directory, say my_modules/. In the program where you need to access your data files, you can import the sys module and append the path of my_modules/ to the sys.path variable and finally import the my_paths module.

```
import sys
sys.path.append('/Users/kate/my_modules/')
import my_paths
```

---

[*] See www.seapine.com/exploreagile/ for an industrial best-practice guide.
[†] Kristian Rother et al. , A toolbox for developing bioinformatics software, *Briefings in Bioformatics* 13:2 (2012), 244–257.

This way, when you change the location of your data files, you do not have to remember in which programs you have used them and modify the paths in all programs one by one, but you can simply change the paths in the my_paths module.

Using import and the dot syntax, you can access each of the data items in the modules separately from the rest of your program.

Of course, you could put the same data into a list, dictionary, or class as well. But if you want to access the same data from different places, importing a module is easier, especially when the data change, and you would have to update it in several places otherwise.

Example 15.2 Creating Your Own Package

A package in Python is a folder with Python modules. You can group together modules into packages by storing them in the same place. To make a package importable, you also need to add a file __init__.py, which may be empty. You can use import to use single modules or the entire package. For instance, if you have a directory neuroimaging/ with three Python files

```
neuron_count.py
shrink_images.py
__init__.py
```

all of the following import statements work:

```
import neuroimaging
from neuroimaging import neuron_count
from neuroimaging.shrink_images import *
```

The __init__.py file is imported automatically by all three commands. Python has a list of default directories where it looks for modules and packages. These include the current directory and the site-packages/ folder (the location of which depends on your operating system and installation). As mentioned earlier, you can see and modify the complete list of directories for modules and packages using sys.path:

```
import sys
print sys.path
```

Alternatively, you can add Python directories to the PYTHONPATH environment variable (see Appendix D, Section D.3.5).

Q & A: WHAT DOES PYTHON DO IF MODULES IMPORT EACH OTHER IN A CIRCULAR MANNER?

Python can handle simple circular imports (A imports B, B imports C, C imports A again). However, more complex imports of this kind can lead to problems. For debugging, it is easier to arrange your modules rather as an acyclic graph where there is a clear hierarchy. When you find yourself thinking about how to arrange modules in a smart way, you should read about Design Patterns (e.g., at http://sourcemaking.com/).

15.5 TESTING YOURSELF

Exercise 15.1 Data Types

In the skeleton program for the aaRS project, what data types would you use for the variables seq, directory, filename, and sequences?

Exercise 15.2 Prediction of Transmembrane Helices

Write down *requirements* for the following project description: You have a table containing the relative frequencies of amino acids in transmembrane helices. Generally, the nonpolar amino acids are more frequent than the polar ones. On the basis of the data, develop a simple predictor for transmembrane helices. The program should read a protein sequence from a FASTA file and run a sliding window (see Chapter 2) over the sequence. For each subsequence of length N, sum up the frequencies from the table. If the sum is above a given threshold, the program should print a message that a transmembrane helix has been found, along with the sequence and position. Test the program using the protein sequences of bacteriorhodopsin and lysozyme. Determine a threshold parameter for which both proteins are clearly distinguished.

Exercise 15.3 Create a Scaffolding

Implement a program scaffolding for the transmembrane helix predictor. Write down the three most important questions that you would have to ask about the project before starting to write the full code.

Exercise 15.4 Implement the Program

Implement the transmembrane helix predictor.

Exercise 15.5 Run pylint

Run the pylint program on one of your programs. Improve the formatting to achieve a score of at least 9.0.

In Part III, you learned all aspects of modular programming. You know now how to write your own functions and the several advantages of using them. In Chapter 10, you learned how to call functions, how to pass them arguments, and how to return results from a function. In Chapter 11, you have been introduced to classes. These are flexible objects that make it possible to structure and reuse your code. Although classes are abstract and it may be difficult to use them at first, grouping data and methods in the same place can help you to manage very complex tasks. In Chapter 12, you learned how to handle programming errors. You read that errors are normal in programming but that if you structure your programs well, you may be able to spot and manage them when they occur without frequent program crashes. Python provides try...except: blocks where you can monitor code for the occurrence of errors and let your program react specifically. Chapter 13 was aimed at illustrating the use of external modules in Python. The external module described in this chapter is RPy2, the Python interface to R, the most frequently used software for statistical computing and graphical analyses. Once you learn the principles underlying the use of one external module, you will be able to use any external modules, provided you spend some time reading their documentation. In any case, R is very useful in biological data analyses per se. You will encounter more external modules in Part IV and Part V. Chapter 14 showed how to build pipelines, i.e., how to connect programs to each other and how to pass parameters to a program from the UNIX command line you use to run the program. Pipelines are very useful when you have a program whose input is the output of another program. Instead of manually running the programs one after the other, you can write a pipeline that contains the command lines for running the external programs in a Python script. Finally, in Chapter 15, you found several best practices to write well-structured, good programs. Good programs not only have aesthetic purposes but also make it possible to easily detect errors; can be better shared, maintained, and extended; and often show higher performance.

IV

Data Visualization

INTRODUCTION

Having strong visual representations of your data is as important for good science as writing good text. This is what Part IV of this book focuses on. Creating scientific images with Python is technically not difficult. In contrast to Part III, where you learned more complex programming techniques, in Part IV you will take a step back in complexity. Here you will learn to handle three big software packages that can help you very much. You saw in Chapter 13 how to create plots both interactively and noninteractively with the Python interface to R. Here you will see more.

Pictures speak to us. In Chapter 16, you will learn to turn data into diagrams with `matplotlib`. The library consists of a handful of commands to create bar plots, line plots, scatterplots, pie charts, and the like. Functions for scaling and exporting diagrams are explained, as are extras such as labels, error bars, and legend boxes. In Chapter 17, you will learn to create images of 3D molecules with PyMOL. PyMOL contains a powerful scripting interface that allows you to create customized, reproducible molecular graphics. You will be able to render protein, DNA, and RNA structures in high resolution. Chapter 18 shows how you can directly manipulate image files. You will learn how to assemble drawings from lines, boxes, circles, and text with the Python Imaging Library. You will learn how to combine many pictures into one and how to make a big picture smaller. Whether you use your skills to add subtitles to pictures or shrink the size of a disk full of photographs is up to you. By the end of Part IV, you will have pushed the doors wide open toward becoming a skilled image wizard.

Creating Scientific Diagrams

L EARNING GOAL: You can use `matplotlib` to generate high-resolution plots.

16.1 IN THIS CHAPTER YOU WILL LEARN

- How to create a bar plot

- How to create a scatterplot

- How to create a line plot

- How to create a pie chart

- How to draw error bars

- How to plot a histogram

16.2 STORY: NUCLEOTIDE FREQUENCIES IN THE RIBOSOME

In 2009, the Nobel Prize in Chemistry went to Venkatraman Ramakrishnan, Thomas Steitz, and Ada Yonath for their studies on the structure and function of the ribosome. The ribosome is among the biggest known molecular machines, made up of three RNA components (in prokaryotes the 23S, 16S, and 5S rRNA) and many proteins. The RNA consists of the four basic ribonucleotides and a few modified nucleotides that fine-tune the ribosomal function.

TABLE 16.1   Nucleotide Numbers in the 23S
Ribosomal Subunit.

Species	A	C	G	U
T. thermophilus	606	759	1024	398
E. coli	762	639	912	591

16.2.1 Problem Description

In Table 16.1, you find the exact number of nucleotides for the 23S subunit of the *Thermus thermophilus* ribosome resolved by Ramakrishnan et al. and the corresponding data for *Escherichia coli*. In this chapter, you will learn how to display these data in a more attractive way.

The following program creates a bar plot using the `matplotlib` library. First, the *y*-values and bar labels are put into list variables. Second, the `figure()` function creates an empty diagram. Third, various `matplotlib` functions like `bar()` draw components of the diagram. Finally, `savefig()` saves the diagram to a `.png` file.

16.2.2 Example Python Session

```
from pylab import figure, title, xlabel, ylabel,\
    xticks, bar, legend, axis, savefig
nucleotides = ["A", "G", "C", "U"]
counts = [
    [606, 1024, 759, 398],
    [762, 912, 639, 591],
    ]
figure()
title('RNA nucleotides in the ribosome')
xlabel('RNA')
ylabel('base count')

x1 = [2.0, 4.0, 6.0, 8.0]
x2 = [x - 0.5 for x in x1]

xticks(x1, nucleotides)

bar(x1, counts[1], width=0.5, color="#cccccc", label="E.coli 23S")
bar(x2, counts[0], width=0.5, color="#808080", label="T. \
    thermophilus 23S")

legend()
axis([1.0, 9.0, 0, 1200])
savefig('barplot.png')
```

Source: Adapted from code published by A.Via/K.Rother under the Python License.

16.3 WHAT DO THE COMMANDS MEAN?

16.3.1 The `matplotlib` Library

The program uses `matplotlib`, a Python library for scientific diagrams. It contains many functions to generate drawings from data, and some of them have a lot of options. Instead of explaining every option exhaustively, we focus on creating standard diagrams from straightforward data such as *x*- and *y*-values. This chapter provides you with ready-made scripts for different kinds of plots that you can then customize.

Q & A: HOW CAN I EXECUTE THE EXAMPLE?

To use `matplotlib`, you need to install it separately from Python. On all systems this can be done by the single terminal command

```
easy_install matplotlib
```

This requires installation of the `easy _ install` program.
    On Ubuntu Linux, you also can use

```
sudo apt-get install python-matplotlib
```

For Mac OS X 10.6 or higher, .dmg files are provided. On Windows, you need to download and install Scientific Python (http://scipy.org/) first.
    The program in Section 16.2.2 creates a file `barplot.png` (see Figure 16.1) in the same directory in which the Python script has been executed.
    You can use `matplotlib` to create a diagram in just four steps:

```
from pylab import figure, plot, savefig
xdata = [1, 2, 3, 4]
ydata = [1.25, 2.5, 5.0, 10.0]
figure()
plot(xdata, ydata)
savefig('figure1.png')
```

Source: Adapted from code published by A.Via/K.Rother under the Python License.
    First, the library is imported. Second, you start a new figure. Third, you plot some data. Fourth, you save the figure to a .png file. This gives you a plot that you can use right away to analyze your data. Below, more options are presented that you can use to create high-quality plots.

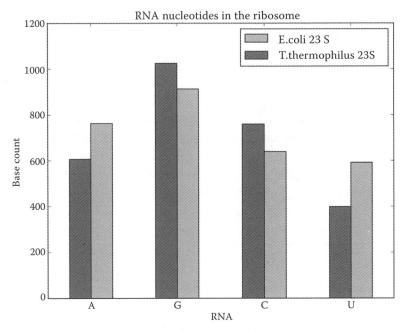

FIGURE 16.1   Bar plot.

Q & A: Can I Create Multiple Figures in the Same Program?

Each time you call the `figure()` function, the background is cleared, and you start creating a new plot. While `matplotlib` is capable of creating multi-panel figures, there are several good reasons to create separate images. First, you can use separate figures for different purposes (posters, presentations, publications) easier than you can use a single multi-panel image. Second, you might want to customize or annotate parts of the figure in a separate program. Finally, single figures are easier to analyze. Creating figures with multiple panels should be left for the stage of manuscript preparation when all your results have been analyzed and confirmed.

16.3.2 Drawing Vertical Bars

The `bar()` function simply draws bars. In the simplest version, it takes as parameters two lists: one for the *x*-axis values and one for the bar heights:

```
bar([1, 2, 3], [20, 50, 100])
```

The first list [1, 2, 3] indicates where the bars should start on the *x*-axis; the second list [20, 50, 100] indicates how high each bar should be.

Similarly, you can create horizontal bars with the `barh()` function:

```
barh([1, 2, 3], [20, 50, 100])
```

The example program in Section 16.2.2 draws four groups with two bars each (see Figure 16.1). If you want to nicely align the bars horizontally with their labels, you need to carefully prepare the *x*-values. There are two lists of *x*-positions in the program: x1 is used for the label of the right bar of each group, and x2 is used for the label of the left bar of each group (shifted by 0.5 to the left with respect to x1). The final `bar()` command has three additional parameters for the bar widths, color, and label that is to appear in the legend box:

```
bar(x1, counts[1], width = 0.5, color = "#cccccc", \
    label = "E.coli 23S")
```

16.3.3 Adding Labels to an *x*-Axis and *y*-Axis

Every scientific diagram that has an *x*-axis and *y*-axis needs to have a concise description on each axis, including the unit used. The `xlabel()` and `ylabel()` functions draw a text on the respective axis:

```
xlabel('protein concentration [mM]')
```

It is possible to use mathematical symbols, subscripts, and superscripts in the labels:

```
xlabel('protein concentration [$\muM$]')
```

A full list of symbols can be found at http://en.wikipedia.org/wiki/Help:Formula.

16.3.4 Adding Tick Marks

Equally important as labeling the axes is adding numerical or text marks along the axis. `matplotlib` adds numbers by itself, but sometimes the result is not what you want. The `xticks()` and `yticks()` functions draw customized ticks:

```
xticks(xpos, bases)
```

writes each element of the `bases` list of strings at the positions `xpos` on the *x*-axis. The `yticks` function works similarly.

16.3.5 Adding a Legend Box

The legend() function takes the labels of all data sets that were plotted so far and writes them to a legend box in the order of appearance. It does not need any argument:

```
legend()
```

16.3.6 Adding a Figure Title

The title() function simply adds text on the top of the diagram, like in the example program in Section 16.2.2:

```
title('RNA bases in the ribosome')
```

16.3.7 Setting the Boundaries of the Diagram

matplotlib automatically chooses the extent of the diagram in such a way that all data will be visible. Sometimes, though, this is not what you want. For instance, if you have a large set of data and want to zoom in on the lower 10%, you can use the axis() function:

```
axis([lower_x, upper_x, lower_y, upper_y])
```

axis() takes a list of four values: the lower and upper boundaries for both the x-axis and the y-axis. In the bar plot program, axis() is used to adjust the margins on the left, right, and top of the canvas to balance the diagram optically:

```
axis([1.0, 9.0, 0, 1200])
```

16.3.8 Exporting an Image File in Low Resolution and High Resolution

The savefig() function writes the entire diagram to an image file in .png format:

```
savefig('barplot.png')
```

By default, a 600 × 600 pixel image with 100 dpi will be created. You can create higher resolution images by adding a precise dpi value:

```
savefig('barplot.png', dpi = 300)
```

You can also export to .tif and .eps formats directly:

```
savefig('barplot.tif', dpi = 300)
```

16.4 EXAMPLES

Example 16.1 How to Plot a Function

matplotlib can create line plots from *x/y* data. The plot() function requires two lists of values that have the same length. The following example plots a sine function (see Figure 16.2):

```
from pylab import figure, plot, text, axis, savefig
import math

figure()

xdata = [0.1 * i for i in range(100)]
ydata = [math.sin(j) for j in xdata]

plot(xdata, ydata, 'kd', linewidth=1)
text(4.8, 0, "$y = sin(x)$",\
     horizontalalignment='center', fontsize=20)
axis([0, 3 * math.pi, -1.2, 1.2])

savefig('sinfunc.png')
```

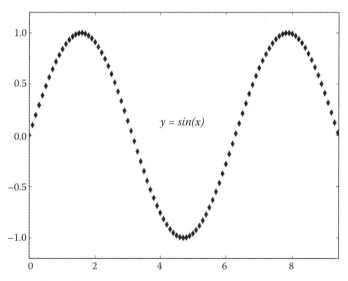

FIGURE 16.2   Plot of a sine function.

First, the program creates a list of equidistant *x*-values using the range() function and a list comprehension (explained in Chapter 4). The *y*-values are created by a list comprehension that calculates sin(x) for each *x*-value. The third parameter in the plot function, 'kd', indicates the color and style of the line. The first character stands for the color (k: black, r: red, g: green, b: blue; others exist). The second character indicates the symbol used for drawing (o: circles, s: squares, v and ^: triangles, d: diamonds, +: crosses, -: lines, :: dotted lines, .: dots). The text() function adds a text marked by the enclosing $ symbols at a given *x/y* position using the mathematical TeX notation (see http://en.wikipedia.org/wiki/Help:Formula). When you limit the *x*-right boundary to 3 * math.pi in the axis() function, the y = sin(x) function begins and ends at the same height (for purely aesthetic reasons).

Q & A: When Using the plot() Function, I See Two Symbols in the Legend Box. How Can I Get Just One?

For a line plot or scatterplot, if you add the legend() function to the code, you will see the symbol used for drawing (a diamond in the example) twice in the legend box. It is possible to fix this within matplotlib, but it is complicated. We recommend leaving the legend out or as it is and edit the legend box manually in a graphics program when all your figures are finished.

Example 16.2 Drawing a Pie Chart

A pie chart like the one in Figure 16.3 can be drawn with the pie() function:

```
from pylab import figure, title, pie, savefig
nucleotides = 'G', 'C', 'A', 'U'
count = [1024, 759, 606, 398]
explode = [0.0, 0.0, 0.05, 0.05]

colors = ["#f0f0f0", "#dddddd", "#bbbbbb", "#999999"]
```

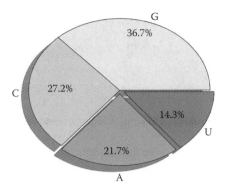

FIGURE 16.3   Pie chart: Bases in 23S RNA from *T. thermophilus*.

```
def get_percent(value):
    '''Formats float values in pie slices to percent.'''
    return "%4.1f%%" % (value)

figure(1)
title('nucleotides in 23S RNA from T.thermophilus')
pie(count, explode=explode, labels=nucleotides, \
    shadow=True, colors=colors, autopct=get_percent)
savefig('piechart.png', dpi=150)
```

Source: Adapted from code published by A.Via/K.Rother under the Python License.

If you want to emphasize that the amount of G and C in the *T. thermophilus* ribosome is more than half of the overall nucleotides, you can use a pie chart. The pie() function in matplotlib works in a similar way as bar() or plot(): you supply the numbers, labels, and colors as lists of items. By the values in the explode list, you can move pie slices out from the center (a 0.0 means it attaches to the other slices). In addition to the labels, you can also write text into each slice. The autopct parameter is a function (the function name is used like a variable here, this is why no parentheses are needed) that gets the size fraction of a slice (from 0.0 to 1.0) and returns a string. The get_percent function converts the number to a string and adds a percent symbol (see Chapter 3 for string formatting).

Q & A: Why Are There Four Percent Symbols in the get_ percent Function?

Each of the four percent symbols in get_percent() is necessary, but they all have a different meaning:

```
return "%4.1f%%" % (value)
```

In string formatting, a single percent character is interpreted as the start of a formatting character (%s for a string, %i for an integer, %f for a float). The first symbol %4.1f is the placeholder for a float inside the string. The double percent character (%%) is the placeholder for a normal percent symbol, because a single percent would be interpreted as another formatting character. Finally, the fourth percent symbol connects the formatting string to the tuple with values to be inserted.

Example 16.3 Adding Error Bars

Error bars can be added to both scatterplots and bar plots (see Figure 16.4). In both cases, you need a third list of numbers for the size of the error bars. A scatterplot with error bars is created by the errorbar() function (which works very similar to plot()), whereas for a bar plot the parameters yerr and ecolor can be added to the bar() function.

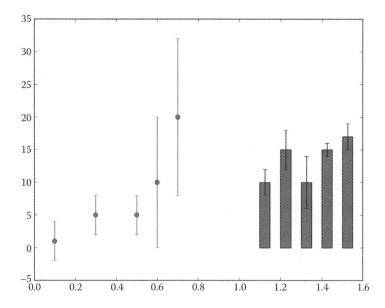

FIGURE 16.4   Error bars in scatterplots and bar plots.

```
from pylab import *
figure()
from pylab import figure, errorbar, bar, savefig
figure()

# scatterplot with error bars
x1 = [0.1, 0.3, 0.5, 0.6, 0.7]
y1 = [1, 5, 5, 10, 20]
err1 = [3, 3, 3, 10, 12]
errorbar(x1, y1, err1 , fmt='ro')

# barplot with error bars
x2 = [1.1, 1.2, 1.3, 1.4, 1.5]
y2 = [10, 15, 10, 15, 17]
err2 = (2, 3, 4, 1, 2)
width = 0.05
bar(x2, y2, width, color='r', yerr=err2, ecolor="black")

savefig('errorbars.png')
```

Source: Adapted from code published by A.Via/K.Rother under the Python License.

Example 16.4 Drawing a Histogram

The following code takes a list of integers and plots their absolute frequencies in five bins (see Figure 16.5):

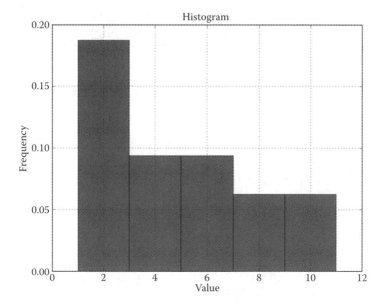

FIGURE 16.5   Histogram.

```
from pylab import figure, title, xlabel, ylabel, \
    hist, axis, grid, savefig
data = [1, 1, 9, 1, 3, 5, 8, 2, 1, 5, 11, 8, 3, 4, 2, 5]
n_bins = 5

figure()
num, bins, patches = hist(data, n_bins, normed=1.0, \
    histtype='bar', facecolor='green', alpha=0.75)
title('Histogram')
xlabel('value')
ylabel('frequency')
axis()
grid(True)
savefig('histogram.png')
```

Source: Adapted from code published by A.Via/K.Rother under the Python License.

The `hist()` function creates a bar plot but groups the data into the given number of bins first. The `grid()` function switches on a grid in the background of the diagram.

Q & A: WHERE CAN I FIND HELP TO DRAW FURTHER DIAGRAM TYPES?

The quickest method to find your way around `matplotlib` is to borrow from working examples. You can check the `matplotlib` gallery web page (http://matplotlib.org/gallery.html) for an example diagram that is the closest to what you want to do. Then copy the source code for the example and start manipulating it (see Exercise 16.4).

16.5 TESTING YOURSELF

Exercise 16.1 Create a Bar Plot

Display the neuron lengths from Chapter 3 as a plot with horizontal bars.

Exercise 16.2 Create a Scatterplot

Plot the number of bases in *T. thermophilus* versus the number of bases in *E. coli* from Table 16.1.

Exercise 16.3 Draw a Histogram

Read a FASTA file with multiple sequences. Plot a histogram of the sequence lengths.

Exercise 16.4 Use the `matplotlib` Gallery

Find out how to create a box-and-whisker plot in the `matplotlib` gallery. Copy and paste the example to a Python file and execute it.

Exercise 16.5 Create a Series of Plots

Write a program that parses a multiple sequence FASTA file, counts the frequencies of the 20 amino acids in each sequence, and creates a separate pie chart for each sequence, showing the top five frequencies and summarizing the rest as "other."

## CHAPTER 17

Creating Molecule Images with PyMOL

L EARNING GOAL: You can create high-quality images of molecules.

17.1 IN THIS CHAPTER YOU WILL LEARN

- How to display a molecule in seven steps

- How to create publication-quality images of molecules

- How to use the PyMOL command line

17.2 STORY: THE ZINC FINGER

Zinc fingers are one of the most abundant three-dimensional motifs by which proteins bind DNA. A single zinc finger consists of two helices held in a specific conformation by a zinc atom so that one helix fits into the major groove of the DNA. If you want to explain to someone how zinc finger proteins bind DNA and what the role of the zinc atom is, you may quickly reach a point where words fail. Discussions about molecular structures cry for illustrations. How do we make a good (i.e., journal-quality) picture of a molecule then?

In photography, you cannot simply aim at a beautiful person, hit the trigger of your camera, and expect to have a shot for the title page of *Cosmopolitan*. A lot of work goes into your model getting dressed, doing the makeup, adjusting the lighting, and postprocessing the image. The same is true for three-dimensional (3D) models of biomolecules.

Fortunately, you can do most with a single program: PyMOL. Before you start, it is worth it to think about what message the picture should convey. Such a message could be as follows: "The image displays how the zinc ions in a zinc finger protein relate to the DNA and protein components, so that the function of zinc finger proteins can be explained." The outcome could look like the one in Figure 17.1. In this chapter, you will learn to create such pictures with the PyMOL software.

17.2.1 What Is PyMOL?

PyMOL is a picture machine for molecules (www.pymol.org). You can use it to create high-resolution images of 3D structures for publications, presentations, and websites. PyMOL has functions for visualizing coordinates of structures and for analyzing their chemical properties. You can use all functions of PyMOL via the graphical interface and a scripting language, or you can combine both. With the graphical interface, it is often more comfortable to quickly create a snapshot. On the other hand, creating quality images is complicated with the graphical interface

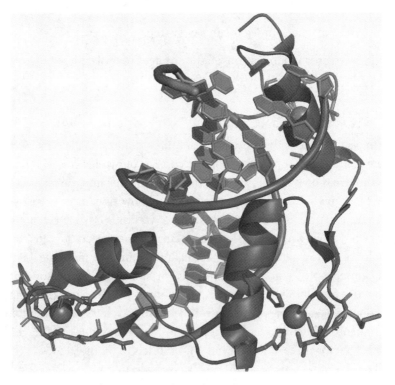

FIGURE 17.1   A zinc finger protein bound to DNA.

alone. The scripting interface allows you to make the generation of your figures fully reproducible, so you won't have to start over when you find out you would like to change a few details. Taken together, the graphical and scripting methods complement each other.

PyMOL was written by the late Warren L. DeLano using Python, while the time-critical parts have been coded in the C language. It runs equally well on Windows, Linux, and Mac OS. There is a free and commercial version of PyMOL available. The free version can be used without restrictions, with all functions fully available. The commercial version, currently maintained by the Schrödinger company, offers extra features like improved rendering and integration with PowerPoint. You can find binary versions for all operating systems at http://sourceforge.net/projects/pymol/ under 'Files' → 'Legacy'.

The following PyMOL script creates the complete zinc finger image from a single protein structure file. The commands used in the following zinc finger script are not Python but commands specific for PyMOL. However, you can use them in a similar way as you use the Python shell: you enter commands in the text console in PyMOL or write them into a script file. The command line is useful if you want to support your work on the graphical interface, to use it as the main way of controlling PyMOL, or simply to try functions of the program. See Boxes 17.1 and 17.2 for a basic overview of the graphical interface and help functions.

BOX 17.1   THE GRAPHICAL USER INTERFACE (GUI)

The PyMOL graphical user environment consists of two windows, a big one and a medium one (see Figure 17.2). The following lists the most important functions:

- *3D screen (center of the big window).* The 3D molecules appear here. Alternatively, a text console can be toggled on and off by pressing ESC.
- *Object list (top right of the big window).* Loaded molecules and user-defined selections appear here. Each of them can be toggled on and off by clicking its name. Also, individual display modes for each object can be selected using the S (show), H (hide), L (label), and C (color) buttons.
- *Command line (bottom of the big and medium windows).* Here, text commands can be entered. Partially written commands can be expanded by pressing TAB. By pressing the up arrow you can see

previously entered commands. To navigate directories, use the commands cd, dir, ls, and pwd from the PyMOL command line just as in the UNIX or Windows shell. On Windows, cd is especially important, because PyMOL starts by default in its own directory, not in the "Desktop" or the "My Documents" folders.

- *Mouse controls (bottom right of the big window).* The functions of the mouse buttons in combination with the keyboard are summarized here. A click into this control box toggles between two modes for selecting and visualizing atoms.
- *Display menu (top of the medium window).* This menu tweaks the quality of the current scene.
- *Settings menu (top of the medium window).* Colors and details of the display modes can be adjusted here.
- *Wizard and Plugin menus (top of the medium window).* This menu contains advanced tools and some demos.

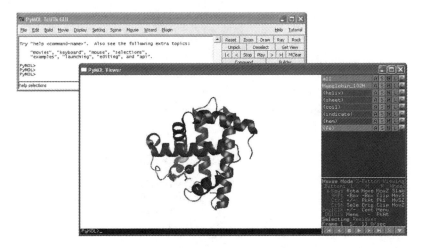

FIGURE 17.2 The PyMOL graphical interface. *Note:* The foreground image shows the big window. The background image represents the medium window.

BOX 17.2 GETTING HELP

PyMOL has excellent built-in documentation. It gives you a list of all available commands and descriptions of each command. To get to the main help page, you can type

```
help
```

This displays available help topics in the main window. By pressing ESC you get back to the graphical screen. To get help with a specific command (e.g., orient), you can type

```
help orient
```

In addition, a few specific help topics are available. The one on selections is particularly useful:

```
help selections
```

17.2.2 Example PyMOL Session

```
delete *
load 1aay.pdb

hide everything
bg_color white

# protein
select zinc_finger, chain a
show cartoon, zinc_finger
color blue, zinc_finger

# DNA
select dna, chain b or chain c
select dna_backbone, elem P
show cartoon, dna
set cartoon_ring_mode, 3
color green, dna
color forest, dna_backbone

# zinc
select zinc, resn zn
show spheres, zinc
color gray, zinc

# binding residues
select atoms_pocket, zinc around 5.0 and not zinc
select pocket, byres atoms_pocket
show sticks, pocket
set valence, 1
color marine, pocket

set_view (\
      0.385022461,     -0.910319746,     -0.151902989,\
     -0.748979092,     -0.212032005,     -0.627752066,\
      0.539247334,      0.355471820,     -0.763447404,\
```

```
      0.000005471,       0.000029832,      -134.466125488,\
      1.499966264,      12.841400146,        50.074134827,\
    100.975906372,     167.958770752,         0.000000000 )
ray 800,600
png zinc_finger.png
```

Source: Adapted from code published by A.Via/K.Rother under the Python License.

Q & A: HOW DO YOU RUN THE SCRIPT?

First, copy the commands into a text file (e.g., `zinc_finger.pml`). Second, start PyMOL. Finally, move to the directory where you have put the `zinc_finger.pml` file, choose the `File->Run` option, and select your script file. On Windows, it might be enough to double-click the file. On Linux, you can use the text console of PyMOL and type `@/home/your_login/Desktop/zinc_finger.pml` (this is less convenient on Windows because folder names are long and complicated). On Mac OS X, you can type `@zinc_finger.pml` in the directory where you have put the file.

17.3 SEVEN STEPS TO CREATE A HIGH-RESOLUTION IMAGE

The script creates the zinc finger image in Figure 17.1. To create similar scripts, you can follow a seven-step pattern:

1. Create a PyMOL script file.

2. Load the molecule.

3. Select atoms and residues.

4. Choose representations for each selection.

5. Color the molecule and background.

6. Set the camera position.

7. Export a high-resolution image.

All seven steps can be done using the graphical interface of PyMOL (see Figure 17.2 and Box 17.1). In this chapter, however, the scripting interface will be explained, because it allows you to reproduce, customize, and improve the way an existing image has been created.

17.3.1 Writing PyMOL Script Files

The commands that the PyMOL text console understands can be stored in script files. PyMOL scripts are plain text files with one command in each line. Usually the text files have the ending .pml. A script can be invoked from the PyMOL "File" menu or by an @ followed by the filename:

```
@zinc_finger.pml
```

PyMOL scripts should always have the .pml extension.

The advantage of using a script is that your picture is fully reproducible. You can easily adapt a script to other tasks, even after months or years. If you want to create a series of pictures, it saves time to have a script like the one at the beginning of the chapter as a starting point and modify it. For your first steps in PyMOL, you can simply copy the zinc finger script and adjust it to your own molecules.

17.3.2 Loading and Saving Molecules
Loading Molecules
To start, you need a file with the 3D structure of your molecule, for instance, one you downloaded from the PDB (www.pdb.org) or PubChem (http://pubchem.ncbi.nlm.nih.gov/) databases. You can load a molecule file using the load command followed by the name of the structure file:

```
load 1aay.pdb
```

The molecule name appears in the panel at the top-right corner: the object list (see Figure 17.2). This is the location where PyMOL shows everything that is loaded or otherwise defined at any given moment. You can assign a name to the loaded molecule with

```
load 1aay.pdb, zinc_finger
```

PyMOL can read many file formats commonly used for molecules: .ent, .pdb, .mol, .mol2, .xplor, .mmod, .ccp4, .r3d, .trj, and .pse.

Q & A: My Molecule Does Not Load. What Went Wrong?

Most probably, PyMOL is looking in the wrong directory. When you enter pwd in the PyMOL command line, PyMOL prints which directory you are in (press ESC to see the result). Use cd <folder name> to change to a different directory, and finally use ls to list all files. If you are in the right place and your file is listed there, the command load <filename> should load the structure. If it still goes wrong, try a different file or open the file in a text editor to check if it really contains 3D coordinates.

Saving Molecules

In the same way, you can store molecules to files. PyMOL recognizes file types by their ending. For protein, DNA, and RNA structures, the .pdb format is most common, because it can be read by many other programs (and most people you might ask for help). By default, PyMOL saves all molecules currently loaded with the save command followed by a filename with the .pdb extension:

```
save everything.pdb
```

You can also save only one object from the list in the top-right corner:

```
save my_molecule.pdb, zinc finger
```

If you use a filename with the .pse extension, PyMOL allows you to save the entire PyMOL session in a single file, including colors and the camera perspective:

```
save my_session.pse
```

When you load a .pse session, everything you had loaded previously gets deleted. Thus, it is not possible to combine two sessions. In that situation scripts are more useful. PyMOL sessions can also be loaded and used in scripts. Do not confuse the ending .pml for script files with that of PyMOL sessions .pse.

Q & A: Can I Use PyMOL Sessions to Save My Progress?

Yes, generally you can write sessions with different filenames while you are working on a structure. This is fine for saving your progress from power failure or other accidents. However, when you try to load your session files with an older or newer PyMOL version, they may be incompatible, or the display may be distorted. Writing scripts circumvents this problem, because you have full control over what is in the script.

17.3.3 Selecting Parts of Molecules

To create a decent picture of your molecule, you will find it worthwhile to first neatly lay out all parts you want to use. For the zinc finger protein, this means defining the protein, the DNA, the zinc, and the binding residues as separate objects. In PyMOL, such molecule parts are called *selections*. Selections appear in the object list in the top-right corner. You can define selections using the *sequence ruler* or the `select` command.

Using the Sequence Ruler

The graphical interface has a box that allows you to pick a few amino acids or nucleotides. When you press the little s button in the bottom-right corner (see Figure 17.2), a selection box with the sequence of your molecule appears on the top of the display. Whatever residues you click in the sequence, they will be immediately represented in the object list as an entry labeled (`sele`). If you want to save the selected residues for later, it helps to copy the selection under a different name using

```
select my_copy, sele
```

in the text console.

You can also create residue selections simply by clicking on individual atoms in the structure. The selection will be indicated by purple squares and represented in the object list as an entry labeled (`sele`). From there, you can store your selection in the same way as described previously. If you want the purple squares to disappear, you can type

```
indicate none
```

or

```
disable sele
```

The `select` Command

The `select` command is the most powerful (and complex) command in PyMOL. You will probably spend most of your time learning PyMOL on this command. To make this time a little shorter, you can start with the examples given here. An overview of the selection algebra is available by typing help selections.

The select command always follows the following pattern:

```
select <selection>,<expression>
```

<selection> is the name for the subset you want to define (easy). <expression> is a rule describing what you want to select (more difficult). Whenever you select something, it appears as a selection in the object list at the top-right corner. Selections are a subset of the atoms in your molecules that you can manipulate separately. For instance, in the zinc finger script in Section 17.2.2, the select command is used four times to select chains, residues, and single atoms. Many different expressions for selection are possible.

Selecting Chains

```
select zinc_finger, chain a
```

This chooses the entire chain A from the structure file. In the object panel, an extra line zinc_finger will appear:

```
select dna, chain b or chain c
```

The DNA consists of two chains, which in this structure are labeled B and C (apart from some exceptions, PyMOL is case insensitive). Both are included in the same selection by the logical or.

Selecting Residues

```
select zinc, resn zn
```

Finally, the three individual zinc atoms are chosen as entire residues and put into a single selection (zinc). Analogously, you can select amino acid residues replacing zn with ala, cys, val, etc. Since the zinc residue consists of a single atom, you could also select the zinc atoms with

```
select zinc, elem ZN
```

instead of selecting a residue (the ZN atom name must be uppercase).

Selecting Atoms

```
select dna_backbone, elem P
```

The phosphates in the DNA backbone are the only phosphates present in the structure. This command simply grabs all of them and puts them into a selection named dna_backbone. More examples of select commands can be found in Table 17.1.

Q & A: How Do I Know Which Chains, Residues, or Atoms My Molecule Contains?

There are three ways to achieve this. First, open the PDB file in a text editor (this may be inconvenient if you are not familiar with the format). Second, look at the summary of your structure at www.pdb.org (given that your structure is from there). Third, left-click on an atom of the chain you want to select. In the text window below the menu bar, PyMOL displays a line similar to

```
You clicked/1aay//C/DC`56/O4'
```

This means you clicked the O4' atom in the molecule 1aay, chain C, residue number 56, which is a deoxycytidine (abbreviated DC). Alternatively, you can right-click on the atom you want to select. A pop-up window with options will appear.

TABLE 17.1   Selection Commands in PyMOL.

Selection	Command
Select an object or selection	select sel1, 1aay
Duplicate an object or selection	create clone1, 1aay
Select a chain	select dna, chain A
Select residues by name	select aromatic, resn phe+tyr+trp
Select residues by number	select sel2, resi 1-100
Select atoms by name	select calpha, name CA
Select atoms by element	select oxygen, elem O
Select by protein secondary structure (helices and sheets)	select helix, ss h select sheet, ss s
Combine selections by "or"	select sel3, resi 1-100 or resi 201-300
Combine selections by "and"	select sel4, resn trp and name ca
Select all oxygens from chain A but no waters	select sel5, elem O and chain A and not resn HOH
Select atoms around a ligand in 5.0 Ångstroms diameter	select sel6, resn HEM around 5.0
Select entire residues around a ligand in 5.0 Ångstroms diameter	select sel7, br. resn HEM around 5.0

Combining Conditions

You can create selections using more than one condition, combining them with any of the keywords and, or, or not and with parentheses. This works in a similar way to the Boolean operators and, or, and not that are used with the if condition in Python (see Chapter 4, Section 4.3.1).

```
select sel02, resi 1-100 or resi 201-300
select sel03, resn trp and name ca
select sel04, ss h and not (resn ile+val+leu)
```

You can use these operators to create selections from other selections.

```
select aa, resi 1-100
select bb, resi 201-300
select cc, aa or bb
```

These expressions can be written in a single line:

```
select cc, resi 1-100 or resi 201-300
```

Selections versus Objects

These two concepts need further clarification. An *object* is the primary representation of a molecule with all atoms, bonds, colors, and display modes in the PyMOL memory. A *selection* is a pointer to a defined set of atoms in one or several objects. It follows that each atom can belong to more than one selection but only to one object. In the PyMOL object list (upper-right corner of the graphical window), the names of the selections are indicated in brackets. Practically, the biggest difference between them is that clicking an object name will hide/show it completely! There is a default selection, named "(all)," that contains all atoms loaded to PyMOL. The difference between selections and objects is demonstrated in an example:

```
load lysozyme.ent, lysozyme
select calpha_sel, lysozyme and name ca
create calpha_obj, lysozyme and name ca
```

The first command creates an object named lysozyme, as it loads new atoms to the memory. The second defines a selection, calpha_sel, which contains *pointers* to all alpha-carbons of lysozyme. The third command creates an object named calpha_obj, which contains the *coordinates* of the alpha-carbons. Effectively, these atoms are duplicated in memory by the create command. When, e.g., the color of both lysozyme and calpha_sel are altered, the changes will override each other. In contrast, calpha_obj remains unaffected, as it has its own atoms. When lysozyme is removed, the calpha_sel selection will also disappear, but the duplicate calpha_obj object will remain.

17.3.4 Choose Representations for Each Selection

For each part of your molecule, you need to decide how you want it to look. You can pick display modes by clicking the S (show) buttons in the object list (see Figure 17.2). Proteins and nucleic acids often look best as cartoons, while smaller molecules and details are better visible as sticks or balls and sticks. Grooves and electrostatic potentials can be visualized by sphere and surface representations.

On the command line, the commands show and hide set the display mode of a given selection or object. The available modes include lines, sticks, cartoon, spheres, and surface. Typical examples of the show and hide commands are as follows:

```
hide all
```

This command disables all representations. The loaded molecules disappear.

```
show cartoon
```

This command displays all molecules in cartoon mode (protein and nucleic acids; for single atoms and small molecules, there is no cartoon mode).

```
show cartoon, zinc_finger
```

This command does the same thing as the previous one but for the zinc_finger selection only.

```
hide cartoon, zinc_finger
```

This one switches the cartoon representation for the zinc finger off.

There are a few tricks that help display proteins, nucleic acids, small molecules, and single atoms better.

Displaying Protein Cartoons

For protein cartoons, you can change the type of cartoon with the cartoon command, e.g.,

```
cartoon arrow
```

Instead of `arrow`, you can use `loop`, `rect`, `oval`, `tube`, and `dumbbell`. The `cartoon` mode can be set specifically for certain secondary structural elements:

```
cartoon tube, ss h
cartoon rect, ss s or ss l
```

Finally, you can adjust a lot of detailed settings for cartoons, e.g.,

```
set cartoon_loop_radius, 2.0
```

Check the 'Settings' → 'Edit All...' menu for all the available settings. They are not documented, but they can be understood by trial and error.

Displaying DNA and RNA Cartoons

Similar to proteins, nucleic acids can be displayed as cartoons. The most typical mode for DNA and RNA is set by

```
show cartoon, dna
set cartoon_ring_mode, 3
```

The `cartoon_ring_mode` number accepts values from 1 to 6. Different `cartoon_ring_mode` values correspond to different representations.

Displaying Small Molecules

The most common display mode for small molecules (or side chains of macromolecular residues) is "sticks." One thing to keep in mind is that the valence of single and double bonds needs to be displayed properly. For the zinc-binding pocket, this is done with the following commands:

```
show sticks, pocket
set valence, 1
```

Displaying Ions and Other Single Atoms

The zinc atoms in the zinc finger are displayed as simple balls:

```
show spheres, zinc
```

The same can be applied to water molecules that frequently occur in PDB files. If the water molecules are unwanted, you can hide them:

```
hide nonbonded
```

Or you can delete them altogether:

```
remove resn hoh
```

17.3.5 Setting Colors

Colors can do a lot to support the message of your image. Before you start thinking about particular colors, it is worth considering basic constraints: What background do you want to have? Are you preparing for an impressive presentation or for print media? Are you going to print color or black-and-white figures? Deciding these things up front can save you a lot of trouble later. If you want to postpone the decision, having a PyMOL script for your scene keeps your options flexible. When coloring your scenery, please bear in mind that 9% of the male population has deuteranopia and may have trouble telling red and green apart. However, when you print your research article in black and white, the percentage of researchers for whom red and green are indistinguishable grows to 100%.

Setting the Background Color
The background color determines all other colors. In live presentations, a black background often makes grasping the depth easier. For printed pages, a white background generally looks better (and is cheaper to print), unless it is printed on expensive glossy paper. You can change the background to white with

```
bg_color white
```

Setting the Colors of Molecules
The `color` command changes the color of the entire object or a given selection:

```
color red
color red, zinc_finger
```

PyMOL has a predefined set of colors such as `firebrick`, `forest`, `teal`, `salmon`, `marine`, and `slate` that you can readily use with the `color` command. A list of colors is available from the "Settings" menu or the `C` (color) button in the object list (the last column in the graphical panel, top right; see Figure 17.2).

Color by Element

If you want to color a molecule and assign different colors to elements, you can use a special function:

```
util.cbag('zinc_finger')
```

This function sets carbon atoms to green. Analogous functions assigning different colors to carbon atoms are util.cbab (blue), util.cbac (cyan), util.cbak (light magenta), util.cbam (magenta), util.cbao (orange), util.cbap (dark magenta), util.cbas (salmon), util.cbaw (white), and util.cbay (yellow).

Defining Your Own Colors

You can use the set_color command to define custom colors:

```
set_color leaves, [0.2, 0.8, 0.0]
color leaves, dna
```

The three numbers in the list correspond to red, green, and blue, respectively. Each color number ranges between 0.0 and 1.0, and you can play with them to obtain your favorite color. For example,

```
set_color leaves, [0.1, 0.0, 0.0]
color leaves, dna
```

will color dna in red.

Q & A: What Do "RGB" and "CMYK" Stand For?

There are two kinds of color schemes: RGB (red-green-blue) for screen display and CMYK (cyan-magenta-yellow-black) for printing. Some RGB colors look terrible when printed, but PyMOL has a built-in CMYK translation that should be enabled when creating print images. You can find it in the 'Display' → 'Color Space' menu. The actual look of CMYK colors is to some extent device dependent. Most of the default colors in PyMOL are CMYK safe.

17.3.6 Setting the Camera Position

Finding a good camera position is not always easy. Often, portions of a structure hide part of what you want to display, or when you show multiple sites, it is difficult to get them all into view at the same time. Positioning the camera may involve trade-offs. A practical solution is to try a series of positions and pick the best image later (photographers do the same!). To get the camera position into your PyMOL script, you first need to position the molecule using the mouse. Once you've found a good orientation, you can transfer the camera position to a script by typing

```
get_view
```

The command prints a set_view() command with lots of numbers in the brackets. The numbers represent the exact camera position, zoom, depth, etc. In practice, you do not need to worry about what the numbers mean. You can simply copy the entire command into your PyMOL script.

Q & A: How Can I Copy Text from the PyMOL Window?

You can select the text from the text field in the smaller PyMOL menu bar window on the top using your mouse and copy it using Ctrl-C.

Centering Your Molecule
If your molecule has slipped out of sight while you are moving it around or the rotation center is weird, you can center the molecule (or parts of it) by typing

```
orient
orient dna
```

17.3.7 Exporting a High-Resolution Image

PyMOL has a built-in ray tracer that can be used to create high-quality light effects. The final image of the zinc finger protein is generated by these commands:

```
indicate none
set fog, 0
ray 800, 600
png zinc_finger.png
```

The first command hides the small purple squares by which selected atoms are indicated. The second command turns the fog (more distant atoms being a little blurred) off. In the 'Setting' → 'Rendering' menu, a few related options such as picture quality, antialiasing, and shadows can be adjusted. The ray command generates the high-quality image. You can simply type

```
ray
```

to display an image of the size of the PyMOL graphical window. Otherwise, you can explicitly give the size in pixels, as in the previous example (see Box 17.3). The execution of the ray command can take some time depending on image complexity and size. Finally, the png command writes an image file in .png format. The calculated image should be immediately stored using the png command, as it is lost as soon as the scenery is changed.

BOX 17.3 HOW MANY PIXELS DO I NEED FOR 300 DPI?

When you export your image for a journal, it is important to know how big the final printed image is going to be. If, for instance, you are preparing a 7-inch-wide color figure with 300 dpi, you need a 7 × 300 = 2,100-pixel-wide image. If you want to leave room for labels or a margin, you can create a slightly smaller image and paste it into an empty canvas in Photoshop or GIMP. Just never, *under any circumstances*, shrink or enlarge an image you want to submit to a journal. Quality almost inevitably goes down the drain that way.

Labels

Your image will become a masterpiece when you add text labels and symbols (circles, arrows, etc.). PyMOL can display labels for atoms, residues, etc., but they are close to worthless for publications. Therefore, it is better to use an external program (e.g., Photoshop, GIMP, or PowerPoint) to add graphical elements.

Final Advice

In photography it is said that to make good pictures, you need to make many pictures. The same applies to molecules. Often the best way to obtain your image is to save your ray-traced scenery with several representations, different colors, and many camera perspectives in a row and choose the best shot afterward. You can find more on PyMOL in the resources given in Box 17.4.

BOX 17.4 PyMOL WEBSITES

Some experienced PyMOL users have written plugins for PyMOL that are available on their personal websites (if not already on the PyMOL wiki):

- The site www.pymol.org contains the PyMOL manual (150+ pages). It explains the mouse navigation and menus in detail and contains a reference for all commands.
- The PyMOL wiki (www.pymolwiki.org) contains user-contributed documentation. The wiki covers mostly intermediate to advanced topics in a high quality and has the advantage that example scripts are quite often given. The main index is reached via the "Top level of contents" point on the main page.
- Robert Campbell's crystallography tools (http://adelie.biochem. queensu.ca/~rlc/work/pymol/).
- A tutorial that explains many things not explained here has been written by Gareth Stockwell (www.ebi.ac.uk/~gareth/pymol/).

17.4 EXAMPLES

Example 17.1 Creating Ball-and-Stick Representation

PyMOL has no direct ball-and-stick representation. However, it can be simulated by combining spheres and sticks:

```
show sticks, pocket
show spheres, pocket
set stick_radius, 0.1, pocket
set sphere_scale, 0.25, pocket
color marine, pocket
```

Source: Adapted from code published by A.Via/K.Rother under the Python License.

The two set commands change the size of the sticks and the size of the spheres only for the pocket selection. The results are small balls connected by sticks.

Q & A: What Other Settings Are There?
Check "Setting" and "Edit all" in the main menu to see what other parameters you can change.

Example 17.2 Transparent Surfaces

It is often helpful and impressive to have a part of the molecule invisible. PyMOL can make surfaces partially transparent:

```
hide all
show surface
show cartoon
set transparency, 0.5
```

If you want to shade the cartoons differently to the surface, you can load the same molecule twice:

```
load 1aay.pdb, zf_surface
load 1aay.pdb, zf_cartoon
hide all
show cartoon, zf_cartoon
show surface, zf_surface
set transparency, 0.5
```

Source: Adapted from code published by A.Via/K.Rother under the Python License.

This way, two copies of the same structure are created, but they look different. Using separate files for parts of your structure, you can create separate surfaces for, e.g., a protein and its ligand.

Pearl Effect

You can use transparency to create a pearl-like effect. You duplicate the atom and create one with a solid sphere and one with a slightly larger, translucent sphere (corona):

```
create zinc2, zinc
set sphere_transparency, 0.4, zinc2
set sphere_scale, 1.05, zinc2
ray
```

Source: Adapted from code published by A.Via/K.Rother under the Python License.

The create command generates extra atoms from the given selection. The duplication of molecules or parts of them is sometimes useful to achieve more complicated effects.

Example 17.3 Highlighting Distances between Atoms

Thin lines connecting atoms and the corresponding distance can be shown by the `distance` command. It takes two atoms as arguments, specifying either the full identifier, as in the example below, or two separate selections containing one atom each.

```
distance dist = (/1aay//C/DA' 58/OP2),(/1aay//B/DG' 10/OP2)
color black, dist
```

Source: Adapted from code published by A.Via/K.Rother under the Python License.

17.5 TESTING YOURSELF

Exercise 17.1 Create a High-Resolution Image

Create a high-resolution image of the hemoglobin molecule that shows how the iron atom in the heme group is kept in position by the two histidine residues (see Figure 17.3). You can use the structure with the PDB code 2DN2.

Exercise 17.2 Select a Hetero Group

Write a PyMOL command that selects the entire heme group from the hemoglobin structure and nothing else.

Exercise 17.3 Select Specific Residues

Write one or more PyMOL commands that select the two histidines attached to the heme group and nothing else.

Exercise 17.4 Draw Ball-and-Stick Mode

Write a PyMOL script that displays the heme group in ball-and-stick mode but gives the sticks a color different from that of the balls.

Hint: You will need to load the molecule twice using different names.

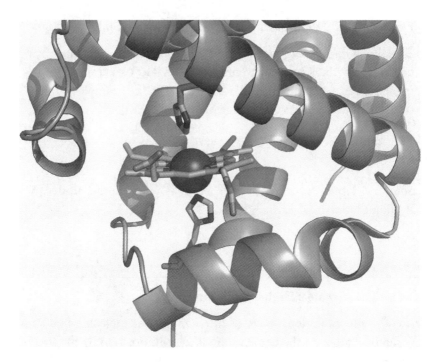

FIGURE 17.3   A high-resolution image of the hemoglobin molecule. *Note:* The iron atom in the hemoglobin heme group is kept in position by two histidine residues.

Exercise 17.5 Create a Movie

Use PyMOL to generate a molecular movie highlighting structural features of the hemoglobin structure: the fold, the heme-binding site, and functional amino acids. You can use the emovie.py plugin (www.weizmann.ac.il/ISPC/eMovie.html) or write your own. In any case, you need to generate a long sequence of .png images from which the movie can be assembled. On Windows you can use MEncoder to assemble them to a movie. The command for running MEncoder in the directory with all your .png files is

```
mencoder "mf://*.png" -mf fps = 25 -o output.avi -ovc
lavc -lavcopts vcodec = mpeg4
```

Manipulating Images

LEARNING GOAL: You can compose images from geometrical shapes and text.

18.1 IN THIS CHAPTER YOU WILL LEARN

- How to draw images with the Python Imaging Library

- How to draw a schematic image of a plasmid

- How to draw geometrical shapes

- How to add text to an image

- How to combine several pictures to a single image

- How to resize an image

- How to convert a color image to a black-and-white image

18.2 STORY: PLOT A PLASMID

One of the first artificial plasmids was constructed in 1977 by Francisco Bolivar and Raymond Rodriguez. It consists of 4,361 DNA base pairs, a replication origin, a gene for ampicillin resistance, and one gene for tetracycline resistance. The plasmid contains many restriction sites and thus serves as a base for constructing genetic vectors.

18.2.1 Problem Description

Schematic diagrams are at the core of visualizing scientific content. Imagine phylogenetic trees without seeing the tree, imagine a metabolic pathway without reaction arrows, and imagine explaining a cell without actually drawing it. The same is true for the structure of genes and proteins.

In this chapter, you are going to draw the pBR322 plasmid. An image of the plasmid would show it as a ring-shaped structure. Regions for the replication origin and both resistance genes need to be marked by different colors. An exemplary cleavage site needs to be indicated, and text labels should be added. And of course, everything must be drawn at the positions corresponding to the correct nucleotide positions.

Using Photoshop or GIMP, it is difficult to manually get the proportions right. Software for drawing vector graphics such as Inkscape or CorelDraw is a better choice but involves a lot of manual work. A multitude of plasmid drawing programs are available on the web, all having advantages and disadvantages. Because none of these solutions are perfect, you are going to build your own. With Python you are in full control of what will be drawn. You are going to draw a precise schematic diagram of the pBR322 plasmid using the *Python Imaging Library* (PIL) (see Box 18.1). PIL contains powerful tools to manipulate images, from moving and rotating parts over graphical elements of the drawing to using complex filters that change the entire image. PIL is one of the most popular Python libraries; for instance, the images on Euro coins were created using PIL. The following example imports two modules from PIL, assigns some constants to variable names, activates drawing tools, and then defines and calls functions for the actual drawing of the plasmid.

BOX 18.1   HOW TO INSTALL THE PYTHON IMAGING LIBRARY

On Linux you can install PIL by typing

```
sudo apt-get install python-imaging
```

On Windows, you need to download a binary distribution from www.pythonware.com/products/pil/ and install it by double-clicking. In both cases you have succeeded when you open a Python shell after installing and the following import works:

```
>>> import PIL
```

18.2.2 Example Python Session

```python
from PIL import Image, ImageDraw
import math

PLASMID_LENGTH = 4361
SIZE = (500, 500)
CENTER = (250, 250)

pBR322 = Image.new('RGB', SIZE, 'white')
DRAW = ImageDraw.Draw(pBR322)

def get_angle(bp, length=PLASMID_LENGTH):
    """Converts base position into an angle."""
    return bp * 360 / length

def coord(angle, center, radius):
    """Return (x,y) coordinates of a point in a circle."""
    rad = math.radians(90 - angle)
    x = int(center[0] + math.sin(rad) * radius)
    y = int(center[1] + math.cos(rad) * radius)
    return x, y

def draw_arrow_tip(start, direction, color):
    """Draws a triangle at the given start angle."""
    p1 = coord(start + direction, CENTER, 185)
    p2 = coord(start, CENTER, 160)
    p3 = coord(start, CENTER, 210)
    DRAW.polygon((p1, p2, p3), fill=color)

TET_START, TET_END = get_angle(88), get_angle(1276)
AMP_START, AMP_END = get_angle(3293), get_angle(4153)
ORI_START, ORI_END = get_angle(2519), get_angle(3133)

# drawing the plasmid
BOX = (50, 50, 450, 450)
DRAW.pieslice(BOX, 0, 360, fill='gray')
DRAW.pieslice(BOX, TET_START, TET_END, fill='blue')
DRAW.pieslice(BOX, AMP_START, AMP_END, fill='orange')
DRAW.pieslice(BOX, ORI_START, ORI_END, fill='darkmagenta')

DRAW.pieslice((80, 80, 420, 420), 0, 360, fill='white')
draw_arrow_tip(TET_END, 10, 'blue')
draw_arrow_tip(AMP_START, -10, 'orange')
draw_arrow_tip(ORI_START, -10, 'darkmagenta')

pBR322.save('plasmid_pBR322.png')
```

Source: Adapted from code published by A.Via/K.Rother under the Python License.

18.3 WHAT DO THE COMMANDS MEAN?

As mentioned earlier, Section 18.2.2 uses PIL (see Box 18.1). The most important parts of PIL are imported by

```
from PIL import Image, ImageDraw
```

At the heart of PIL is the `Image` module. Practically everything you can do to an image uses this module. For instance, when you read a picture from a file, you get an `Image` object. When you draw graphical elements, they are drawn on an `Image` object. When you write text—you got the idea. The `ImageDraw` module is a collection of tools to draw things into images.

The plasmid diagram in Figure 18.1 is composed of one big gray circle with colored pie slices for the three marked regions. After you draw these four parts of the plasmid, you draw a smaller white circle in the middle to cut out the center part of the plasmid. Only afterward do you add the arrow tips.

The steps in the program include creating an empty image, drawing various geometrical shapes and lines, adding text, and finally saving the

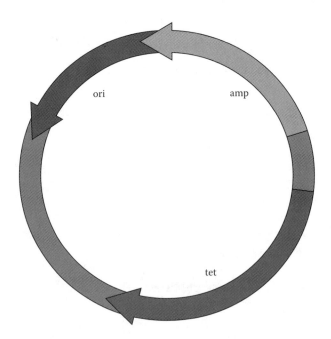

FIGURE 18.1   A plasmid diagram.

image to a .png file. The program also defines three helper functions: angle(), coord(), and draw_arrow_tip(). The angle() function helps calculate angles from base pair numbers so that you can define constants for all regions in the plasmid:

```
TET_START, TET_END = angle(88), angle(1276)
```

Here, 88 and 1276 are base positions that the angle() function converts to degrees. The other two functions are explained below.

18.3.1 Creating an Image

An empty Image object can be created by Image.new(), which takes three arguments:

```
pBR322 = Image.new('RGB', SIZE, 'white')
```

The string 'RGB' indicates that a red-green-blue color scheme shall be used for the image, which is appropriate for most images. The tuple SIZE containing (500, 500) indicates the *x* and *y* size of the image in pixels. Finally, the string 'white' sets the background color to white.

ImageDraw.Draw activates the drawing tools (lines, circles, text, etc.) for the plasmid image. Drawing tools are activated on the pBR322 Image object, by passing it as argument to the ImageDraw.Draw() function. The result is assigned to the DRAW variable:

```
DRAW = ImageDraw.Draw(pBR322)
```

The DRAW variable is used later throughout the script.

18.3.2 Reading and Writing Images

The PIL library can read and write practically all image formats (see Box 18.2). You can write an image to a file with

```
image.save('plasmid_pBR322.png')
```

as in the last line of the script in Section 18.2.2. You can read the same image file later with

```
image = Image.open('plasmid_pBR322.png, 'r')
```

BOX 18.2   HOW DO THE COMMON IMAGE FORMATS DIFFER?

If images were stored on computers as a simple table of color values, the files would be awfully big. This is why compression procedures have been invented. Most image formats differ in the way they compress information and whether they allow for quality losses.

- *BMP:* This format is actually a simple table of pixels. This is why files are awfully big.
- *PNG:* This format preserves the color of every single pixel. When you convert an image to PNG, you can be sure that no information is lost. PNG images can be partially transparent.
- *GIF:* GIF is similar to PNG but is older. GIFs can be animated (this used to be popular in the early days of the web but has gone out of fashion).
- *JPG:* This strongly compressed format saves space by blurring colors of adjacent pixels a little. It is great for photographs but ruins the precision of line art drawings.
- *TIF:* TIF has an accurate pixel format, and it makes files a lot bigger than PNG. The LZW compression is popular. This format is often used in the context of layout work and other print media.

---

and work on it, e.g., by creating a new `ImageDraw.Draw` tool set:

```
d1 = ImageDraw.Draw(image)
```

18.3.3 Coordinates

Whenever you want to change something in a specific location in your image, you need to specify coordinates. In the examples in this chapter, two kinds of coordinates are used: *points* and *rectangles*. First, points are written as (x, y) tuples. For instance,

```
point = (100, 100)
```

is a point 100 pixels from the left border and 100 pixels from the top of the image. The top-left corner has the coordinates $(0, 0)$. Second, rectangular shapes are written as tuples of four values (x, y, x', y') describing the upper-left (x, y) and lower-right (x', y') corner positions. For instance,

```
box = (100, 100, 150, 150)
```

defines a square-shaped box that starts 100 pixels from the top and left borders and is 50 pixels wide. The 150, 150 are the (x', y') coordinates of the bottom-right corner of the box.

In the script in Section 18.2.2 the `coord()` function creates tuples with point coordinates on a circle with a given angle and center point.

18.3.4 Drawing Geometrical Shapes

As stated earlier, the `ImageDraw.Draw` command activates the drawing tools. For any image, you need to use it to draw anything. Call the `ImageDraw.Draw()` function with an `Image` object as the argument and assign the result to a variable. Then, use the dot syntax on this variable to call the `ImageDraw.Draw` methods for drawing objects. Here, we cover only the most common methods and show how to draw circles, rectangles, polygons, and lines. In all examples, we use the variable `DRAW` for the tool object.

Drawing Circles

The `d.pieslice()` function draws circles and parts thereof. The simplest variant is a full circle:

```
BOX = (50, 50, 450, 450)
DRAW.pieslice(BOX, 0, 360, fill = 'grey')
```

BOX is a tuple of four numbers with the coordinates of the top-left and bottom-right corner of the box enclosing the circle. Alternatively, you could write the tuple directly in the command:

```
DRAW.pieslice((50, 50, 450, 450), 0, 360, fill = 'grey')
```

Sections of circles can be drawn by giving two angle values in degrees specifying the start and end angles of the pie slice:

```
DRAW.pieslice(BOX, TET_START, TET_END, fill = 'blue')
```

Or more explicitly:

```
DRAW.pieslice(BOX, 7, 105, fill = 'blue')
```

This will draw a filled slice of a circle starting at 7 degrees and ending at 105 degrees with the zero angle at 3:00.

A thin border can be added by the extra `outline` option:

```
DRAW.pieslice(BOX, 0, 360, fill = 'white', outline = 'black')
```

If you want to draw a circle outline (one that is not filled), you can use the `arc` function:

```
DRAW.arc(BOX, 0, 360, fill = 'black')
```

Drawing Rectangles

Rectangles and squares are fairly easy to draw. The coordinates and colors work in the same way as for circles, but no angles need to be given:

```
DRAW.rectangle(BOX, fill = 'lightblue', outline = 'black')
```

To draw rectangles with sides not parallel to the *x*-axis or *y*-axis, draw a polygon, as explained next.

Drawing Polygons

The arrow tips in the plasmid are drawn as triangles with the `polygon()` method. Instead of a box, the first parameter is a tuple or list of (*x*, *y*) points:

```
DRAW.polygon((point1, point2, point3), fill = 'lightblue',
outline = 'blue')
```

The `draw_arrow_tip()` function defined in Section 18.2.2 first calculates the coordinates of the three points (`point1`, `point2`, `point3`) using the `coord()` function and then passes the coordinates to the `DRAW.polygon()` function. `coord(angle, center, radius)` calculates points on circles. The first argument is the angle, the second is the center of the circle, and the third is the radius. The `direction` argument of `draw_arrow_tip()` determines in which direction the arrow tip should point, and `color` specifies its color:

```
def draw_arrow_tip(start, direction, color):
    """Draws a triangle at the given start angle."""
    p1 = coord(start + direction, CENTER, 185)
    p2 = coord(start, CENTER, 160)
    p3 = coord(start, CENTER, 210)
    DRAW.polygon((p1, p2, p3), fill=color)
```

The polygon() function can also take the outline option as an argument. For instance, to draw a square standing on a corner, you could write the following:

```
DRAW.polygon([(100, 50), (50, 100), (100, 150), (150, 100)],
fill = 'blue', outline = 'black')
```

Drawing Lines

Lines can be drawn as a list of points similar to polygons. For instance, you could add a line to the plasmid script to mark the EcoRI restriction site, as follows:

```
ECOR1 = angle(4359)
p1 = coord(ECOR1, CENTER, 160)
p2 = coord(ECOR1, CENTER, 210)
DRAW.line((p1, p2), fill = 'black', width = 3)
```

Notice that when you draw a line over more than two points, the end points are not automatically connected. Thus, to draw a square at the same location as the previous polygon, you need to add the starting point a second time as the last item of the first argument of DRAW.line:

```
DRAW.line([(100, 50), (50, 100), (100, 150), (150, 100),
(100, 50)], fill = 'black', width = 3)
```

18.3.5 Rotating an Image

You can rotate any image by any number of degrees using the rotate() method:

```
pBR322 = pBR322.rotate(45)
```

This method returns a new image object that you need to put into a variable. Usually, rotating by angles other than multiples of 90 will blur fine structures of the image a little. Therefore, if possible, add text only after rotating.

18.3.6 Adding Text Labels

Text can be added to any location in the picture to label parts of the image.

```
DRAW.text((370, 240), "EcoR1", fill = "black")
DRAW.text((320, 370), "TET", fill = "black")
DRAW.text((330, 130), "AMP", fill = "black")
DRAW.text((150, 130), "ORI", fill = "black")
```

Here (370, 240) is the (*x*, *y*) position at the beginning of the text, and fill is specified to set the text color. The coordinates were determined by trial and error. Although it would not be difficult to determine some coordinates for the text elements automatically, it is very difficult to make the image look good that way. The default font of PIL is small and not very beautiful. Fortunately, PIL can handle any font for which you have a True Type Font (TTF) file (this includes practically all fonts on your computer; you can search for .ttf files on your hard disk or on the web). To use a TTF file in PIL, you need to load it first:

```
from PIL import ImageFont
arial16 = ImageFont.truetype('arial.ttf',16)
DRAW.text((370, 240), "EcoR1", fill = 'black',font = arial16)
```

The output of the code is visible in Figure 18.1.

Caveat: To use TTFs on Windows, you need an additional library that is difficult to install. On Windows, we recommend adding labels in a drawing program or sticking with the default font.

18.3.7 Colors

PIL has 140 colors that you can write as strings ('red', 'lightred', 'magenta', etc.). Alternatively, you can specify the precise red-green-blue (RGB) values ranging from 0 to 255. In RGB, red is (255, 0, 0), green is (0, 255, 0), and blue is (0, 0, 255). You can use decimal or hexadecimal values: 'white' can be written as either (255, 255, 255) or '#ffffff', and 'black' can be written as (0, 0, 0) or '#000000'.

Q & A: WHERE DO I FIND ALL AVAILABLE COLOR NAMES?

The 140 named colors in PIL are known as the *X11 color names* (http://en.wikipedia.org/wiki/X11_color_names). Take note that some color names have spaces that are not allowed in Python; i.e., to use the color Peach Puff in your programs, you need to write it as "peachpuff" or "PeachPuff."

TABLE 18.1   Python Commands for Image Drawing and Manipulation.

Command	Purpose
`from Image import Image, ImageDraw`	Import the PIL library
`i = Image.new(mode, size,` `bg_color)`	Create an empty image
`i = Image.read(filename)`	Read an image from a file
`i.write(filename)`	Write an image to a file
`d = ImageDraw.Draw(i)`	Activate the drawing tools for an image
`d.pieslice(box, angle1, angle2,` `fill)`	Draw a circle or part thereof
`d.arc(box, angle1, angle2, fill)`	Draw the outline of a circle
`d.rectangle(box, fill)`	Draw a rectangle
`d.polygon(points, fill)`	Draw a polygon from a list of points
`d.line((p1, p2), fill, width)`	Draw a line between two points
`i2 = i.rotate(angle)`	Rotate an entire image
`d.text(pos, text, fill, font)`	Write text

`from PIL import ImageFont`
`f = ImageFont.truetype(ttf,size)`
`d.text(pos, text, fill, font)`
	Write text using a custom TrueType font (Linux and Mac OS only)

`i2 = i.resize(new_size)`	Create a shrunken or enlarged image
`i.size`	Tuple containing the x/y size of an image

Table 18.1 reports a summary of Python commands to manipulate images and the corresponding actions.

18.3.8 Helper Variables

Before the script starts to draw anything, some values are stored in variables:

```
PLASMID_LENGTH = 4361
SIZE = (500, 500)
CENTER = (250, 250)
```

The first, PLASMID_LENGTH, is the total number of base pairs in the entire plasmid. This number helps calculate the angles in the circular structure. The second, SIZE, is the size of the entire image. It is stored as a tuple of (x, y) coordinates. The third, CENTER, is the center point of the plasmid circle, which is in the center of the image. The whole point of defining these constants is to make the program code easier to read and to edit.

18.4 EXAMPLES

Example 18.1 Combining Several Pictures into One Image

Imagine you have a series of many images and you would like to add the same element, e.g., a color scale or a legend, to all of them. The following script adds a small label to a bigger image:

```
from PIL import Image
image = Image.open('color.png', 'r')
label = Image.open('label.png', 'r')
image.paste(label, (40, 460))
image.save('combined.png')
```

Source: Adapted from code published by A.Via/K.Rother under the Python License.

The `image.paste()` function takes the second image to be pasted as an argument and the coordinates of the top-left corner where the second image is pasted. When you paste images, their size stays the same. So, if the color image in the example would be too small to accommodate the label image, the label would be cropped at the borders. You can use the `image.paste()` function to create multi-panel figures from a few diagrams. First, you load each of the individual diagrams. Then you create a bigger, empty image. Use the `paste` function to copy each diagram into the bigger one. Finally, add text to the merged image and save it under a new filename.

Example 18.2 Shrinking the Size of Images

With PIL it is easy to change the size of a picture. For instance, you can create a 100 × 100 pixel image from a bigger one:

```
import Image
image = Image.open('big.png')
small = image.resize((100, 100))
small.save('small.png')
```

Source: Adapted from code published by A.Via/K.Rother under the Python License.

However, often it would be great to shrink not just one but a series of images. For instance, you might have a series of photographs that you want to make smaller so that they take less disk space. You can run a `for` loop over all files from a directory using `os.listdir()` (see also Chapter 14). If the images do not have the same size and you'd like to preserve their proportions, you need to find out the exact size of the original images before you shrink them. The following script shrinks all .png images in the current directory to half their size in pixels:

```
import Image
import os
for filename in os.listdir('.'):
    if filename.endswith('.png'):
        im = Image.open(filename)
        x = im.size[0] / 2
        y = im.size[1] / 2
        small = im.resize((x, y))
        small.save('small_'+filename)
```

Source: Adapted from code published by A.Via/K.Rother under the Python License.

Caveat: Shrinking images with original experimental data is dangerous. When you shrink an image containing, e.g., a western blot, gel, scanned photograph, etc., it might cause blurry regions to emerge or disappear, which can suggest a different interpretation. While it may be necessary to change the size of an image for means of presentation or publication, you obviously should preserve a copy of the original image file in order to follow good scientific practice.

Example 18.3 Making a Black-and-White Image out of a Color Image

One common reason to create black-and-white images or color ones is that you want to check how your figures look when printed. A simple approach is to desaturate all colors in PIL and save a black-and-white image:

```
from PIL import Image
image = Image.open('color.png', 'r')
bw_image = Image.new('LA', image.size, (255, 255))
bw_image.paste(image, (0, 0))
bw_image.save('black_white.png')
```

Source: Adapted from code published by A.Via/K.Rother under the Python License.

First, a color image is read. Then an empty black-and-white image is created (the `'LA'` indicates monochrome mode). When the color image is pasted into the black-and-white image, the colors are automatically converted to grayscale. There are more sophisticated procedures in many graphics programs that optimize contrast (e.g., in photographs). However, for technical drawings and diagrams, this procedure is sufficient.

18.5 TESTING YOURSELF

Exercise 18.1 Draw Another Plasmid

The plasmid pUC19 is a successor of pBR322, which is easier to handle in the lab. It consists of 2,686 base pairs, has the Amp gene in position 2486-1626, a LacZ domain in position 469-146, and the replication origin from 1466-852. Draw an image of the plasmid.

Exercise 18.2 Draw an Analog Clock

Write a script that draws the image of an analog clock with two hands indicating hours and minutes.

Hint: You can get the current time from the time library:

```
import time
local = time.localtime()
hour = local.tm_hour
minutes = local.tm_min
```

Hint: You can draw the hands of the clock as lines from the center to an *x-y* position on a circle. To calculate the exact coordinates, you can adapt the `coord()` function from the plasmid script.

Exercise 18.3 Shrink Your Photographs

Go to a folder with digital camera images. To save disk space, reduce the width of all images in the folder to 50%. Test the program on a backup copy first.

FIGURE 18.2   A schematic image of protein domains.

Exercise 18.4 Assemble a Multi-panel Figure

Write a script that creates a multi-panel figure of the four diagrams you created in Chapter 16 and label them A, B, C, and D.

Hint: The script should create a big empty image, load the four panels as separate images, and paste them into the big one. Save the big image after adding text labels.

Exercise 18.5 Plot Protein Domains

Write a program that creates schematic images of protein domains along the sequence (see Figure 18.2 for an example). The program should have the start and end positions of several domains as variables. Add labels to the image.

Hint: You can find examples of protein domain architectures in the InterPro database (www.ebi.ac.uk/interpro).

Hint: Search a protein containing an SH3 domain and retrieve the domain boundaries from Interpro or from the sequence entry in Uniprot.

In Part IV, you learned how to visualize your data. Different types of data need different ways to be visualized. If you want to represent the correlation between two sets of points, you can draw a scatterplot; if you want to compare frequencies, a histogram will fit your needs better. However, if you want to prepare a figure displaying the three-dimensional structure of the trypsin catalytic triad for a paper or to draw a plasmid with restriction sites for a slide presentation, you have to use completely different visualization tools. Chapter 16 is about scientific diagrams. You learned how to draw a bar plot, a scatterplot, a line plot, a pie chart, and a histogram. You also learned how to add error bars in a plot, labels to the x-axis and y-axis, tick marks, a legend box, and a figure title. In Chapter 17, you learned how to create the 3D image of a molecule with PyMOL. You can use the graphical interface of PyMOL and type the commands in the PyMOL text console, but if you want to create complex reproducible images or customize and improve existing ones, you should write them into a PyMOL script file, which is a text file with the .pml extension, listing PyMOL commands. PyMOL scripts can be easily run from the PyMOL graphical interface. Finally, in Chapter 18, you learned how to manipulate images using the Python Imaging Library (PIL). PIL, which is one of the most popular Python libraries, contains tools to move, rotate, and modify images. The library makes it possible to create an Image object and provides a large number of methods to manipulate it. You saw how to draw different types of geometrical shapes, such as circles, polygons, and lines, and how to rotate, add text labels and colors to, and shrink an image. You also learned how to combine several pictures into one image and how to make a black-and-white image out of a color image.

At this stage, you have acquired most of what you need to create and manipulate nearly any kind of image. With this background, you will be able to easily make even more elaborate figures.

V

Biopython

INTRODUCTION: USING A BIG PROGRAMMING LIBRARY

Biopython (http://biopython.org/wiki/Main_Page) is a collection of modules for computational molecular biology, which allows performing many of the basic tasks required in a bioinformatics project.

The most common tasks that can be performed using Biopython include the following:

- parsing (i.e., extracting information from) file formats for gene and protein sequences, protein structures, PubMed records, etc.;

- downloading files from repositories such as NCBI, ExPASy, etc.;

- running (locally or remotely) popular bioinformatics algorithms such as BLAST, ClustalW, etc.; and

- running Biopython implementations of algorithms for clustering, machine learning, data analysis, and data visualization.

It also provides classes that you can use to handle data (such as sequences) and methods to perform operations on them (such as translation, complement, etc.). The *Biopython Tutorial and Cookbook* (http://biopython.org/DIST/docs/tutorial/Tutorial.html) is a good starting point to get a grasp of what Biopython can do for you. More advanced tutorials can be found for specific packages (e.g., "The Biopython Structural Bioinformatics FAQ," http://biopython.org/DIST/docs/cookbook/biopdb_faq.pdf).

YOU CAN USE BIOPYTHON IN SEVERAL WAYS

Biopython is not a program itself; it is a collection of Python tools for bioinformatics programming. You can build a research pipeline solely based on Biopython, or you can write new code for some specific tasks and use Biopython for more standard operations, or you can modify Biopython open-source code to better adapt it to your own needs.

Write your own code when

- the algorithm implementation or coding is interesting to you,

- Biopython data structure mapping is too complex for your task,

- Biopython does not provide tools for your specific task, or

- you want to have fine-tuned control on what your code does.

Use Biopython when

- its modules and/or methods fit your needs,

- your task is unchallenging or boring (e.g., Why are you wasting your time? Don't "re-invent the wheel" unless you're doing it as a learning exercise), or

- your task will take you a lot of writing effort.

Extend Biopython (i.e., modify the Biopython source code) when

- the Biopython modules and/or methods almost do what you need but do not exactly fit your need.

- Note that this might be challenging for a beginner. It can be difficult to read and understand someone else's code.

Remember the following:

- When managing biological data, keep Biopython in mind and have a look at the available tools.

- Browse the documentation and become familiar with its capabilities.

- Use `help()`, `type()`, `dir()`, and other built-in features to explore Biopython modules.

INSTALLING BIOPYTHON

Biopython is not part of the official Python distribution and, as such, must be downloaded (from http://biopython.org/wiki/Download) and installed independently. A prerequisite for Biopython is the installation of the NumPy package (downloadable from http://numpy.scipy.org/).

While Biopython installation is usually pain free (follow the instructions at http://biopython.org/DIST/docs/install/Installation.html), NumPy can be more problematic, especially on Macs, for which specially prepared installers exist (at http://stronginference.com/scipy-superpack/). Be careful: Biopython and NumPy versions are coordinated, meaning that specific Biopython releases must be installed for specific NumPy releases.

Additional packages can optionally be installed to allow Biopython to perform additional tasks (mainly for graphical output and plots of various kinds).

OTHER SIMILAR RESOURCES

Biopython is the result of a large, international, collaborative effort and is currently the most used resource written in Python for biological computation. However, we would like to mention that it is not the only project available for managing biological data and resources. PyCogent (http://pycogent.org/) is another valuable software library for genomic biology. It has many features similar to those of Biopython, even though it is more focused on RNA and phylogeny. Even though PyCogent is not thoroughly described in this book, some scripts are presented that make use of its tools (see Recipe 1).

LET'S GET STARTED

To start using Biopython, you have to tell your computer where you have installed the package. Unless you have added the Biopython installation directory to your PATH environment variable, you have to add it to the `sys.path` Python variable (this latter solution is shown in the following example). Then you have to import the module `Bio`, which is the main Biopython module. All the other (sub)modules will be imported from `Bio`.

```
>>> import sys
>>> sys.path.append("/Users/home/kate/source/biopython-1.57")
>>> import Bio
>>> dir(Bio)
['BiopythonDeprecationWarning',
'MissingExternalDependencyError',
'MissingPythonDependencyError', '__builtins__',
'__doc__', '__file__', '__name__', '__package__',
'__path__', '__version__']
>>> Bio.__version__
'1.60'
>>> help(Bio)
NAME
    Bio - Collection of modules for dealing with biological
data in Python.

FILE
    /Users/home/kate/source/biopython-1.57/Bio/__init__.py

DESCRIPTION
    The Biopython Project is an international association of
developers
    of freely available Python tools for computational
molecular biology.

    http://biopython.org

PACKAGE CONTENTS
PACKAGE CONTENTS
    Affy (package)
    Align (package)
    AlignIO (package)
    Alphabet (package)
    Application (package)
    Blast (package)
    ...
```

All other modules will be imported from `Bio` as follows:

```
>>> from Bio import Seq
>>> from Bio.Alphabet import IUPAC
>>> from Bio.Data import CodonTable
```

In Chapter 19, we present tools to work with nucleic acid and protein sequences. You will see how to read a sequence and build sequence objects owning many features and methods acting on them. You will learn how to write sequences to files in specific formats (FASTA, GenBank) and how to work with functions designed to parse different kinds of sequence records. In Chapter 20, you will basically learn tools to retrieve records from web resources such as NCBI. This chapter deals not only with nucleic and protein sequence records but also with PubMed records and describes how to search them by, e.g., keywords. Finally, Chapter 21 describes a powerful Biopython module that makes it possible to easily work with PDB protein structures. In Chapter 10, you saw how difficult it could be to work with the PDB file format. In Chapter 21, you will see a Biopython structure that will allow you to extract information from PDB files without pain.

## CHAPTER 19

Working with Sequence Data

L EARNING GOAL: You can manipulate DNA, RNA, and protein sequences using Biopython.

19.1 IN THIS CHAPTER YOU WILL LEARN

- How to create a sequence object

- How to reverse and transcribe a DNA sequence

- How to translate an RNA sequence into a protein sequence

- How to create a sequence record

- How to read sequence files in different formats

- How to write formatted sequence files

19.2 STORY: HOW TO TRANSLATE A DNA CODING SEQUENCE INTO THE CORRESPONDING PROTEIN SEQUENCE AND WRITE IT TO A FASTA FILE

19.2.1 Problem Description

In Chapter 4 you learned how to parse sequence files using elementary operations on strings. For example, you learned that the ">" symbol in FASTA formatted files at the first position of a row starts the header of the record and contains the sequence ID and some concise annotations.

You also learned that you can transcribe a DNA sequence by replacing the Ts by Us and translate it using a dictionary for the genetic code (see Chapter 5, Section 5.3.1). Here, these and other actions are accomplished using Biopython modules and methods. Biopython gives you a shortcut for accomplishing these tasks by providing tools to manipulate sequences and sequence files, thus making it very simple to work with different file formats, to annotate sequence records, to write them to files, etc. In the following Python session, four modules from the Bio package are used: Seq is needed to create a sequence object; IUPAC is used to assign a biological alphabet (e.g., DNA or protein) to a sequence object; SeqRecord allows creating a sequence record object equipped with ID, annotation, description, etc.; and SeqIO provides methods to read and write formatted sequence files.

19.2.2 Example Python Session

```
from Bio import Seq
from Bio.Alphabet import IUPAC
from Bio.SeqRecord import SeqRecord
from Bio import SeqIO
# read the input sequence
dna = open("hemoglobin-gene.txt").read().strip()
dna = Seq.Seq(dna, IUPAC.unambiguous_dna)
# transcribe and translate
mrna = dna.transcribe()
protein = mrna.translate()
# write the protein sequence to a file
protein_record = SeqRecord(protein,\
    id='sp|P69905.2|HBA_HUMAN',\
    description="Hemoglobin subunit alpha, human")
outfile = open("HBA_HUMAN.fasta", "w")
SeqIO.write(protein_record, outfile,"fasta")
outfile.close()
```

Source: Adapted from code published by A.Via/K.Rother under the Python License.

19.3 WHAT DO THE COMMANDS MEAN?

In Section 19.2.2, the DNA coding sequence from the hemoglobin subunit alpha is read from a plain text file (hemoglobin-gene.txt; the sequence is identical to the one in Appendix C, Section C.2, "A Single Nucleotide Sequence File in FASTA Format"). It is then transcribed into an mRNA sequence, translated into a peptide sequence, and written to the HBA_HUMAN.fasta output file. Let's look at the objects imported in the Python session one by one.

19.3.1 The Seq Object

The first imported name is Seq, which is a module within the Bio library. The Seq.Seq class creates a *sequence* object, i.e., a sequence associated with an alphabet attribute, which specifies the kind of sequence stored in the object. You can create Seq objects with or without specifying an alphabet:

```
>>> from Bio import Seq
>>> my_seq = Seq.Seq("AGCATCGTAGCATGCAC")
>>> my_seq
Seq('AGCATCGTAGCATGCAC', Alphabet())
```

Biopython contains a set of precompiled alphabets that cover all biological sequence types. IUPAC-defined alphabets (http://www.chem.qmw.ac.uk/iupac) are the most frequently used. If you want to use alphabets, you have to import the IUPAC module from the Bio.Alphabet module (as in Section 19.2.2). It contains the alphabets IUPACUnambiguousDNA (basic ACGT letters), IUPACAmbiguousDNA (includes ambiguous letters), ExtendedIUPACDNA (includes modified bases), IUPACUnambiguousRNA, IUPACAmbiguousRNA, IUPACProtein (IUPAC standard amino acids), and ExtendedIUPACProtein (includes selenocysteine, X, etc.). The dna variable defined in Section 19.2.2 is a sequence object characterized by the IUPAC.unambiguous_dna alphabet.

Transcribing and Translating Sequences

Methods of Seq objects are designed specifically for biological sequences; for example, you can obtain the transcription of a DNA sequence using the transcribe method, as shown in Section 19.2.2:

```
>>> my_seq.transcribe()
Seq('AGCAUCGUAGCAUGCAC', RNAAlphabet())
```

The transcribe method just changes the Ts to Us and sets the alphabet to RNA. It assumes that the input DNA sequence is the coding strand.

If you have a template strand and want to perform the transcription, you need to get the reverse complement first and then transcribe it:

```
>>> dna = Seq.Seq("AGCATCGTAGCATGCAC", IUPAC.unambiguous_dna)
>>> cdna = dna.reverse_complement()
>>> cdna
Seq('GTGCATGCTACGATGCT', IUPACUnambiguousDNA())
>>> mrna = codingStrand.transcribe()
>>> mrna
Seq('GUGCAUGCUACGAUGCU', IUPACUnambiguousRNA())
```

Or in a single command line:

```
>>> dna.reverse_complement().transcribe()
Seq('GUGCAUGCUACGAUGCU', IUPACUnambiguousRNA())
```

These methods are available for sequences assigned to protein alphabets as well, but their execution will raise errors.

A DNA or RNA sequence object can also be translated into a protein sequence. To this aim, a number of genetic codes are available through the `CodonTable` module of the `Bio.Data` module:

```
>>> from Bio.Data import CodonTable
```

You can access the tables through the dictionaries available in the `CodonTable` module (the list of dictionaries available can be visualized using the `dir()` function). For example, the `unambiguous_dna_by_name` dictionary makes it possible to access the set of DNA codon tables by name (e.g., `"Standard"`, `"Vertebrate Mitochondrial"`, etc.).

```
>>> from Bio.Data import CodonTable
>>> standard_table = \
... CodonTable.unambiguous_dna_by_name["Standard"]
```

In contrast, the `unambiguous_dna_by_id` uses a numerical identifier (1 corresponds to the `"Standard"` codon table, 2 to the `"Vertebrate Mitochondrial"`, etc.). All NCBI-defined alphabets are available and identified by the NCBI table identifier (see www.ncbi.nlm.nih.gov/Taxonomy/Utils/wprintgc.cgi). By default, Biopython translation will use the standard genetic code (corresponding to the NCBI table ID 1).

If you print the `standard_table` variable, you will get the codon table from Figure 19.1. Codon table objects have other useful attributes as well, such as start and stop codons:

```
>>> standard_table.start_codons
['TTG', 'CTG', 'ATG']
>>> standard_table.stop_codons
['TAA', 'TAG', 'TGA']
>>> mito_table = \
... CodonTable.unambiguous_dna_by_name["Vertebrate Mitochondrial"]
>>> mito_table.start_codons
['ATT', 'ATC', 'ATA', 'ATG', 'GTG']
>>> mito_table.stop_codons
['TAA', 'TAG', 'AGA', 'AGG']
```

The `translate` method translates an RNA or DNA sequence using either the default or a specified genetic code and returns a `Seq` object, the alphabet of which will contain additional information. In the following

	T	C	A	G	
T	TTT F	TCT S	TAT Y	TGT C	T
T	TTC F	TCC S	TAC Y	TGC C	C
T	TTA L	TCA S	TAA Stop	TGA Stop	A
T	TTG L(s)	TCG S	TAG Stop	TGG W	G
C	CTT L	CCT P	CAT H	CGT R	T
C	CTC L	CCC P	CAC H	CGC R	C
C	CTA L	CCA P	CAA Q	CGA R	A
C	CTG L(s)	CCG P	CAG Q	CGG R	G
A	ATT I	ACT T	AAT N	AGT S	T
A	ATC I	ACC T	AAC N	AGC S	C
A	ATA I	ACA T	AAA K	AGA R	A
A	ATG M(s)	ACG T	AAG K	AGG R	G
G	GTT V	GCT A	GAT D	GGT G	T
G	GTC V	GCC A	GAC D	GGC G	C
G	GTA V	GCA A	GAA E	GGA G	A
G	GTG V	GCG A	GAG E	GGG G	G

FIGURE 19.1 The DNA unambiguous codon table.

example, the output alphabet contains information related to the presence of stop codons in the translated sequence:

```
>>> mrna = \
... Seq.Seq('AUGGCCAUUGUA AUGGGCCGCUGAA AGGGAUAG',\
... IUPAC.unambiguous_rna)
>>> mrna.translate(table = "Standard")
Seq('MAIVMGR*KG*', HasStopCodon(IUPACProtein(), '*'))
>>> mrna.translate(table = "Vertebrate Mitochondrial")
Seq('MAIVMGRWKG*', HasStopCodon(IUPACProtein(), '*'))
```

By default, all stop codons encountered during the translation of the nucleic acid sequence are returned as stars ("*"). Notice that there are two stop codons in the mRNA sequence if the standard codon table (table = "Standard") is used (UGA and UAG) and only one if the vertebrate mitochondrial codon table is used (table = "Vertebrate Mitochondrial"). In fact, the UGA codon is recognized as a tryptophan (W) in the mitochondrion. You can impose the translation to stop at the first encountered stop codon:

```
>>> mrna.translate(to_stop = True, table = 1)
Seq('MAIVMGR', IUPACProtein())
>>> mrna.translate(to_stop = True, table = 2)
Seq('MAIVMGRWKG', IUPACProtein())
```

19.3.2 Working with Sequences as Strings

You can manipulate sequence objects in Biopython in the same way as strings. For example, you can index, slice, split, convert the sequence to uppercase or lowercase, count occurrences of characters, and so on:

```
>>> from Bio import Seq
>>> my_seq = Seq.Seq("AGCATCGTA GCATGCAC")
>>> my_seq[0]
'A'
>>> my_seq[0:3]
Seq('AGC', Alphabet())
>>> my_seq.split('T')
[Seq('AGCA', Alphabet()), Seq('CG', Alphabet()),
    Seq('AGCA', Alphabet()), Seq('GCAC', Alphabet())]
>>> my_seq.count('A')
5
>>> my_seq.count('A') / float(len(my_seq))
0.29411764705882354
```

Notice that when you slice a Seq object, or you split it, the methods return not just strings but other Seq objects. Sequence objects can also be concatenated, but only if their alphabets are compatible (or are generic alphabets):

```
>>> my_seq = Seq.Seq("AGCATCGTAGCATGCAC", IUPAC.unambiguous_dna)
>>> my_seq_2 = Seq.Seq("CGTC", IUPAC.unambiguous_dna)
>>> my_seq + my_seq_2
Seq('AGCATCGTAGCATGCACCGTC', IUPACUnambiguousDNA())
```

You can search the sequence for the occurrence of specific substrings using the `find` method. If the subsequence is not found, Python will return -1; if the subsequence is found, the position of the leftmost matching character in the target sequence is returned:

```
>>> my_seq = Seq.Seq("AGCATCGTAGCATGCAC", IUPAC.unambiguous_dna)
>>> my_seq.find("TCGT")
4
>>> my_seq.find("TTTT")
-1
```

It is also possible to search for patterns represented by regular expressions using the Python `re` module or using the Biopython module `Bio.Motif` (see the Biopython tutorial).

Finally, Biopython provides a number of functions, such as `transcribe()` or `translate()`, that can be used on strings directly:

```
>>> my_seq_str = "AGCATCGTAGCATGCAC"
>>> Bio.Seq.transcribe(my_seq_str)
'AGCAUCGUAGCAUGCAC'
>>> Bio.Seq.translate(my_seq_str)
'SIVAC'
>>> Bio.Seq.reverse_complement(my_seq_str)
'GTGCATGCTACGATGCT'
```

19.3.3 The `MutableSeq` Object

`Seq` objects behave similarly to Python strings, in the sense that they are immutable. Therefore, if you try to reassign a residue in a sequence object, you will get an error message. Biopython provides the `MutableSeq` object to create mutable sequence objects:

```
>>> my_seq = Seq.Seq("AGCATCGTAGCATG", IUPAC.unambiguous_dna)
>>> my_seq[5] = "T"
Traceback (most recent call last):
    File "<stdin>", line 1, in <module>
```

```
TypeError: 'Seq' object does not support item assignment
>>> my_seq = \
... Seq.MutableSeq("AGCATCGTAGCATG", IUPAC.unambiguous_dna)
>>> my_seq[5] = "T"
>>> my_seq
MutableSeq('AGCATTGTAGCATG', IUPACUnambiguousDNA())
```

Source: Adapted from code published by A.Via/K.Rother under the Python License.

You cannot use methods such as reverse() or remove() on Seq objects, but you can use them on MutableSeq objects.

Finally, it is possible to convert an immutable Seq object into a mutable one, and vice versa, using the tomutable() method of Seq objects and the toseq() method of MutableSeq objects:

```
>>> my_mut_seq = my_seq.tomutable()
>>> my_mut_seq
MutableSeq('AGCATCGTAGCATGCAC', IUPACUnambiguousDNA())
>>> my_seq = my_mut_seq.toseq()
>>> my_seq
Seq('AGCATCGTAGCATGCAC', IUPACUnambiguousDNA())
```

Because MutableSeq objects can change (much like lists or sets), they cannot be used as dictionary keys, while Seq objects can.

19.3.4 The SeqRecord Object

The SeqRecord class provides a container for a sequence and its annotation. In the Python session in Section 19.2.2, the protein sequence in the protein_seq variable, obtained by translating the mRNA sequence object, is converted into a SeqRecord object:

```
protein_record = \
SeqRecord(protein_seq,id = 'sp|P69905.2|HBA_HUMAN', \
description = "Hemoglobin subunit alpha, Homo sapiens")
```

The arguments passed to SeqRecord to create the object are a Seq object (stored in the protein_seq variable), an id and a description, which must be both strings. SeqRecord objects allow associating features to a sequence object, such as the identifier or a description. The available features are as follows:

- seq: This is a biological sequence, typically in the form of a Seq object.

- id: This is the primary ID used to identify the sequence.

- `name`: This is a "common" molecule name.

- `description`: This is a description of the sequence/molecule.

- `letter_annotations`: This is a dictionary with per-residue annotations. Keys are the type of annotation (e.g., "secondary structure"), and values are Python sequences (lists, tuples, or strings) having the same length as the sequence, where each element is a per-residue annotation (e.g., secondary structure type indicated with a single character: S = strand, H = helix, etc.). It is useful for assigning quality scores, secondary structure or accessibility preferences, etc. to residues.

- `annotations`: This is a dictionary of additional information about the sequence. The keys are the type of information, and the information is contained in the value.

- `features`: This is a list of `SeqFeature` objects, with more structured information about sequence features (e.g., position of genes on a genome, or domains on a protein sequence; see following text).

- `dbxrefs`: This is a list of database cross-references.

Such features can be manually created by the user or imported from a database record (e.g., a GenBank or SwissProt file; see also Chapter 20). In Section 19.2.2, a `SeqRecord` object associated with a `Seq` object was created with an ID and a description. Both features can be retrieved directly:

```
>>> protein_record.id
'sp|P69905.2|HBA_HUMAN'
>>> protein_record.description
'Hemoglobin subunit alpha, Homo sapiens'
```

Features can also be assigned on the fly:

```
>>> protein_record.name = "Hemoglobin"
```

The `annotation` attribute is an empty dictionary that can be used to store all kinds of information that do not fall into the categories already provided by `SeqRecord`:

```
>>> protein_record.annotations["origin"] = "human"
>>> protein_record.annotations["subunit"] = "alpha"
>>> protein_record.annotations
{'origin': 'human', 'subunit': 'alpha'}
```

Similarly, the `letter_annotations` attribute is an empty dictionary the values of which must be strings, lists, or tuples of exactly the same length of the sequence:

```
>>> protein_record.letter_annotations[\
..."secondary structure"] = \
...'HHHHHHHHHHHHHHHHHHHHHHHHHHHHHHHHHHHHHHHHHHHSSSSSS...'
```

Converting `SeqRecord` Objects to File Formats

Once you have set attributes for your sequence, you can convert it to some of the most popular storage formats for sequences by using the `format` method:

```
>>> print protein_record.format("fasta")
>sp|P69905.2|HBA_HUMAN Hemoglobin subunit alpha, Homo sapiens
MVLSPADKTNVKAAWGKVGAHAGEYGAEALERMFLSFPTTKTYFPHFDLSHGSAQVKGHG
KKVADALTNAVAHVDDMPNALSALSDLHAHKLRVDPVNFKLLSHCLLVTLAAHLPAEFTP
AVHASLDKFLASVSTVLTSKYR*
```

See what happens if you use the "genbank" format:

```
>>> print protein_record.format("genbank")
```

Finally, you can slice a `SeqRecord` object: where possible, the annotation will be sliced accordingly (e.g., `letter_annotations`), but some features (e.g., `dbxrefs`) will not be extended to the sliced object. Two `SeqRecord` objects can be also concatenated by adding them into a new `SeqRecord`. The new object will inherit some of the features that are identical in the two parent `SeqRecord` objects (e.g., `id`), whereas some others are not inherited in any case (such as `annotations`).

19.3.5 The `SeqIO` Module

In Chapter 4, we introduced procedures to parse files using standard Python commands. Here, you will see how to read and write sequence files using Biopython. The `SeqIO` module is very useful to parse many common file formats and write annotated sequences to standard file formats. In Section 19.2.2, `protein_record` (a `SeqRecord` object) is written to the `HBA_HUMAN.fasta` output file in FASTA format.

The Biopython `SeqIO` module provides parsers for many common file formats. These parsers extract information from an input file (either local or retrieved from a database) and automatically convert it into a `SeqRecord` object. `SeqIO` also provides a method to write `SeqRecord` objects to conveniently formatted files.

Parsing Files

There are two methods for sequence file parsing: `SeqIO.parse()` and `SeqIO.read()`; both of them require two mandatory arguments and one optional argument:

1. a file (mandatory; also called a "handle" object) that specifies where the data must be read from (could be a filename, a file opened for reading, data downloaded from a database using a script, or the output of another piece of code);

2. a string indicating the format of the data (mandatory; e.g., `"fasta"` or `"genbank"`; a full list of supported formats is available at http://biopython.org/wiki/SeqIO);

3. an argument that specifies the alphabet of the sequence data (optional).

The difference between the two methods `SeqIO.parse()` and `SeqIO.read()` is that `SeqIO.parse()` returns an iterator that produces `SeqRecord` objects from an input file of several records. You can use the iterator like a list in `for` or `while` loops. See Example 19.2. If you have a file containing a single record, you have to use `SeqIO.read()` instead. It returns a `SeqRecord` object. While `SeqIO.parse()` can process any number of records in the input handle, `SeqIO.read()` parses only single-record files by first checking whether there is only one record in the handle and raises an error if this condition is not met.

Q & A: WHAT IS AN ITERATOR?

An iterator is a data structure that produces a series of entries (e.g., `SeqRecord` objects). It can be used like a list in loops, but technically it is not a list. An iterator has no length, and it cannot be indexed and sliced. You can only request the next object from it. When you do it, the iterator looks to see if there are more records available. This way, the iterator does not need to keep all records in the memory all the time.

Parsing Large Sequence Files

The usage of iterators is a way to parse large files without consuming large amounts of memory. For a big number of records, you can use `SeqIO.index()`, a method that needs two arguments: a record filename and a file format. The `SeqIO.index()` method returns a dictionary-like object that gives you access to all records without keeping all data in the memory. The dictionary keys are the IDs of the records, and the values contain the entire record, which can be accessed using the attributes `id`, `description`, etc. When a particular record is accessed, the record content is parsed on the fly. This method allows you to manipulate huge files, with a little cost in flexibility and speed. Notice that these dictionary-like objects are read-only, meaning that once they are created, no records can be inserted or removed.

Writing Files

The `SeqIO.write()` method writes one or more `SeqRecord` objects to a file in the format specified by the user. The method requires three arguments: one or more `SeqRecord` objects, a handle object (i.e., a file opened with the `"w"` modality) or a filename to write to, and a sequence format (e.g., `"fasta"` or `"genbank"`).

The first argument can be a list, an iterator, or an individual `SeqRecord`, as shown in Section 19.2.2. When writing GenBank files, the alphabet must be set for the sequence.

Concluding Remarks

In some cases, you may prefer to use traditional programming, e.g., if you want a customized parser or when you have a nonstandard format that Biopython fails to parse. In other cases, you may find it more convenient to use the `SeqIO` module, e.g., when you have to index large files. In both cases, you have to be aware that file formats change occasionally and that they may contain unexpected characters, lines, and exceptions, which could break even the best-designed parser.

19.4 EXAMPLES

Example 19.1 Using the `Bio.SeqIO` Module to Parse a Multiple Sequence FASTA File

In the following example, the multiple sequence FASTA file shown in Appendix C, Section C.4, "A Multiple Sequence File in FASTA Format" is parsed:

```
from Bio import SeqIO
fasta_file = open("Uniprot.fasta","r")
for seq_record in SeqIO.parse(fasta_file, "fasta"):
    print seq_record.id
    print repr(seq_record.seq)
    print len(seq_record)
fasta_file.close()
```

Source: Adapted from code published by A.Via/K.Rother under the Python License.

The code writes the identifiers, sequences, and lengths for all three entries in the FASTA file:

```
sp|P03372|ESR1_HUMAN
Seq('MTMTLHTKASGMALL HQIQGNELEPLNRPQLKIPLER
PLGEVYLDSSKPAVYNY...ATV', SingleLetterAlphabet())
595
sp|P62333|PRS10_HUMAN
Seq('MADPRDKALQDYRK KLLEHKEIDGRLKELREQLKELT
KQYEKSENDLKALQSVG...KPV', SingleLetterAlphabet())
389
sp|P62509|ERR3_MOUSE
Seq('MDSVELCLPESFS LHYEEELLCRMSNKDRHIDSSCSS
FIKTEPSSPASLTDSVN...AKV', SingleLetterAlphabet())
458
```

Since the handle is a file, it is a good habit to close it when the processing is done. Remember that the iterator "empties" the file, meaning that to scan the records another time, the file must be closed, then opened again, and then used again as the handle argument of SeqIO.parse(). You can also use SeqIO.parse() by omitting the explicit creation of the handle and directly passing a filename or a complete path to SeqIO.parse. For example,

```
>>> for seq_record in SeqIO.parse("Uniprot.fasta", "fasta"):
... print seq_record.id
sp|P03372|ESR1_HUMAN
sp|P62333|PRS10_HUMAN
sp|P62509|ERR3_MOUSE
```

You can also parse records one by one using the next() method of the iterator:

```
>>> uniprot_iterator = SeqIO.parse("Uniprot.fasta","fasta")
>>> uniprot_iterator.next().id
```

```
'sp|P03372|ESR1_HUMAN'
>>> uniprot_iterator.next().id
'sp|P62333|PRS10_HUMAN'
```

When all records have been read, the next() method will return either None or a StopIteration exception (depending on your Biopython version).

Example 19.2 Using the SeqIO Module to Parse a Record File and Store Its Content in a List or a Dictionary

You can easily store all records from a file in a list:

```
from Bio import SeqIO
uniprot_iterator = SeqIO.parse("Uniprot.fasta", "fasta")
records = list(uniprot_iterator)
print records[0].id
print records[0].seq
```

Source: Adapted from code published by A.Via/K.Rother under the Python License.

This code generates the output:

```
sp|P03372|ESR1_HUMAN
MTMTLHTKASGMALLHQIQGNELEPLNRPQLKI…
```

Alternatively, you can use a dictionary, the keys of which are the record IDs and the values of which contain the record information:

```
uniprot_iterator = SeqIO.parse("Uniprot.fasta", "fasta")
records = SeqIO.to_dict(uniprot_iterator)
print records['sp|P03372|ESR1_HUMAN'].id
print records['sp|P03372|ESR1_HUMAN'].seq
```

Source: Adapted from code published by A.Via/K.Rother under the Python License.

This code generates the same output as above.

Example 19.3 Using SeqIO.index()to Parse a Big File

The usage of the index method helps process large sequence files that don't fit into the memory at the same time. In the example, the file from Example 19.1 will be read.

```
records = SeqIO.index("Uniprot.fasta","fasta")
print records.keys()
print len(records['sp|P03372|ESR1_HUMAN'].seq)
```

Source: Adapted from code published by A.Via/K.Rother under the Python License.

This code produces the output:

```
['sp|P03372|ESR1_HUMAN', 'sp|P62333|PRS10_HUMAN',
'sp|P62509|ERR3_MOUSE']
595
```

Example 19.4 Converting between Sequence File Formats

You can convert sequence file formats by combining the Bio.SeqIO. parse() and Bio.SeqIO.write() methods. The following script converts a GenBank file to a FASTA file:

```
from Bio import SeqIO
genbank_file = open ("AY810830.gbk", "r")
output_file = open("AY810830.fasta", "w")
records = SeqIO.parse(genbank_file, "genbank")
SeqIO.write(records, output_file, "fasta")
output_file.close()
```

Source: Adapted from code published by A.Via/K.Rother under the Python License.

Notice that if you do not close the output file, the writing cannot be completed.

19.5 TESTING YOURSELF

Exercise 19.1 Parse a Single Sequence Record

Read a single record from a FASTA formatted file, extract its ID and its sequence, and print them.

Exercise 19.2 Build and Write a SeqRecord Object to a File

Use the sequence ID and the sequence from Exercise 19.1 to create a SeqRecord object. Manually add a customized description. Write the SeqRecord object to a file in FASTA format and to a second file in GenBank format. Note that for GenBank format an alphabet must be assigned when creating the Seq object.

Exercise 19.3

Parse a multiple-record file and write to a file only the IDs of all records.

Exercise 19.4 Write GenBank Sequences to Separate Files

Parse a multiple-record file in GenBank format and write each record to a separate file in FASTA format. Use the IDs of the entries to create filenames.

Hint: You can manually create the input file by going to the GenBank website.

Exercise 19.5 Format Conversion

Try to convert the *protein* sequence FASTA formatted file of Example 19.1 (or a similar one) into GenBank format. What happens?

## CHAPTER 20

Retrieving Data from Web Resources

L EARNING GOAL: You can search and fetch database records from NCBI via Biopython.

20.1 IN THIS CHAPTER YOU WILL LEARN

- How to read sequence files from the web

- How to submit PubMed queries

- How to submit queries to the NCBI nucleotide database

- How to retrieve Uniprot records and write them to a file

20.2 STORY: SEARCHING PUBLICATIONS BY KEYWORDS IN PUBMED AND DOWNLOADING AND PARSING THE CORRESPONDING RECORDS

20.2.1 Problem Description

In the previous chapter, you used Biopython to manipulate local sequence files (e.g., FASTA and GenBank files). In this chapter, you will use Biopython to access online NCBI databases, such as PubMed and GenBank, and Expasy resources, such as Uniprot, and retrieve and parse their contents. The following Python session shows how to find publications about PyCogent, a Python library complementary to Biopython. First, PubMed entries containing the keyword "PyCogent" need to be found and retrieved, and the

---

BOX 20.1   DOCUMENTATION AND SAMPLE QUERIES

Documentation

www.ncbi.nlm.nih.gov/books/NBK25500/

Searching for Papers in PubMed

http://eutils.ncbi.nlm.nih.gov/entrez/eutils/esearch.fcgi?db=pubmed&term= thermophilic,packing&rettype=uilist

Retrieving Publication Records in Medline Format

http://eutils.ncbi.nlm.nih.gov/entrez/eutils/efetch.fcgi?db=pubmed&id=1174 8933,11700088&retmode=text&rettype=medline

Searching for Protein Database Entries by Keywords

http://eutils.ncbi.nlm.nih.gov/entrez/eutils/esearch.fcgi?db=protein&term=c ancer+AND+human

Retrieving Protein Database Entries in FASTA Format

http://eutils.ncbi.nlm.nih.gov/entrez/eutils/efetch.fcgi?db=protein&id=1234 567&rettype=fasta

Retrieving Protein Database Entries in GenBank Format

http://eutils.ncbi.nlm.nih.gov/entrez/eutils/efetch.fcgi?db=protein&id=1234 567&rettype=gb

Retrieving Nucleotide Database Entries

http://eutils.ncbi.nlm.nih.gov/entrez/eutils/efetch.fcgi?db=nucleotide&id=979 0228&rettype=gb

---

resulting records need to be parsed. Since PubMed is one of the NCBI databases (www.ncbi.nlm.nih.gov/), it is connected to the Entrez data retrieval system (www.ncbi.nlm.nih.gov/Entrez). See Box 20.1 for sample queries to the NCBI server. The Biopython module to access NCBI web services is called `Entrez` as well. The `Entrez` module is needed to access and download NCBI database records. To further parse publication records, you need a specialized parser from the `Bio.Medline` module.

20.2.2 Python Session

```
from Bio import Entrez
from Bio import Medline
keyword = "PyCogent"
```

```
# search publications in PubMed
Entrez.email = "my_email@address.com"
handle = Entrez.esearch(db="pubmed", term=keyword)
record = Entrez.read(handle)
pmids = record['IdList']
print pmids
# retrieve Medline entries from PubMed
handle = Entrez.efetch(db="pubmed", id=pmids,\
    rettype="medline", retmode="text")
medline_records = Medline.parse(handle)
records = list(medline_records)
n = 1
for record in records:
    if keyword in record["TI"]:
        print n, ')', record["TI"]
        n += 1
```

Source: Adapted from code published by A.Via/K.Rother under the Python License.

The code produces the output:

```
['22479120', '18230758', '17708774']
1 ) Abstractions, algorithms and data structures for
    structural bioinformatics in PyCogent.
2 ) PyCogent: a toolkit for making sense from sequence.
```

20.3 WHAT DO THE COMMANDS MEAN?

20.3.1 The `Entrez` Module

`Entrez` provides a connection to the `esearch` and `efetch` tools on the NCBI servers. You can list methods and attributes available in the `Entrez` module by typing

```
>>> from Bio import Entrez
>>> dir(Entrez)
```

In the output, you will notice the `mail` attribute and the `esearch()` and `efetch()` functions used in Section 20.2.2. The `mail` attribute tells your email address to NCBI. This is not mandatory, but NCBI wants to be able to contact users in case of problems, and supplying your email address is fair. You can also provide your email address with each single access to NCBI by including `email = "my_email@address.com"` in the list of arguments of `Entrez.esearch()`.

`Entrez.esearch()`

The `Entrez.esearch()` conducts searches in NCBI databases using a query text. The function takes two mandatory arguments: db, the database to search (default is `pubmed`), and `term`, the query text.

In Section 20.2.2, the term to search is "PyCogent." If you want to search more than one keyword you can use "AND" or "OR", as you would do in an online search. You can also use keyword specifications such as [Year], [Organism], [Gene], etc. (see Example 20.2). `Entrez.esearch()` returns a list of database identifiers in the form of a "handle," which can be read using the `Entrez.read()` function. The latter returns a dictionary with the keys that include, among others, "IdList" (its value is a list of IDs matching the text query) and "Count" (its value is the total number of IDs). In the case of the PubMed query shown in Section 20.2.2, the PMIDs are contained in the `record['IdList']` value of the `record` dictionary.

You can use the optional parameter `retmax` (maximum retrieved) to set how many entries matching the query text are to be retrieved (see Example 20.2). Other useful optional arguments are `datetype`, `reldate`, `mindate`, and `maxdate` (both in the form "YYYY/MM/DD"). `datetype` can be used to choose a type of date ("mdat": modification date, "pdat": publication date, "edat": Entrez date) to limit your search. `reldate` must be an integer n and tells the `esearch()` method to return only the IDs of the records matching `datetype` within the last n days. `mindate` and `maxdate` specify a date range that can be used to limit the search by the date type specified by `datetype`. For instance, the following query returns only papers in the period during which this book was written:

```
>>> handle = Entrez.esearch(db = "pubmed", \
... term = "Python", datetype = "pdat", \
... mindate = "2011/08/01", maxdate = "2013/10/28")
>>> record = Entrez.read(handle)
>>> record['Count']
'3'
```

To count matches, you can also use the `Entrez.egquery()` method, which returns the number of matches of the search term in each of the Entrez databases. It takes a single mandatory argument, which is the term to be searched:

```
>>> handle = Entrez.egquery(term = "PyCogent")
>>> record = Entrez.read(handle)
```

```
>>> for r in record["eGQueryResult"]:
...    print r["DbName"], r["Count"]
pubmed 3
pmc 49
mesh 0
...
```

Entrez.efetch()

So far, you have seen how to identify the IDs of records for one or more search terms. If you want to download these records from the NCBI server, you can use the `Entrez.efetch()` tool. In Biopython, the `Entrez.efetch()` function takes as arguments the database from which the records are to be retrieved and one ID or the list of IDs of the records you want to download. In the Python session in Section 20.2.2, the list of PMIDs to be downloaded is saved in the `pmids` variable. It is subsequently used as an ID list for `efetch()`.

The `Entrez.efetch()` function has many optional arguments: `retmode` specifies the format of the record(s) retrieved (text, HTML, XML), and `rettype` specifies what types of records are shown. It depends on the database you are accessing. For PubMed the `rettype` value can be, for example, `abstract`, `citation`, or `medline`. For Uniprot you can set `rettype` to `fasta` to retrieve the sequence of a protein record (see Example 20.4); `retmax` is the total number of records to be retrieved (up to a maximum of 10,000).

`Entrez.fetch()` returns a handle that "contains" your records. You can read the raw data from the handle like you would read an open Python file (with a `for` loop) or parse them using specialized functions.

20.3.2 The `Medline` Module

To parse the PubMed records you downloaded with `Entrez. efetch()`, you have to import the Biopython `Medline` module, which provides the `Medline.parse()` function. The result of this function can be conveniently converted into a list. This list contains `Bio.Medline.Record` objects that work like dictionaries. The most common keys are `TI` (Title), `PMID`, `PG` (pages), `AB` (Abstract), and `AU` (Authors). Not all keys are present in each dictionary. For example, if there is no abstract available for a PubMed record, the `AB` key will be missing in the dictionary. The keys available for a given record can be visualized by typing the following:

```
>>> handle = Entrez.efetch(db = "pubmed", \
... retype = "medline", id = ['22479120', \
... '18230758', '17708774'], \
... retmode = 'text')
>>> records = Medline.parse(handle)
>>> list(records)[0].keys()
['STAT', 'IP', 'DEP', 'DA', 'AID', 'CRDT', 'DP', 'OWN',
    'PT', 'LA', 'FAU', 'JT', 'PG', 'PMC', 'TA', 'JID',
    'AB', 'VI', 'IS', 'TI', 'AU', 'MHDA', 'PHST', 'EDAT',
    'SO', 'PMID', 'PST']
```

Or if you also want to display the corresponding values:

```
>>> for record in records:
...     for k, v in record.items():
...         print k, v
```

Notice that if you have to parse a single record, you can use the `Medline.read()` function instead of `Medline.parse()`.

20.4 EXAMPLES

Example 20.1 What Are the Available Entrez Databases?

If you want to get information about the Entrez databases, you can use the function `Entrez.einfo()`. If you use it without arguments, you will get a dictionary with a single `key:value` pair where the value is a list of available databases in Entrez. If you pass a given database name as argument to `Entrez.einfo()`, you will get information about that database:

```
from Bio import Entrez
handle = Entrez.einfo()
info = Entrez.read(handle)
print info
raw_input('... press enter for a list of fields in PubMed')
handle = Entrez.einfo(db="pubmed")
record = Entrez.read(handle)
print record.keys()
print record['DbInfo']['Description']
print record['DbInfo']
```

Source: Adapted from code published by A.Via/K.Rother under the Python License.

The program generates the output:

```
{u'DbList': ['pubmed', 'protein', 'nuccore', 'nucleotide',
    'nucgss', 'nucest', 'structure', 'genome', 'assembly',
    'genomeprj', 'bioproject', 'biosample', 'blastdbinfo',
    'books', 'cdd', 'clinvar', 'clone', 'gap', 'gapplus',
    'dbvar', 'epigenomics', 'gene', 'gds', 'geoprofiles',
    'homologene', 'medgen', 'journals', 'mesh', 'ncbisearch',
    'nlmcatalog', 'omia', 'omim', 'pmc', 'popset', 'probe',
    'proteinclusters', 'pcassay', 'biosystems', 'pccompound',
    'pcsubstance', 'pubmedhealth', 'seqannot', 'snp', 'sra',
    'taxonomy', 'toolkit', 'toolkitall', 'toolkitbook',
    'unigene', 'unists', 'gencoll']}
```

... and a long list of fields after pressing enter.

Example 20.2 Searching PubMed with More Than One Term, Combining Keywords with AND/OR

```
from Bio import Entrez
handle = Entrez.esearch(db="pubmed", term="PyCogent AND RNA")
record = Entrez.read(handle)
print record['IdList']
handle = Entrez.esearch(db="pubmed", term="PyCogent OR RNA")
record = Entrez.read(handle)
print record['Count']
handle = Entrez.esearch(db="pubmed", \
    term="PyCogent AND 2008[Year]")
record = Entrez.read(handle)
print record['IdList']
handle = Entrez.esearch(db="pubmed", term= \
    "C. elegans[Organism] AND 2008[Year] AND Mapk[Gene]")
record = Entrez.read(handle)
print record['Count']
```

Source: Adapted from code published by A.Via/K.Rother under the Python License.

The program writes the lists of PMIDs and respective paper counts for the four queries. The optional parameter retmax (maximum retrieved items) makes it possible to set the maximum number of retrieved matches of the query text:

```
handle = Entrez.esearch(db = "pubmed", \
    term = "PyCogent OR RNA", retmax = "3")
record = Entrez.read(handle)
print record['IdList']
```

which results in:

```
['23285493', '23285311', '23285230']
```

Example 20.3 Retrieving and Parsing Nucleotide
Database Entries in GenBank Format

This procedure is nearly identical to the procedure to retrieve and parse PubMed records. The main difference is that the IDs you need to fetch are the GI numbers of the sequences. Multiple IDs must be passed in the form of a string of comma-separated GI numbers instead of a list, and the file format (`retmode`) must be set to `xml`.

```python
from Bio import Entrez
# search sequences by a combination of keywords
handle = Entrez.esearch(db="nucleotide", \
    term="Homo sapiens AND mRNA AND MapK")
records = Entrez.read(handle)
print records['Count']
top3_records = records['IdList'][0:3]
print top3_records
# retrieve the sequences by their GI numbers
gi_list = ','.join(top3_records)
print gi_list
handle = Entrez.efetch(db="nucleotide", \
    id=gi_list, rettype="gb", retmode="xml")
records = Entrez.read(handle)
print len(records)
print records[0].keys()
print records[0]['GBSeq_organism']
```

Source: Adapted from code published by A.Via/K.Rother under the Python License.

At time of writing, this code generates the output:

```
1053
['472824973', '433282995', '433282994']
472824973,433282995,433282994
3
[u'GBSeq_moltype', u'GBSeq_comment', u'GBSeq_feature-table',
    u'GBSeq_primary', u'GBSeq_references', u'GBSeq_locus',
    u'GBSeq_keywords', u'GBSeq_secondary-accessions', u'GBSeq_
    definition', u'GBSeq_organism', u'GBSeq_strandedness',
    u'GBSeq_source', u'GBSeq_sequence',
```

```
    u'GBSeq_primary-accession', u'GBSeq_accession-version',
    u'GBSeq_length', u'GBSeq_create-date', u'GBSeq_division',
    u'GBSeq_update-date', u'GBSeq_topology', u'GBSeq_other-
    seqids', u'GBSeq_taxonomy']
Homo sapiens
```

For a single gi, you can also use the "text" (retmode = "text") format:

```
handle = Entrez.efetch(db = "nucleotide", \
    id = "186972394", rettype = "gb", retmode = "text")
record = handle.read()
```

Example 20.4 Searching for NCBI Protein Database Entries by Keywords

This procedure is very similar to that shown for PubMed and nucleotide records (Section 20.2.2 and Example 20.3, respectively):

```
from Bio import Entrez
# search IDs of protein sequences by keywords
handle = Entrez.esearch(db="protein", \
    term="Human AND cancer AND p21")
records = Entrez.read(handle)
print records['Count']
id_list = records['IdList'][0:3]
# retrieve sequences
id_list = ",".join(id_list)
print id_list
handle = Entrez.efetch(db="protein", \
    id=id_list, rettype="fasta", retmode="xml")
records = Entrez.read(handle)
rec = list(records)
print rec[0].keys()
print rec[0]['TSeq_defline']
```

Source: Adapted from code published by A.Via/K.Rother under the Python License.

This code creates the output:

```
920
229577056,131890016,113677036
[u'TSeq_accver', u'TSeq_sequence', u'TSeq_length',
    u'TSeq_taxid', u'TSeq_orgname', u'TSeq_gi',
    u'TSeq_seqtype', u'TSeq_defline']
CDC42 small effector protein 2 [Danio rerio]
```

Example 20.5 Retrieving SwissProt Database Entries and Writing Them to a File in FASTA Format

Biopython provides a module (called ExPASy) to access the SwissProt database and other Expasy resources (http://www.expasy.org/). The get_sprot_raw() method of the ExPASy module returns a handle, which can be read using the SeqIO.read() method (see Chapter 19). It is therefore necessary to import the SeqIO module first. As you learned in Chapter 19, the object returned by SeqIO.read() is a SeqRecord object and, as such, has id, Seq, and description attributes and can be written to a FASTA formatted file using the SeqIO.write() method.

```
from Bio import ExPASy
from Bio import SeqIO
handle = ExPASy.get_sprot_raw("P04637")
seq_record = SeqIO.read(handle, "swiss")
out = open('myfile.fasta','w')
fasta = SeqIO.write(seq_record, out, "fasta")
out.close()
```

Source: Adapted from code published by A.Via/K.Rother under the Python License.

Notice that if you want to do this for several SwissProt ACs, you have to retrieve and parse them one by one:

```
ac_list = ['P04637', 'P0CQ42', 'Q13671']
records = []
for ac in ac_list:
    handle = ExPASy.get_sprot_raw(ac)
    record = SeqIO.read(handle, "swiss")
    records.append(record)
out = open('myfile.fasta','w')
for rec in records:
    fasta = Bio.SeqIO.write(rec, out, "fasta")
out.close()
```

This code creates a local multiple FASTA file.

20.5 TESTING YOURSELF

Exercise 20.1 Search PubMed by Keywords

Use Entrez.esearch() to retrieve a list of PMIDs of papers about tRNA aminoacylation from 2008, using as search terms *trna, aminoacylation, "2008"[Publication Date]*. How many papers do you find?

Exercise 20.2 Get Paper Information and Save It to a File

Use `Entrez.efetch()` to get information from papers retrieved in Exercise 20.1 in the Medline format and save it to a file. How many lines does the file have?

Exercise 20.3 Fetch a Nucleotide Sequence

Use `Entrez.efetch()` to download the nucleotide sequence with the GI 433282994 and write it to a file in FASTA format.

Exercise 20.4 Search for Protein Sequences by Keyword

Use `Entrez.esearch()` to find protein sequences for the bacteriorhodopsin protein. Retrieve the first 20 sequences and save them to a file in GenBank format.

Exercise 20.5

Write a small program that performs a function similar to that of EndNote or Mendeley. The program should read a text with PMIDs in square brackets (e.g., [23285311]) and replace them by an increasing number in square brackets (e.g., [1]), plus a formatted reference at the end of the document, e.g.,

```
[1] Cieslik M, Derewenda ZS, Mura C. Abstractions, algorithms
    and data structures for structural bioinformatics in
    PyCogent. J Appl Crystallogr. 2011 Apr 1;44:424-428.
```

## CHAPTER 21

Working with 3D Structure Data

L EARNING GOAL: You can use Biopython to work with macromolecular 3D structures.

21.1 IN THIS CHAPTER YOU WILL LEARN

- How to parse PDB files with Biopython

- How to access chains, residues, and atoms

- How to superimpose structures onto corresponding residues

21.2 STORY: EXTRACTING ATOM NAMES AND THREE-DIMENSIONAL COORDINATES FROM A PDB FILE

The PDB format is the most popular file format for 3D coordinates of mac-
romolecules in structural bioinformatics. In Chapter 10, you learned how to
extract information from a PDB file writing your own scripts. You may have
thought that it is a pretty difficult task, which is true. Once you are capable
of writing a PDB parser by yourself (i.e., without using Biopython), you can
claim you have become a good programmer! However, PDB files contain a
lot more than just coordinates. In particular, if you want to fully represent
the structure of atoms, residues, chains, etc., you need to write a lot more
code. Fortunately, the Biopython developers have done that for you.

The `Bio.PDB` package is a powerful tool to retrieve macromolecular
structures from the web, to read and write PDB files, to calculate distances

and angles between atomic coordinates, and to superimpose structures. In this chapter, you will see how to parse PDB files using Biopython and how `Structure` objects in Biopython can be used.

21.2.1 Problem Description

In this example, a PDB file is downloaded from the Protein Data Bank and read into a `Structure` object. The `Structure` object is a container for the structural information in PDB entries, organized in a matryoshka-like hierarchy: a structure contains models, which contain chains, which contain residues, which contain atoms. In the following, this hierarchy will be abbreviated as "SMCRA" (Structure → Model(s) → Chain(s) → Residues → Atoms). In the following Python session, models, chains, residues, and atoms are extracted from a structure. Each object (`model`, `chain`, `residue`, `atom`) is in turn a container for additional information. For example, each atom object contains the atom name, element, and spatial coordinates. Here, you will see how the SMCRA hierarchy works and how to use it to manipulate PDB structures.

21.2.2 Example Python Session

```
from Bio import PDB
pdbl = PDB.PDBList()
pdbl.retrieve_pdb_file("2DN1")

parser = PDB.PDBParser()
structure = parser.get_structure("2DN1", "dn/pdb2dn1.ent")

for model in structure:
    for chain in model:
        print chain
        for residue in chain:
            print residue.resname, residue.id[1]
            for atom in residue:
                print atom.name, atom.coord
```

Source: Adapted from code published by A.Via/K.Rother under the Python License.

21.3 WHAT DO THE COMMANDS MEAN?

21.3.1 The `Bio.PDB` Module

To start working with Biopython, first you have to import the PDB module:

```
from Bio import PDB
```

The script consists of three parts. The first paragraph downloads a PDB structure file from the web and writes it to a file. The second paragraph reads it into a `Structure` object. The third paragraph walks through the hierarchical SMCRA structure and prints chains, residues, and atoms.

Q & A: IS THE SMCRA STRUCTURE A TREE?

Yes, it is a tree because there is one object (`Structure`) that contains all others over several parent–child relationships, and each child object has exactly one parent. In the SMCRA tree, all objects originate from the same Python class (they are subclasses of the `Entity` class). This structure is also called a *composite*.

Downloading PDB Structures

In the first line of the script, the `PDB` module from Biopython (`Bio.PDB`) is imported. Then the `PDB.PDBList` class is called. This class provides access to the PDB structure codes (e.g., "2DN1") on the RCSB server. The status of each structure (modified or obsolete) is also specified. Updated lists are released each week. The instruction

```
pdbl.retrieve_pdb_file("2DN1")
```

is the one that downloads the file and saves it in a directory with a name that corresponds to the two alphabetical characters of the PDB code (e.g., dn/in the case of 2DN1). This directory is created in the current directory (i.e., where you are running the script), and the PDB file is stored with the pdb2dn1.ent name.

Parsing PDB Structures

The following two instructions create a `PDB.PDBParser` object

```
parser = PDB.PDBParser()
structure = parser.get_structure("2DN1", "dn/pdb2dn1.ent")
```

and use the `get_structure()` method to parse the PDB file returning the `Structure` object mentioned previously, which is at the top of the

SMCRA hierarchy. In Section 21.2.2, the `Structure` object is assigned to the `structure` variable. The `get_structure()` method takes two arguments: a textual identifier for the `Structure` object (it does not make a difference whether you call it "my_PDB" or "Jim") and the PDB filename (and location) you want to parse. Notice that you also can parse files that are already present on your computer; you don't need to download them each time. If the `get_structure()` function fails to return an SMCRA object, most probably there is something wrong with the PDB file. This may happen for some old or modified PDB files that do not correspond to the PDB standard by the letter, in which case you may need to parse the PDB file with your own script or at least standardize it.

Parsing mmCIF Files

The `PDB.PDBParser` class provides tools to parse files in the PDB format (usually ending with .pdb or .ent). If you want to parse files in the mmCIF format (see www.ebi.ac.uk/pdbe/docs/documentation/mmcif. html), you have to download files in that format and use a different parser:

```
parser = MMCIFParser()
structure = parser.get_structure("2DN1","2DN1.cif")
```

All the rest works in the same way.

21.3.2 The SMCRA Object Hierarchy

The first level of the hierarchy is the `Structure`, which contains only one `Model` for most structures (NMR structures usually have more). The `Chain` level contains all macromolecular chains in a given model. Each `Chain` is composed of `Residue` objects, which in turn are composed of `Atom` objects. Each object of the hierarchy has its own set of attributes and methods, although they share a common pool because they all derive from the `Entity` class.

Generally, each object in the SMCRA has an identifier. Chains and atoms have string identifiers, while models and residues have numbers. Using identifiers, you can access the child objects within the hierarchy like you would use a dictionary. This way of accessing child objects holds for all objects of the SMCRA hierarchy. For example,

```
model = structure[0]
chain = model['A']
residue = chain[2]
atom = residue['CA']
```

Methods of `Structure` *Objects*

You can visualize the methods and attributes of `Structure` objects using the `dir()` function, passing as argument the variable name you have chosen for the `Structure` object (`structure` in the Python session in Section 21.2.2). When you type

```
>>> from Bio import PDB
>>> parser = PDB.PDBParser()
>>> structure = parser.get_structure("2DN1", \
... "dn/pdb2dn1.ent")
>>> dir(structure)
```

all methods and attributes will be displayed. Here are some examples of `Structure` functions and attributes and how they work:

```
>>> structure.get_id()
'2DN1'
>>> structure.get_level()
'S'
```

`'S'` stands for "structure level."

```
>>> structure.child_list
[<Model id = 0>]
```

The `child_list` attribute returns a list of the children of the structure, i.e., a list of objects on the next level of the hierarchy: the models. This structure has a single model (in fact it is an X-ray structure). You will obtain the same result with the following:

```
>>> list(structure)
[<Model id = 0>]
```

As you will see, the `child_list` attribute and the `list()` function can be applied to all objects of the hierarchy. You can use them to snoop inside objects and see what is contained in models, chains, and residues.

```
>>> structure.header
{'structure_method': 'x-ray diffraction','head':
'oxygen storage/transport',...}
```

The `header` is a dictionary (here visualized only in part) collecting information from the PDB header. You can extract individual fields using the keys of the dictionary:

```
>>> structure.header['structure_method']
'x-ray diffraction'
```

Methods of Model Objects

To use a model that is the child of a Structure object, you will find it is convenient to assign it to a variable first:

```
>>> model = structure.child_list[0]
```

Alternatively, you can just run a variable (e.g., model) over the Structure object, even without performing actions in the loop. Such a variable will take the values of the Model objects belonging to the Structure object:

```
>>> for model in structure: pass
...
```

When the loop is exited, the model variable will contain a Model object. Then you can explore methods and attributes of the Model object, applying the dir() function to the model variable:

```
>>> dir(model)
```

Among others, you have the following attributes:

```
>>> model.child_list
[<Chain id = A>, <Chain id = B>]
>>> list(model)
[<Chain id = A>, <Chain id = B>]
>>> model.child_dict
{'A': <Chain id = A>, 'B': <Chain id = B>}
>>> model.level
'M'
```

You can see that the 2DN1 structure has only one model and two chains: A and B.

Methods of Chain Objects

You can obtain the chain identifiers using the id attribute of a Model object (assigned to the model variable):

```
>>> for chain in model:
...     print chain.id
```

```
...
A
B
```

If you want to know what attributes and methods the chain objects have, again, assign one of the chains to a variable and then use the dir() function:

```
>> chain = model.child_list[0]
```

or

```
>>> for chain in model: pass
...
>>> dir(chain)
```

You can see which residues the chain object contains using either the child_list attribute or the list() function:

```
>>> list(chain)
[<Residue LEU het = resseq = 2 icode = >,...]
```

This list has as many elements as there are residues in the chain.

Methods of Residue Objects

Similar to Structure, Model, and Chain objects, Residue objects, which are the children of the Chain objects, have an id attribute:

```
>>> for residue in chain:
...     print residue.id
...
(' ', 2, ' ')
(' ', 3, ' ')
...
```

The Residue object id is a tuple of three elements for each residue instead of a single character as in case of Chain objects. The three elements are (1) a flag for nonpolymer atoms (e.g., water, ions, and ligands, also called "heteroatoms"), which consist of H_XXX, where XXX is the name of the hetero residue (e.g., H_OXY), or W in the case of water; (2) the sequence identifier of the residue along the chain; and (3) the "insertion code," i.e., a code that is used to represent, e.g., insertion mutants.

You can retrieve the type of a residue via the `resname` attribute of Residue objects:

```
>>> for residue in chain:
...     print residue.resname, residue.id
...
HIS (' ', 2, ' ')
LEU (' ', 3, ' ')
THR (' ', 4, ' ')
...
```

Keep in mind that the residue type is also visible when printing the Residue object:

```
>>> for residue in chain:
...     print residue
...
<Residue HIS het = resseq = 2 icode = >
<Residue LEU het = resseq = 3 icode = >
<Residue THR het = resseq = 4 icode = >
...
```

Q & A: WHAT DO I HAVE TO DO TO ACCESS A SPECIFIC RESIDUE?

You can use the three-element tuple `id` or simply use the residue number as a dictionary key of a `Chain` object:

```
>>> residue = chain[((' ', 2, ' '))]
>>> residue
<Residue HIS het = resseq = 2 icode = >
>>> residue = chain[2]
>>> residue
<Residue HIS het = resseq = 2 icode = >
```

Then, you can retrieve the residue type using `resname`:

```
>>> chain[2].resname
'HIS'
>>>
```

Notice that in case of Residue objects, the `child_list` attribute and the `list()` function will return a list of the atoms of each residue:

```
>> list(residue)
[<Atom N>, <Atom CA>, <Atom C>, <Atom O>]
>>> residue.child_list
[<Atom N>, <Atom CA>, <Atom C>, <Atom O>]
```

Running over all residues of a chain and printing their `resname` and `child_list` attributes will display residues and their constituent atoms:

```
>>> for residue in chain:
...    print residue.resname, residue.child_list
...
HIS [<Atom N>, <Atom CA>, <Atom C>, <Atom O>]
LEU [<Atom N>, <Atom CA>, <Atom C>, <Atom O>, <Atom CB>,
    <Atom CG>, <Atom CD1>, <Atom CD2>]
THR [<Atom N>, <Atom CA>, <Atom C>, <Atom O>, <Atom CB>,
    <Atom OG1>, <Atom CG2>]
...
```

Q & A: DO THE RESIDUES ALWAYS APPEAR IN THE SAME ORDER AS THEY ARE IN THE SEQUENCE?

Generally they do. However, when a chain contains modified residues that may be represented by heteroatoms, these will usually appear after all normal residues. For instance, modified nucleotides in tRNA structures sometimes appear at the end of the structure. As a general rule, the PDB file order is kept in the corresponding `Structure` object. However, if you depend on the order of residues, it is safer to sort them explicitly.

Methods of `Atom` Objects

As you may have noticed, the atom `id` (or atom name) is not just the element symbol but an abbreviation of the atom position within a residue. For instance, "CA" is the atom `id` of all C-alpha atoms in protein structures. Each `Atom` object has plenty of methods to extract the information available in the PDB files. For a given residue, you can easily obtain the `id`, serial number, and coordinates of each of its atoms:

```
>>> residue = chain[2]
>>> for atom in residue:
... print atom.get_id(), atom.get_serial_number(), \
... atom.get_coord()
...
N 1064 [14.4829998 15.80900002 11.95800018]
CA 1065 [14.82299995 15.06900024 13.15100002]
C 1066 [14.93500042 15.76099968 14.47500038]
O 1067 [15.99600029 15.7159996 15.13099957]
>>>
```

`Atom` objects have methods to retrieve the following data: the isotropic B factor (`atom.get_bfactor()`), the anisotropic B factor (`atom.get_anisou()`), the atom occupancy (`atom.get_occu-pancy()`), and a `Vector` object (`atom.get_vector()`) that facilitates some calculations with coordinates.

The Python session in Section 21.2.2 basically runs over all models (1), all chains (A and B), all residues of each chain, and all atoms of each residue. For each atom, it prints the atom `name` and its 3D coordinates.

Q & A: WHAT ARE THE B-FACTOR AND OCCUPANCY?

The B-factor is a statistical measure that describes the reliability of the coordinates of a given atom. In X-ray structures, it represents the Brownian motion of the atom and inaccuracy of measurement. The higher the B-factor is, the more uncertain the position of the atom. The occupancy is used to add weights to alternative atom positions. For any given atom, the occupancies should always add up to 1.0.

21.4 EXAMPLES

Example 21.1 Distance between Atoms

To calculate a distance between two atoms in Biopython, you can use the difference operator on two `Atom` objects:

```
from Bio import PDB
parser = PDB.PDBParser()
structure = parser.get_structure("2DN1", \
    "dn/pdb2dn1.ent")
atom1 = structure[0]['A'][2]['CA']
atom2 = structure[0]['A'][3]['CA']
dist = atom_1 - atom_2
print dist
```

Source: Adapted from code published by A.Via/K.Rother under the Python License.

This code generates the output:

```
3.76608
```

Note that unlike in the examples in Section 21.2.2 and Section 21.3.2, the levels of the hierarchy are all traversed by a single line for each atom here (e.g., `structure[0]["A"][2]["CA"]`).

Example 21.2 Extracting the Sequence from a Structure

This task can be done using the polypeptide builder (a class named PPBuilder) from the Polypeptide module of the Bio.PDB module. You first create a PPBuilder instance from the class. A Structure object can be parsed using the build_peptides() method of PPBuilder(), thus generating a list of "peptide objects." The sequence of each polypeptide chain can be obtained using the get_sequence() method of these peptide objects:

```
from Bio import PDB
from Bio.PDB.Polypeptide import PPBuilder
parser = PDB.PDBParser()
structure = parser.get_structure("2DN1", \
    "dn/pdb2dn1.ent")
ppb = PPBuilder()
peptides = ppb.build_peptides(structure)
for pep in peptides:
    print pep.get_sequence()
```

Source: Adapted from code published by A.Via/K.Rother under the Python License.

This code generates the output:

```
LSPADKTNVKAAWGKVGAHAGEYGAEALERMFLSFPTTKTYFPHFDLSH
GSAQVKGHGKKVADALTNAVAHVDDMPNALSALSDLHAHKLRVDPVNFK
LLSHCLLVTLAAHLPAEFTPAVHASLDKFLASVSTVLTSKYR

HLTPEEKSAVTALWGKVNVDEVGGEALGRLLVVYPWTQRFFESFGDLST
PDAVMGNPKVKAHGKKVLGAFSDGLAHLDNLKGTFATLSELHCDKLHVD
PENFRLLGNVLVCVLAHHFGKEFTPPVQAAYQKVVAGVANALAHKYH
```

The same can also be done for a single polypeptide:

```
peptides = ppb.build_peptides(structure)
seq = peptides[0].get_sequence()
```

The seq variable contains:

```
Seq('LSPADKTNVKAAWGKVGAHAGEYGAEALE
RMFLSFPTTKTYFPHFDLSHGSAQV...
KYR', ProteinAlphabet())
```

You will have noticed that the get_sequence() method of peptide objects returns a Biopython Seq() object (see Chapter 19).

Q & A: CAN I GET THE SEQUENCE OF DNA
AND RNA STRUCTURES AS WELL?

The polypeptide builder works for proteins only. Nucleotides can be processed by the ModeRNA library for building comparative models of RNA (http://iimcb.genesilico.pl/moderna/; see Recipe 18).

Example 21.3 Superimposition of Two Structures

To compare two structures visually, it is useful to superimpose them. During superposition, the structures are placed in the same coordinate system such that the distances between pairs of corresponding atoms become minimal. Two structures can be superimposed using at least three pairs of corresponding atoms. To this aim, you can use the `Superimposer` class of the `Bio.PDB` module. You have to define which atoms in each structure you want to superimpose and decide which structure will be fixed in its coordinate system and which will be moving (i.e., rotated and translated). The atoms to build a superposition matrix can be stored in two separate lists and passed as arguments to the `set_atoms()` method of the `PDB.Superimposer()` object. The `set_atoms(list_1, list_2)` function superimposes the atoms listed in `list_2` on the atoms listed in `list_1`. One outcome of the superposition function is a superposition matrix that can be used to move an entire structure. The quality of the fit is given by the root mean square deviation of the atom pairs (RMSD). In this example, the rotation and translation matrix and the RMSD of the superimposed structures are also printed.

```
from Bio import PDB
parser = PDB.PDBParser()
structure = parser.get_structure("2DN1", "dn/pdb2dn1.ent")
atom1 = structure[0]["A"][10]["CA"]
atom2 = structure[0]["A"][20]["CA"]
atom3 = structure[0]["A"][30]["CA"]
atom4 = structure[0]["B"][10]["CA"]
atom5 = structure[0]["B"][20]["CA"]
atom6 = structure[0]["B"][30]["CA"]
moving = [atom1, atom2, atom3]
fixed = [atom4, atom5, atom6]
sup = PDB.Superimposer()
```

```
sup.set_atoms(fixed, moving)
print sup.rotran
print 'RMS:', sup.rms
```

Source: Adapted from code published by A.Via/K.Rother under the Python License.

In this example, three CA atoms per chain of 2DN1 have been superimposed. However, you could do it for the CA atoms of catalytic triads of two different serine proteases. In this case, the triads of atoms will come from two different structures instead of two different chains of the same structure (see Recipe 20).

Q & A: CAN BIOPYTHON DETERMINE THE LISTS OF ATOMS TO SUPERIMPOSE AUTOMATICALLY?

No, sorry. If you have two structures with identical sequences with atoms in the same order, you can simply take the list of all atoms. But often the order between different PDB entries is messed up, and some atoms may be missing in a crystal structure. In practice, using all atoms from two structures only works for comparing a structure with its model obtained, for example, by homology modeling, provided they are both complete and have the same number of residues/atoms.

Example 21.4 Saving a `Structure` Object to a File

The `PDBIO` module writes PDB files. In this example, the `Structure` object obtained from 2DN1 is saved to a new file `'my_structure.pdb'`. This example simply copies the atom coordinates from the input file to an output file. To write a `Structure` object to a file, you can use the `set_structure()` and `save()` methods of `PDBIO()`:

```
from Bio import PDB
from Bio.PDB import PDBIO
parser = PDB.PDBParser()
structure = parser.get_structure("2DN1", "dn/pdb2dn1.ent")
io = PDBIO()
io.set_structure(structure)
io.save('my_structure.pdb')
```

Source: Adapted from code published by A.Via/K.Rother under the Python License.

21.5 TESTING YOURSELF

Exercise 21.1 Download and Parse a tRNA Structure

Retrieve the Phenylalanyl-tRNA structure 1EHZ from the PDB website and parse it to a `Structure` object using Biopython.

Exercise 21.2 Count Residues and Atoms

Calculate the total number of residues and atoms from the `Structure` object of Exercise 21.1. How many residues consist of heteroatoms (i.e., have an "H..." in the first element of the `residue.id` tuple)?

Exercise 21.3 Calculate the Distance between Paired RNA Backbones

In the 1EHZ structure, the residues 1 and 72 form a base pair, as well as 54 and 64. Calculate the distances between the phosphate atoms (name P) of both pairing residues and print them.

Exercise 21.4 Calculate Disulphide Bonds

Disulphide bonds play an important role in the folding and stability of some proteins, usually proteins secreted to the extracellular medium. Disulphide bonds in proteins are formed between the thiol groups of cysteine residues, consisting of a Cβ and an Sγ atom. The structure of a disulphide bond can be described by the dihedral angle between its Cβ- Sγ - Sγ - Cβ atoms, which is usually close to 90 degrees. The distance between Sγ - Sγ is usually in the range between 1.9 Å and 2.1 Å. In Biopython you can calculate a torsion angle with the following:

```
from Bio.PDB import Vector
v1 = atom1.get_vector()
v2 = atom2.get_vector()
v3 = atom3.get_vector()
v4 = atom4.get_vector()
Vector.calc_dihedral(v1, v2, v3, v4)
```

Using this information, write a script that finds all potential disulphide bonds in the PDB structure 1C9X (ribonuclease A). How many disulphide bonds are you able to find?

Hint: In your script you have to verify if the atoms of the cysteine thiol groups meet a condition on the dihedral angle and one on the distance.

Exercise 21.5 Draw a Ramachandran Plot

For the final exercise of the book, write a program that calculates and draws a Ramachandran plot using the ϕ and ψ angles from a protein structure. The ϕ torsion angle is between the N_i, $C\alpha_i$, C_i, and N_{i+1} atoms along the backbone, the ψ torsion angle between C_i, N_{i+1}, $C\alpha_{i+1}$, and C_{i+1}. The plot should be written to a .png file.

To complete this exercise, you will need most of the knowledge and skills you have learned in this book. We wish you good luck. Have fun programming!

V SUMMARY

In Part V, you met Biopython, a comprehensive library for managing biological data and resources. Biopython has objects and methods to deal with sequences and annotated sequence records, to access NCBI resources, to work with PDB structures, and much more. It also makes it possible to run programs such as BLAST and parse PubMed records.

In Chapter 19, several objects and methods to work with protein, DNA, and RNA sequence data are illustrated. Seq objects can be used to manipulate sequences enriched with an alphabet and have a lot in common with strings. MutableSeq objects are similar to Seq objects; they make it possible to modify sequences. A SeqRecord object encodes not only a sequence but also several kinds of annotations such as the sequence ID, its source organism, literature cross-references, etc. In Chapter 19, the SeqIO module is also introduced. It can be used to parse data files, such as sequence or multiple alignment files, and store their content in lists and dictionaries. Moreover, you can use SeqIO to read and write sequence files in a variety of formats (e.g., FASTA).

Chapter 20 is about retrieving data from web resources. The chapter focuses mainly on NCBI resources by introducing the Entrez module and shows how to read nucleotide and protein sequence files from the web and how to submit PubMed queries. It also describes how to retrieve Uniprot records and write them to a file.

Finally, in Chapter 21, you learned how to work with three-dimensional structures using the Biopython PDB module. When you use this module, files from the Protein Data Bank (PDB) can be read to Structure objects, which are organized in a matryoshka-like hierarchy abbreviated as "SMCRA" (Structure $\rightarrow$ Model(s) $\rightarrow$ Chain(s) $\rightarrow$ Residues $\rightarrow$ Atoms). Retrieving specific chains, residues, and even atoms and their properties from a structure object is an easy task. In particular, you can extract atom coordinates and use them to superimpose protein structures and calculate their RMSD.

Biopython is not the only Python library available for these purposes. PyCogent, introduced in Recipe 1, also provides a large set of tools for biological data management.

VI

Cookbook

INTRODUCTION

In this part of the book, we cover a set of typical bioinformatics tasks that are not explicitly described in Parts I–V. In principle, after reading this book, you should be able to write programs to accomplish them on your own. In any case, we believe that it is useful to have ready-to-use recipes. You will find here some particular bioinformatics applications such as running BLAST or building phylogenetic trees, some PyCogent examples, and several parsers (of multiple sequence alignments, of HTML pages, of BLAST XML output, of SBML files, and of RNA Vienna files). Furthermore, several recipes are aimed at working with 3D protein and RNA structures. The recipes cover very diverse topics, and we did our best to include the most interesting and frequent bioinformatics applications. You can use the code as it is or use it as a starting point to customize your own programs. Welcome to our zoo of recipes!

Recipe 1: The PyCogent Library

PyCogent [1] (http://pycogent.org) is a powerful alternative library to Biopython. Many things like creating sequence objects, and reading and writing common sequence formats, work in a similar way as in Biopython (see Chapter 20), although the precise syntax differs. PyCogent includes wrappers for many common bioinformatics applications, provides a lot of functions to work with RNA sequences and secondary structures, and makes it possible to calculate phylogenetic trees. In this recipe, functions in PyCogent designed to work with multiple sequence alignments are presented.

To use PyCogent you need to install it separately (installation on Linux and Mac works with a few commands, listed on the download page; for installation on Windows, see the last section of this recipe). There is vast documentation for PyCogent, including thoroughly tested code examples available on http://pycogent.org.

The following script loads a multiple alignment of protein sequences from a FASTA file and calculates the gap fraction for each column:

```
from cogent.core.alignment import Alignment

fasta_file = open('align.fasta')
ali = Alignment(fasta_file.read())
print ali.toFasta()

for column in ali.iterPositions():
    gap_fraction = float(column.count('-')) / len(column)
    print '%4.2f' % gap_fraction,
print
```

Source: Adapted from code published by A.Via/K.Rother under the Python License.

The first two lines import the `Alignment` class from PyCogent and open the FASTA file for reading. In the third line, an instance of the `Alignment` class is created with the open file as a parameter. `Alignment` objects have a `toFasta()` method that returns a string in FASTA format. You can use it to display the alignment or write the string to a file.

To calculate the gap fraction (the relative amount of `'-'` symbols in one column), use the `iterPosition()` method. It returns the columns of the alignment as separate lists:

```
['L', 'S', '-']
```

From this list the number of gaps can be counted using the `count()` method as in Chapter 2. The gap fraction is the number of `'-'` symbols divided by the length of the column. The last line of the script prints all gap fractions. The comma at the end causes all numbers to be written into one line.

ACCESSING ROWS AND COLUMNS IN `Alignment` OBJECTS

PyCogent allows you to work with alignments as if they were a table. For instance, you can access single and multiple columns by numerical indices:

```
print ali[3]
print ali[5:8]
```

The result of the indexing operation is an `Alignment` object that you can print, convert to FASTA, or combine with other alignments:

```
print ali[5:8] + ali[7:9]
```

Extracting sequences from the alignment is slightly more complicated, because you need to access the `Names` dictionary:

```
print ali.getSeq(ali.Names[1])
```

The `takeSeqs()` method allows you to create a new alignment consisting of only a few selected sequences:

```
ali2 = ali.takeSeqs(ali.Names[2], ali.Names[0])
print ali2
```

Alignment objects have two other methods worth mentioning: degap() and `variablePositions()`. degap() return a collection of sequences where the gaps have been removed:

```
seq_coll = ali.degap()
print seq_coll.toFasta()
```

The second, `variablePositions()`, returns a list of column indices where the sequences are not conserved; i.e., at least one of the sequences differs from the others:

```
print ali.variablePositions()
```

These methods represent only a tiny fraction of what PyCogent is capable of doing. Taken together, the library provides many shortcuts to work with sequences, trees, structures, and other kinds of biological data.

INSTALLING PYCOGENT ON WINDOWS

The installation on Windows may require a few hints unless you use the `easy_install` tool (which may be challenging to install on Windows as well). Before installing PyCogent, make sure you have installed Python 2.6 or 2.7 and Scientific Python. To check whether the installation of Scientific Python was successful, open a Python shell and type

```
>>> import numpy
```

Second, unzip the PyCogent file. Third, run the setup script from the Windows shell:

```
C:\Python26\python.exe setup.py build
```

Normally, you would continue with

```
C:\Python26\python.exe setup.py install
```

However, we have seen this fail on several Windows machines without an easy fix in sight. Alternatively, you can look for the `cogent/` directory in the `build/` directory and copy it to any place where you want to import the modules or move it to `C:\Python2.6\lib\ site_packages\`.

Finally, check whether you can import the PyCogent library from Python:

```
>>> import cogent
```

REFERENCE

[1] R. Knight, P. Maxwell, and A. Birmingham, "PyCogent: A Toolkit for Making Sense from Sequence," *Genome Biology* 8 (2007): R171.

Recipe 2: Reversing and Randomizing a Sequence

S INCE EVERY SEQUENCE PATTERN may occur in a sequence by chance, finding a match for a functional motif in a nucleotide or protein sequence does not ensure that the match is biologically meaningful. In other words, we cannot exclude a priori that an occurrence is a false positive (FP) match; namely, that it occurred by chance. To assess the biological significance of the occurrence of a motif, the examined sequences are commonly compared to a random set. Sequence motifs that are significantly overrepresented in a given set of sequences with respect to a set of randomized sequences are likely to encode a functional property (i.e., are likely to be biologically meaningful). A good set of random sequences will be devoid of biological meaning but have the same amino acid/nucleotide composition as the biological ones. Such sets can be generated in several ways, for example, by reversing the original sequences, by reshuffling them or by creating a random sequence from scratch. This recipe illustrates how to use Python to reverse, shuffle, and randomize a sequence.

REVERSING A SEQUENCE

A sequence is a string of characters. There are at least three alternative ways to reverse a string:

1. Converting a string into a list and back.

```
seq = 'ABCDEFGHIJKLMNOPQRSTUVWXYZ'
seq_list = list(seq)
seq_list.reverse()
rev_seq = ''.join(seq_list)
print rev_seq
```

Source: Adapted from code published by A.Via/K.Rother under the Python License.

The reason for converting to a list is that lists have a `reverse()` method but strings do not. The string method `join()` concatenates the elements of the list and generates a string.

2. Using the `reversed()` built-in function, which applies to an iterable data type (string, list, tuple) and returns an iterator that loops over the elements of the iterable in reverse order.

```
rev_seq = ''
for s in reversed(seq):
     rev_seq = rev_seq + s
print rev_seq
```

Instead of the loop, you can directly convert to a string, similarly to the first example:

```
rev_seq = ''.join(list(reversed(seq)))
```

3. Using the extended slice syntax that applies to sequences of objects (strings, lists, tuples). Given a sequence `seq`, `seq[start:end:step]` returns a slice of `seq` including `seq` elements from `start` to `end` with step `step`. By leaving start and end empty (which corresponds to setting `start = 0` and `end = last element+1`) and setting `step = -1`, you will obtain the reversed sequence:

```
seq = 'ABCDEFGHIJKLMNOPQRSTUVWXYZ'
rev_seq = seq[::-1]
print rev_seq
```

RANDOMIZING A SEQUENCE

The following approaches make use of the random module:

1. *Using* `random.sample()`.

```
import random
seq = 'ABCDEFGHIJKLMNOPQRSTUVWXYZ'
ran_seq = ''.join(random.sample(seq, len(seq)))
print ran_seq
```

The `random.sample()` function returns a list of length = `len(seq)` of elements of seq randomly sampled from seq:

```
print random.sample(seq, len(seq))
```

will return:

```
['I', 'X', 'C', 'A', 'Q', 'Z', 'S', 'B', 'U', 'L', 'P', 'H',
'O', 'T', 'N', 'K', 'D', 'I', 'Y', 'R', 'M', 'E', 'W', 'G',
'V', 'F']
```

2. *Using* random.choice(). The string method join() joins the elements of the list and generates a string. Alternatively, you can use the random.choice() function, which randomly picks up a single character from seq. Then, you can use the Python list comprehension, which provides a concise way to create lists. Finally, you can use the join() string method to generate a string from the list:

```
import random

seq = 'ABCDEFGHIJKLMNOPQRSTUVWXYZ'
ran_seq = ''.join([random.choice(seq) \
    for x in range(len(seq))])
print ran_seq
```

Notice that in both approaches, the frequency of each character in the resulting random sequence is *on average* the same as in the original sequence but, unlike shuffling, may fluctuate randomly.

3. *Shuffling a sequence.* To create a random sequence with the exact composition of a given one, you can use the random.shuffle() function. It takes a list and changes the order of elements randomly:

```
import random
seq = 'ABCDEFGHIJKLMNOPQRSTUVWXYZ'
data = list(seq)
random.shuffle(data)
shuffled_seq = ''.join(data)
print shuffled_seq
```

Recipe 3: Creating a Random Sequence with Probabilities

IN RECIPE 2, you saw how to create random sequences with the same amino acid or nucleotide composition as a given sequence. In this recipe, you will learn how to create a random sequence with specific probabilities for each symbol; e.g, a given GC-content. In the following, DNA sequences are considered. When the probabilities for all nucleotides are identical, you can use the `random.choice()` function described in Recipe 2 to create a random DNA sequence (of, e.g., length 100):

```
import random
nucleotides = list('ACGT')
dna = ''
while len(dna) < 100:
    dna += random.choice(nucleotides)
print dna
```

Now, when the nucleotide frequencies shall differ, two things need to be added to this program. First, the probabilities need to be stored somewhere. Second, the probabilities need to be taken into account when composing the random sequence.

A dictionary is well suited to store pairs of nucleotides and their probabilities:

```
probs = {'A': 0.3, 'C': 0.2, 'G': 0.2, 'T': 0.3}
```

This corresponds to a GC-content of 0.4. When you use a dictionary with probabilities, it is important to make sure that the numbers are well balanced, especially if you edit them manually or there are many of them (e.g., amino acids). In that case it helps to check the values automatically. For instance, the sum of the probabilities should be exactly 1.0. The following lines check whether the sum of the probabilities is 1.0 and terminate the program if something is wrong:

```
if sum(probs.values()) != 1.0:
    raise Exception('Sum of probabilities is not 1.0!')
```

In Python this can be written as a shorter assert statement:

```
assert sum(probs.values()) == 1.0
```

You could also check, e.g., whether the probabilities for A+T and C+G are identical, but in biological sequences this is not necessarily given to the last decimal place.

To take the probabilities into account, the program "rolls dice" to decide whether to accept or reject a nucleotide. First, a nucleotide is chosen with random.choice() (with equal probabilities). Second, a random float number is created with random.random(). Third, only if the random number is smaller or equal to the probability of the nucleotide, it will be added to the DNA sequence (checked by if dice < probs[nuc]: in the following code). If it is bigger, no nucleotide is added, and the while loop in the program takes an extra round until the sequence has the desired length. The entire program looks like this:

```
import random

nucleotides = list('ACGT')
probs = {'A': 0.3, 'C': 0.2, 'G': 0.2, 'T': 0.3}
assert sum(probs.values()) == 1.0

dna = ''
while len(dna) < 100:
    nuc = random.choice(nucleotides)
    dice = random.random()
    if dice < probs[nuc]:
        dna += nuc
print dna
```

Source: Adapted from code published by A.Via/K.Rother under the Python License.

Recipe 4: Parsing Multiple Sequence Alignments Using Biopython

---

BIOPYTHON PROVIDES a data structure to store multiple alignments (the `MultipleSeqAlignment` class), and the `Bio.AlignIO` module for reading and writing them in various file formats.

Let's consider a multiple sequence alignment (MSA) of the Pfam Globin Family (PF00149), containing 73 protein seed sequences at the moment of writing. The file is in the Stockholm format, which is one of the most popular formats for multiple alignments. To get this file, you need to go to the Pfam website (http://pfam.sanger.ac.uk/) and search for the Globin entry by entering either the accession (PF00042), or the ID (Globin); once on the Globin page, click on the "Alignments" link (left menu). By specifying the MSA format ("Format an alignment" section), and submitting the query ("Generate" button) in the "Alignment" page, you will download the Globin *seed* sequences in Stockholm format. You can save the alignment to a file with the name `PF00042.sth`.

The `Bio.AlignIO` module provides two methods to parse multiple alignments: `Bio.AlignIO.read()`, if you want to parse just one alignment, and `Bio.AlignIO.parse()`, which parses files containing many alignments. Both methods require two mandatory and one optional argument. The mandatory arguments are

- a handle to the multiple alignment that could be either a file object or a filename, and

- the format of the multiple alignment (a full list of available formats can be found at http://biopython.org/wiki/AlignIO).

The required optional argument is

- the alphabet used by the alignment.

`Bio.AlignIO.read()` returns a single `MultipleSeqAlignment` object (or an error if there is more than one alignment), which can be printed to the screen using the `print` statement or written to a file using the `write()` method of the `Bio.AlignIO` module:

```
from Bio import AlignIO, SeqIO
alignment = AlignIO.read("PF00042.sth", "stockholm")
print alignment
handle = open("PF00042.fasta", "w")
AlignIO.write(alignment, handle, "fasta")
handle.close()
```

Source: Adapted from code published by A.Via/K.Rother under the Python License.

The `AlignIO.write()` function takes three arguments: the `MultipleSeqAlignment` object, a handle to a file, and the output format. Importantly, the output format can be different from the format of the original alignment (in the example, the original format is Stockholm and the output format is FASTA). This trick converts the format of an alignment into another. Notice that you can also use the `SeqIO` module (see Chapter 19) to write a `MultipleSeqAlignment` object to a file:

```
from Bio import AlignIO, SeqIO
alignment = AlignIO.read("PF00042.sth", "stockholm")
handle = open("PF00042.fasta", "w")
SeqIO.write(alignment, handle, "fasta")
handle.close()
```

Source: Adapted from code published by A.Via/K.Rother under the Python License.

If you want to extract information about single sequences in the alignment, you can run over an `Alignment` object, which, in this case, will return `SeqRecord` objects containing the sequence, ID, and annotations (see Chapter 19):

```
from Bio import AlignIO, SeqIO
alignment = AlignIO.read("PF00042.sth", "stockholm")
for record in alignment:
    print record.id, record.annotations, record.seq
```

The `AlignIO.parse()` method returns an iterator that runs over several alignments providing `MultipleSeqAlignment` objects, one for each alignment. To see how it works on more than one MSA, you can download a second MSA in Stockholm format from the Pfam website. For example, use the MSA for calcineurin-like phosphoesterases (accession: PF00149, ID: Metallophos, 324 *seed* sequences at the moment of writing) and then copy and paste one record after the other (PF00042.sth and PF00149.sth) to a new file (PF00042-PF00149.sth). Then you can use the PF00042-PF00149.sth filename as an argument of `AlignIO.parse()`:

```
from Bio import AlignIO
alignments = AlignIO.parse("PF00042-PF00149.sth", "stockholm")
for alignment in alignments:
    print alignment
```

This will print the two MSAs one after the other (not the entire record, just the alignments).

And if you want to extract information from each single sequence record in each of the two alignments in the PF00042-PF00149.sth file:

```
from Bio import AlignIO
alignments = AlignIO.parse("PF00042-PF00149.sth", "stockholm")
for alignment in alignments:
    for record in alignment:
        print record.id, record.annotations, record.seq
```

Finally, if you want to write the alignments to a file in FASTA format:

```
from Bio import AlignIO
alignments = AlignIO.parse("PF00042-PF00149.sth", "stockholm")
handle = open("PF00042-PF00149.fasta", "w")
AlignIO.write(alignments, handle, "fasta")
handle.close()
```

Recipe 5: Calculating a Consensus Sequence from a Multiple Sequence Alignment

W HEN YOU HAVE multiple aligned sequences, a common question is what sequence best represents the entire alignment. This is called a consensus sequence, and it expresses the most frequent residue(s) (i.e., the most conserved) in each column of the multiple alignment. To calculate it, you first need to calculate how often each character occurs for each column of the alignment. Second, you choose the most frequent character for each column. The program below calculates a consensus for a set of short DNA sequences:

```
seqs = [
    'ATCCAGCT',
    'GGGCAACT',
    'ATGGATCT',
    'AAGCAACC',
    'TTGGAACT',
    'ATGCCATT',
    'ATGGCACT'
    ]
n = len(seqs[0])
profile = {'A':[0]*n, 'C':[0]*n, 'G':[0]*n, 'T':[0]*n }

for seq in seqs:
    for i, char in enumerate(seq):
        profile[char][i] += 1
```

```
consensus = ""
for i in range(n):
    col = [(profile[nt][i], nt) for nt in "ACGT"]
    consensus += max(col)[1]
print consensus
```

Source: Adapted from code published by A.Via/K.Rother under the Python License.

The sequences are stored in `seqs`, a list of strings. Next, a profile table is created as a dictionary containing a list of zeroes for each character. The length of each list of zeroes is the same as the sequence length. The expression `[0]*n` with n equal to 8 results in `[0, 0, 0, 0, 0, 0, 0, 0]`.

The first of two `for` loops used in the program fills the `profile` table by counting nucleotides in each position of each sequence of the alignment. It uses the `enumerate()` function to obtain the indices corresponding to the positions in a sequence. After this loop, the profile table will look like this:

```
{
    'A': [5, 1, 0, 0, 5, 5, 0, 0],
    'C': [0, 0, 1, 4, 2, 0, 6, 1],
    'T': [1, 5, 0, 0, 0, 1, 1, 6],
    'G': [1, 1, 6, 3, 0, 1, 0, 0]
}
```

This table shows that the first position of the sequences contains five A's, one G, and one T. The final paragraph of the program searches the most frequently occurring nucleotides in each position and builds a sequence consensus out of them. The program goes through all columns of the profile table. The line

```
col = [(profile[nt][i], nt) for nt in "ACGT"]
```

creates a list containing both the nucleotide and its count. For the first position, `col` would contain

```
[(5, 'A'), (0, 'C'), (1, 'G'), (1, 'T')]
```

This allows identification of the most frequent nucleotide by the `max()` function in the second-to-last line. The same could be achieved with the following code:

```
consensus = ""
for i in range(n):
    max_count = 0
    max_nt = 'x'
    for nt in "ACGT":
        if profile[nt][i] > max_count:
            max_count = profile[nt][i]
            max_nt = nt
    consensus + = max_nt
print consensus
```

Source: Adapted from code published by A.Via/K.Rother under the Python License.

Here, two for loops are used, and the respective, most frequent nucleotide is recorded in the max_nt variable.

Using the most frequent nucleotide has a disadvantage though; if two nucleotides have the same frequency, the one preceding in the alphabet is preferred. A balanced consensus would represent both letters with the highest frequency, for example, [AG] in a sequence pattern. This could be achieved with a minor change to the program.

Recipe 6: Calculating the Distance between Phylogenetic Tree Nodes

S EQUENCES FROM PROTEIN FAMILIES are frequently used to construct phylogenetic trees. This recipe presents a few ways to read and analyze a tree with Biopython.

The Newick format is a universal format for representing phylogenetic trees. It can be read by the majority of programs, including MrBayes, GARLI, PHYLIP, TREE-PUZZLE, and many others. Newick trees contain the name of species and ancestral nodes, their relationship, and the values assigned to each node (commonly used for the distance to the parent node). The following program loads a Newick tree and calculates the distance between two nodes (also called clades):

```
from Bio import Phylo
tree = Phylo.read("newick_small.txt", "newick")
Phylo.draw_ascii(tree)
a = tree.find_clades(name = 'Hadrurus_virgo').next()
b = tree.find_clades(name = 'Coleonyx_godlewskii').next()
ancestor = tree.common_ancestor(a, b)
print tree.distance(a, b)
print tree.distance(ancestor, a)
print tree.distance(ancestor, b)
```

Source: Adapted from code published by A.Via/K.Rother under the Python License.

PARSING NEWICK TREES

Biopython can easily parse a Newick file using the `Bio.Phylo` module:

```
from Bio import Phylo
tree = Phylo.read('newick_small.txt', 'newick')
```

The `Phylo.read()` function takes two parameters. First, the filename, and second, the format. Alternatively, you can parse a Newick tree from a string directly:

```
from cStringIO import StringIO
text = '(A:0.1, (B:0.2, C:0.3):0.1, (D:5, E:0.7):0.15)'
handle = StringIO(text)
tree = Phylo.read(handle, 'newick')
```

Parsing the tree results in a `Tree` object. The `Tree` object contains other objects for the nodes, which in turn may contain objects themselves so that the overall data structure is quite complex. Using a `Tree` object is made easier by a couple of methods, which will be explored in the following.

DISPLAYING A TREE

When you print a `Tree` object, you will see the objects contained within each other. To get a first overview of your tree, a minimal graphical representation may be more useful. Biopython can draw a tree to the text console:

```
Phylo.draw_ascii(tree)
```

More sophisticated graphics can be generated with Biopython 1.58 and higher versions when `matplotlib` is installed as well:

```
Phylo.draw(tree)
```

FINDING A SPECIES NODE

You can retrieve one particular species from the tree using the `find_clades()` function:

```
tree.find_clades(name = 'Hadrurus_virgo').next()
```

By default, find_clades() returns an iterator, because it is possible that several nodes with the same name exist. The next() call at the end of the line returns the first occurrence of the name. If no matching nodes are found, the line will result in an exception. If your tree file provides other annotation apart from the name, it is possible to search by that annotation using a different keyword argument than "name".

FINDING A COMMON ANCESTOR

When you have retrieved two nodes from the same tree, you can identify their common ancestor:

```
ancestor = tree.common_ancestor(a, b)
```

CALCULATING THE DISTANCE BETWEEN TWO TREE NODES

Finally, you can calculate the total distance between any two nodes using the distance() method of the Tree object:

```
print tree.distance(a, b)
print tree.distance(ancestor, a)
```

This works for both ancestral and terminal nodes. The respective distance values from the tree file are summed up along the tree.

More examples on using trees with Biopython can be found at http://biopython.org/wiki/Phylo and http://biopython.org/wiki/Phylo_cookbook.

Recipe 7: Codon Frequencies in a Nucleotide Sequence

C ALCULATING THE FREQUENCY of codon occurrences in a nucleotide sequence (e.g., the whole genome of an organism) may be very useful when you want to discover if some codons for a given amino acid have been preferred over others during evolution. Therefore, given a nucleotide sequence (RNA, in this example) and the list of 20 amino acids, you may be interested in the following kind of information:

```
AA        codon       hits         frequency
A         GCU         0            0.000
A         GCC         8            0.471
A         GCA         7            0.412
A         GCG         2            0.118
...
```

and so on for each amino acid. Here, for each amino acid, the frequency of a codon X (e.g., GCC) is measured as the number of occurrences (hits) of X divided by the total number of occurrences of the amino acid in the input sequence. In the previous example, the frequency of GCC is calculated as

$$freq(GCC) = 8.0/(0+8+7+2) = 0.471 \tag{1}$$

where 0, 8, 7, and 2 are the number of occurrences in the input sequence of GCU, GCC, GCA, and GCG, respectively. Remember that to perform this calculation in Python, you need to convert either the numerator or the denominator in (1) to a floating-point number in order to obtain a float value for freq(GCC).

In this example, two dictionaries are defined. One (aa_codon) collects the relationships between the 20 amino acids and their corresponding codons (aa_codon = {'A': ['GCU', 'GCC', 'GCA', 'GCG'], ...}), and the other (codon_count) has the codons as keys and integer numbers as values (codon_count = {'GCU':0, 'GCC':8, 'GCA':7, ...}). These numbers are the count of occurrences of each codon. Their values are initialized to 0. The number n of occurrences of a codon X is increased by one when X is encountered along the nucleotide sequence. When the codon_count dictionary has been filled with the numbers of all occurrences (i.e., when the sequence scanning is over), this is passed as argument to the calc_freq() function, which has two for loops, both running over the keys of the aa_codon dictionary.

In the first for loop, the total number of codon hits *for each amino acid* is calculated and converted into a floating number (in other words, the denominator of (1) is calculated). In the second for loop, the function calculates the frequency of each codon (as shown in (1) for the GCC codon).

The input sequence is read from a FASTA file, and the sequence is stored in a single string, which is read three times starting from the first, second, and third nucleotide, respectively. This makes it possible to determine the codon frequency for the three reading frames separately. If you want to calculate the codon frequency in a DNA sequence (instead of RNA), you have to replace Us with Ts in all the codons in both aa_codon and codon_count dictionaries. Here is the program:

```
# This program calculates the codon frequency in a RNA sequence
# (it could also be an entire genome)

aa_codon = {
'A':['GCU','GCC','GCA','GCG'], 'C':['UGU','UGC'],
'D':['GAU','GAC'],'E':['GAA','GAG'],'F':['UUU','UUC'],
'G':['GGU','GGC','GGA','GGG'], 'H':['CAU','CAC'],
'K':['AAA','AAG'],'I':['AUU','AUC','AUA','AUU','AUC','AUA'],
'L':['UUA','UUG','CUU','CUC','CUA','CUG'],'M':['AUG'],
'N':['AAU','AAC'],'P':['CCU','CCC','CCA','CCG'],
'Q':['CAA','CAG'],'R':['CGU','CGC','CGA','CGG','AGA','AGG'],
'S':['UCU','UCC','UCA','UCG','AGU','AGC',],
'Y':['UAU','UAC'],'T':['ACU','ACC','ACA','ACG'],
'V':['GUU','GUC','GUA','GUG'],'W':['UGG'],
'STOP':['UAG','UGA','UAA']}
```

```
codon_count = {
'GCU':0,'GCC':0,'GCA':0,'GCG':0,'CGU':0,'CGC':0,
'CGA':0,'CGG':0,'AGA':0,'AGG':0,'UCU':0,'UCC':0,
'UCA':0,'UCG':0,'AGU':0,'AGC':0,'AUU':0,'AUC':0,
'AUA':0,'AUU':0,'AUC':0,'AUA':0,'UUA':0,'UUG':0,
'CUU':0,'CUC':0,'CUA':0,'CUG':0,'GGU':0,'GGC':0,
'GGA':0,'GGG':0,'GUU':0,'GUC':0,'GUA':0,'GUG':0,
'ACU':0,'ACC':0,'ACA':0,'ACG':0,'CCU':0,'CCC':0,
'CCA':0,'CCG':0,'AAU':0,'AAC':0,'GAU':0,'GAC':0,
'UGU':0,'UGC':0,'CAA':0,'CAG':0,'GAA':0,'GAG':0,
'CAU':0,'CAC':0,'AAA':0,'AAG':0,'UUU':0,'UUC':0,
'UAU':0,'UAC':0,'AUG':0,'UGG':0,'UAG':0,
'UGA':0,'UAA':0}

# Writes the frequency of each codon to a file
def calc_freq(codon_count, out_file):
    count_tot = {}
    for aa in aa_codon.keys():
        n = 0
        for codon in aa_codon[aa]:
            n = n + codon_count[codon]
        count_tot[aa] = float(n)
    for aa in aa_codon.keys():
        for codon in aa_codon[aa]:
            if count_tot[aa] != 0.0:
                freq =  codon_count[codon] / count_tot[aa]
            else:
                freq = 0.0
            out_file.write('%4s\t%5s\t%4d\t%9.3f\n'% \
                (aa,codon,codon_count[codon], freq))

in_file = open('A06662.1.fasta')
out_file = open('CodonFrequency.txt', 'w')

# Reads the RNA sequence into a single string
rna = ''
for line in in_file:
    if not line[0] == '>':
        rna = rna + line.strip()

# Scans the sequence frame by frame,
# counts the number of occurrences
# of each codon, and stores it in codon_count dictionary.
# Then calls calc_freq()
for j in range(3):
    out_file.write('!!Codon frequency in frame %d\n' %(j+1))
    out_file.write('  AA\tcodon\thits\tfrequency\n')
    prot = ''
```

```
    for i in range(j, len(rna), 3):
        codon = rna[i:i + 3]
        if codon in codon_count:
            codon_count[codon] = codon_count[codon] + 1
    calc_freq(codon_count, out_file)

out_file.close()
```

Source: Adapted from code published by A.Via/K.Rother under the Python License.

Recipe 8: Parsing RNA 2D Structures in the Vienna Format

O<small>NE OF THE MOST</small> important properties of RNA sequences is their ability to fold and form base pairs. Unlike DNA, the base pairing in RNA is not restricted to double helices and can form complex structures. Predicting these base pairs, the secondary structure of RNA, is a common task in RNA bioinformatics. The Vienna package (http://rna.tbi.univie. ac.at/) is a collection of command-line and web tools for basic tasks, such as predicting base pairs from an RNA sequence [1]. These results are often represented in the Vienna format:

```
> two hairpin loops
AAACCCCGUUUCGGGGAACCACCA
((((...)))).((((..)).)).
```

Like in the FASTA format, the first line contains the name of the RNA sequence and the second the sequence itself. The third line contains the RNA secondary structure in dot-bracket notation. Corresponding brackets indicate a base pair, and unpaired bases are represented by dots. For instance, a helix with three pairs and a loop at the end would be (((....))). A helix interrupted by a bulge would be ((((....))..)) and so on.

This recipe explains how to read the Vienna format and extract the positions of all base pairs.

```
class RNAStructure:

    def __init__(self, vienna):
        lines = vienna.split('\n')
        self.name = lines[0].strip()
        self.sequence = lines[1].strip()
        self.basepairs = \
            sorted(self.parse_basepairs(lines[2].strip()))

    def parse_basepairs(self, dotbracket):
        stack = []
        for i, char in enumerate(dotbracket):
            if char == '(':
                stack.append(i)
            elif char == ')':
                j = stack.pop()
        yield j, i

vienna = '''> two hairpin loops
AAACCCCGUUUCGGGGAACCACCA
((((...)))).((((..)).)).
'''

rna = RNAStructure(vienna)
print rna.name
print rna.sequence
for base1, base2 in rna.basepairs:
    print '(%i, %i)'%(base1, base2)
```

Source: Adapted from code published by A.Via/K.Rother under the Python License.

PARSING THE VIENNA FORMAT USING A CLASS

In the first part, the program defines a class, RNAStructure, to structure the data read from the Vienna format. In the final paragraph, the program creates an instance of the RNAStructure class using a string in the Vienna format. Then, three attributes of the RNAStructure instance are printed: rna.name, rna.sequence, and a list of base pairs each having an opening and closing position.

The RNAStructure class has two methods. The constructor __ init__() sets the three attributes from separate lines of the Vienna string. The constructor also calls the second method, parse_basepairs(), to create a list of base pair indices.

PARSING THE BASE PAIRS

In the parse_basepairs() method, pairs of indices are generated for each base pair. The method returns a list of base pair indices. yield is a Python command that works like return, except the method will return an iterator. The method contains a for loop that goes through each character in a dot-bracket structure and analyzes them. Each time a pair of opening "(" and closing ")" brackets is found, its index and the index of the last open bracket are added to the list result as a tuple. When the entire dot-bracket structure has been processed and the end of the for loop is reached, the method also ends. In the constructor, the sorted() function sorts the resulting base pairs and stores them in the rna.basepairs list.

MATCHING BASE PAIRS USING A STACK

The main challenge in parsing base pairs is to find the corresponding pairs of brackets for each base pair. The stack list is used to find the matching brackets. Each time an opening bracket is encountered, its index i is appended to the stack list. Each time a closing bracket is encountered, the position of the last opening bracket is retrieved from that list by pop(). This way, the positions of opening brackets are returned in the reverse order than they were added. The pop() method always returns the last opening bracket for which no closing bracket has been found yet. Such a list to which elements are appended and then retrieved in reverse order is called a stack. A stack is a basic programming tool that is used in many algorithms.

REFERENCE

[1] A.R. Gruber, R. Lorenz, S.H. Bernhart, R. Neuböck, and I.L. Hofacker, "The Vienna RNA Website," *Nucleic Acids Research* 36 (2008): W70–W74.

Recipe 9: Parsing BLAST XML Output

W HEN RUNNING BLAST ON more than one sequence, you obtain a lot of result files. In that situation, you may save time by processing the output data with a program before you do a more thorough manual evaluation. In this recipe, we use BLAST+, which provides many output options. One of them is XML. When running BLAST from the command line, you can get XML output by adding the –outfmt 5 option. For example,

```
blastp -query P05480.fasta -db nr.00 -out blast_output.xml
-outfmt 5
```

XML is a structured format that is easy for computers to parse. Python has a standard module for parsing XML (see Recipe 10). Biopython offers a parser specific for the BLAST output, which translates output files into neat data structures.

This recipe illustrates how to use the Biopython BLAST XML parser to read an XML BLAST output file, filter hits by e-value, and print the corresponding alignment data.

```
xml_file = open("blast_output.xml")
blast_out = NCBIXML.parse(xml_file)

for record in blast_out:
    for alignment in record.alignments:
        print alignment.title
        for hsp in alignment.hsps:
            # filter by e-value
            if hsp.expect < 0.0001:
                print "score:", hsp.score
                print "query:", hsp.query
```

```
print "match:", hsp.match
print "sbjct:", hsp.sbjct
print '#' * 70
print
```

Source: Adapted from code published by A.Via/K.Rother under the Python License.

The NCBIXML.parse() function returns a blast_out instance that contains one-to-many record objects (BLAST results in one record, but, e.g., in PSI-BLAST there is one record per query). Each record consists of one-to-many hits or alignments, which in turn contain one-to-many local alignments (or HSPs). Once the XML output has been parsed, the program goes through these three levels of the hierarchy (records, alignments, HSPs) in three nested for loops. On each level, different attributes are available, as seen in the code. Each of them can be printed or used for filtering, such as the hsp.expect value.

Note that not all HSPs necessarily cover the length of the entire query sequence. The BLAST algorithm can produce both short, high-quality alignments and long, more permissive alignments, depending on the input parameters. Deciding which is better is not trivial, especially since the individual HSPs may complement each other when searching the sequence of a multi-domain protein. It is possible to write a program to construct a full alignment from the HSPs, but discussing the implementation is beyond the scope of this book.

In summary, the Bio.Blast module provides a data structure that allows you to vary the pattern in the recipe to selectively retrieve information from BLAST XML output files.

Recipe 10: Parsing SBML Files

THE SBML (Systems Biology Markup Language) FORMAT is a standard format for storing information about pathways, reactions, and regulatory networks. (See http://sbml.org for plenty of software listed to work with SBML files.) An SBML document annotating three metabolites could contain the following lines:

```
<?xml version="1.0" encoding="UTF-8"?>
<sbml xmlns="http://www.sbml.org/sbml/level2">

    <model name="SBML file with three metabolites">
    <listOfSpecies>
        <species id="M_m78" name="Inosine">
           <p>FORMULA: C10H12N4O5</p>
        </species>
        <species id="M_m79" name="Xanthosine">
           <p>FORMULA: C10H12N4O6</p>
        </species>
        <species id="M_m80" name="Xanthosine">
           <p>FORMULA: C10H12N4O6</p>
        </species>
    </listOfSpecies>

</model></sbml>
```

SBML is an XML dialect. It contains hierarchically organized tags with a special meaning, e.g., everything between a `<species>` tag and a `</species>` tag belongs together. The first line of the file starting with `<?xml.. >` identifies this as an XML document. The first tag after the XML definition, also called the *root node* (starting with `<sbml...>`), specifies the kind of document used. For instance, in SBML it is specifically defined which tag names and attributes are allowed (these definitions can be found at http://www.sbml.org/sbml/level2, the web address in the second line). XML files may also contain comments, starting with <!-- and ending with -->. The XML example contains three compounds (called *species* in SBML), each identified by a *name* (name), an *ID* (id), and a *chemical formula* (FORMULA). The SBML document has been simplified for this book. Usually, SBML stretches over dozens of screen pages and contains many different kinds of annotation. The following example program reads the SBML file, extracts all chemical compounds, and prints their ID, name, and chemical formula as a tab-separated table.

```python
from xml.dom.minidom import parse
document = parse('sbml_example.xml')
species_list = document.getElementsByTagName("species")

for species in species_list:
    species_id = species.getAttribute('id')
    name = species.getAttribute('name')
    p_list = species.getElementsByTagName("p")
    p = p_list[0]
    text = p.childNodes[0]
    formula = text.nodeValue
    print "%-20s\t%5s\t%s"%(name, species_id, formula)
```

Source: Adapted from code published by A.Via/K.Rother under the Python License.

The program uses the `minidom` XML Python parser. The `xml.dom.minidom.parse` function returns a document object that represents the entire tree of SBML tags translated into Python. The program reads the SBML file, extracts all <species> tags using the `document.getElementsByTagName()` method, and stores them in `species_list`. The `for` loop then goes through all <species> tags. For each species, the *name* and *ID* fields are extracted by `species.getAttribute()`. For parsing the chemical formula, it is necessary to parse the <p> tag within the <species> tags as well. The command `species.getElementsByTagName("p")` only returns the <p> tag within the given species. In the same way, multiple levels of the tag hierarchy can be analyzed. The <p> tag contains a single child node, `p.childNodes[0]`, which contains the text with the chemical formula.

To format and print the name, ID, and formula, string formatting parameters are used in the last line of the program. Note that the %-20s causes the text to be left-justified.

The output of the program looks like this:

```
Inosine                  M_m78 FORMULA: C10H12N4O5
Xanthosine               M_m79 FORMULA: C10H12N4O6
Xanthosine               M_m80 FORMULA: C10H12N4O6
```

FUNCTIONS FOR PARSING XML IN PYTHON

The `xml.dom.minidom` module in Python parses all kinds of XML files. The most important functions used in the previous recipe include the following:

1. *Reading a document from an XML file.* The `parse()` function reads and parses an XML file and returns an object corresponding to the root node of the XML tree.

   ```
   document = parse('sbml_example.xml')
   ```

2. *Getting all tags with a particular name.* From any given XML tag object, you can get a list of all tags with a given name with the `getElementsByTagName()` function. The function returns a list of tags. It does not matter whether the tags are directly contained or whether they are sub-sub-tags. For instance, you can extract all <species> tags from the document without retrieving the <listOfSpecies> tag first:

   ```
   species_list = document.getElementsByTagName('species')
   ```

   Also, the `getElementsByTagName()` function is used in the program to find the <p> tag within the <species> tag.

3. *Getting child tags from a tag.* Each tag contains a list of child tags you can use to access the children directly by its `childNodes` attribute.

   If you know that there is only one child, you get it using the [0] index:

   ```
   p = species.childNodes[0]
   ```

4. *Extracting attributes from a tag.* To access the data within a tag, you can use the getAttribute() function:

```
species_id = species.getAttribute('id')
name = species.getAttribute('name')
```

5. *Extracting text values from a tag.* Extracting the text is a little more complicated, because minidom parses the text as the value of an extra child node. The text node works like a regular tag, but it is invisible in the XML file. The nodeValue attribute gives the text that is inside the text node:

```
text = p.childNodes[0]
formula = text.nodeValue
```

Recipe 11:
Running BLAST

Y OU HAVE FIVE BASIC ways to run BLAST, four of which are as follows: (1) locally from the shell command line, (2) locally from a Python script or interactive Python session, (3) locally using Biopython, and (4) through the NCBI web server using Biopython locally. The fifth way, using your browser and the BLAST web page, will not be described here. The other four methods have the advantage of being automatized easily (e.g., if you want to run BLAST for a whole set of query sequences).

What are the advantages of running BLAST locally? First of all, you can search a query sequence in a customized database, e.g., in a newly sequenced genome you are studying or a set of protein sequences of your interest (e.g., only protein kinases). Second, you may want to insert the program in a pipeline, e.g., to search a large number of query sequences instead of a single one. For example, you might wish for each of them to retain and parse the BLAST output only under certain conditions (e.g., if you have at least a match with >50% sequence identity to the query). Finally, only by running BLAST locally will you have full control over the sequence database and, by that, reproducibility of your search.

To run BLAST locally (either from the shell or from a script), you have to download and install the BLAST+ package (http://blast.ncbi. nlm.nih.gov/Blast.cgi?CMD=Web&PAGE_TYPE=BlastDocs&DOC_TYPE=Download). BLAST+ is a new suite of BLAST tools that utilizes the NCBI C++ Toolkit. You can find instructions on how to download and install the package at http://www.ncbi.nlm.nih.gov/books/NBK1762/.

Once the downloaded files are unpacked and the package is installed, you will have to set up two environment variables in order to tell your system where to look for the installed BLAST programs and inform the

BLAST programs in which directory to search for the databases: PATH and BLASTDB. As for the former, you will have to add the path of the BLAST bin directory to the PATH environment variable of your computer (see Appendix D, Section D.3.5). Otherwise, you have to change to the BLAST bin directory on the terminal shell and run BLAST from there.

Q & A: How Can I Know Which Is the Path of the BLAST bin Directory?

If you install the BLAST program from source, you will have to place the downloaded package under a desired directory, e.g., /home/john. When you unpack the package, a BLAST directory will appear in /home/john (e.g., /home/john/ncbi-blast-2.2.23+). You have to add to the PATH environment variable the bin directory under this BLAST directory. In order to do this, under the bash shell you can use the command line:

```
PATH:/home/john/blast-2.2.23+/bin
export PATH
```

Under the tcsh shell:

```
setenv PATH ${PATH}:/home/john/ncbi-blast-2.2.23+/bin
```

If you want to permanently modify the PATH variable, you can add this (these) line(s) to the .bash_profile or .cshrc sturtup files in your home directory, respectively, and restart your computer.
Notice that when you use the dmg disk to install BLAST+ on Mac OS X (10.4 or higher), all BLAST+ programs will be installed under /usr/local/ncbi/blast/bin.

Then you have to modify the BLASTDB environment variable. In order to do this, you have to create the BLAST database directory /blast/db in your home directory:

```
mkdir /home/john/blast/db
```

This is the directory where you will put all the databases (either downloaded from the BLAST website or your custom ones) that you will use with BLAST.

Create a .ncbirc text file in your home directory having the following path specification:

```
; Start the section for BLAST configuration
[BLAST]
; Specifies the path where BLAST databases are installed
BLASTDB=/home/john/blast/db
```

The semicolon at the beginning of the first and third lines indicates a comment.

Save and exit the file. Save at least a database in `/home/john/blast/db`. If you want to download a database from NCBI, go to ftp://ftp.ncbi.nlm.nih.gov/blast/db. Now you should be ready to run BLAST.

RUNNING BLAST LOCALLY FROM THE SHELL COMMAND LINE

You have to choose the alignment program from the BLAST package you want to run. Depending on the type of query and target sequences (nucleotide or protein) and type of search (one sequence against several sequences, or pairwise), you will have different programs and/or options from which to choose. You can find a detailed user manual at http://www.ncbi.nlm.nih.gov/books/NBK1763/. Unless you use a preformatted database downloaded from the NCBI ftp site, you will need to format your custom sequence file. To this aim, the BLAST package provides the `makeblastdb` application, which produces BLAST databases from FASTA files:

```
makeblastdb -in mygenome.fasta -parse_seqids -dbtype prot
```

`-in` is the option for the input file, `-parse_seqids` enables parsing of sequence ids, and `-dbtype` specifies the type of input molecules (`nucl` or `prot`). Notice that `–parse_seqids` will be able to parse the sequence identifiers if they start immediately after the ">" and follow the format described in http://www.ncbi.nlm.nih.go v/books/NBK7183/?rendertype =table&id=ch_demo.T5.

The query sequence can be in FASTA format and this is the structure of the command line:

```
blastProgram -query InputSeq.fasta -db Database -out OutFile
```

For example,

```
blastp -query P05480.fasta -db nr -out blast_output
```

`blastp` aligns protein sequences, `nr` is the name of the BLAST-formatted database, `blast_output` is the name you have chosen for the BLAST output file, and `P05480.fasta` is a file that contains your query sequence in FASTA format. There are several other options you can add to this command line (e.g., if you want to set a threshold on the BLAST e-value). You can obtain a list of the available options by simply typing the program name followed by –h (usage and description) or –help (usage, description, and description of arguments):

```
blastp -help
```

RUNNING BLAST LOCALLY FROM A PYTHON SCRIPT OR INTERACTIVE SESSION

To run BLAST from a Python script, you can translate the previous command to Python code. In this case, you just have to pass the BLAST shell command line as string argument of the system method of the os module, after having imported it:

```
import os
cmd = "blastp -query P05480.fasta -db nr.00 -out blast_output"
os.system(cmd)
```

In this example, the nr.00 database downloaded from the NCBI FTP site is used. To work more flexibly with BLAST via the os.system() method, you can concatenate the command string using variables for the filenames or for other parameters.

RUNNING BLAST LOCALLY USING BIOPYTHON

You can achieve the same results with Biopython. The only difference is that Biopython provides functions to create suitable command lines. Then you have to use os.system() anyway:

```
from Bio.Blast.Applications import NcbiblastpCommandline
import os
comm_line = NcbiblastpCommandline(query = \
    "P05480.fasta", db = "nr.00", out = "Blast.out")
print comm_line
os.system(str(comm_line))
```

Source: Adapted from code published by A.Via/K.Rother under the Python License.

The advantage of this method is that the keyword parameters in NcbiblastpCommandline are a little more explicit than in the concise command previously used (this may make it easier to find out if you misspelled an argument). For example, if you want the output in XML format you can define comm_line as follows:

```
comm_line = NcbiblastpCommandline(query= \
    "P05480.fasta", db="nr.00", out="Blast.xml", \
    outfmt=5)
```

The actual choice may depend on your personal preference, which is why we explain both versions here.

Notice that, the object to import in the case you want to run BLAST+ with nucleotide sequences is NcbiblastnCommandline. For example:

```
cline = NcbiblastnCommandline(query="my_gene.fasta", \
    db="nt", strand="plus", evalue=0.001, \
    out="Blast.xml", outfmt=5)
```

RUNNING BLAST THROUGH THE WEB USING BIOPYTHON

```
from Bio.Blast import NCBIWWW
BlastResult_handle = NCBIWWW.qblast("blastp","nr"," P05480")
BlastOut = open("P05480_blastp.out", "w")
BlastOut.write(BlastResult_handle.read())
BlastOut.close()
BlastResult_handle.close()
```

Source: Adapted from code published by A.Via/K.Rother under the Python License.

When running BLAST through the web, Biopython sends a query to the NCBI server and retrieves the result. The qblast() method of the NCBIWWW module of Bio.Blast runs the BLAST program by taking BLAST parameters, the input sequence, and the target data set as arguments. It returns a "handle" (BlastResult_handle) that can be read like a file using the method read() and finally written to an output file specifically opened in "w" mode. The output file is in XML format, which can be parsed using the BLAST output parser (see Recipe 9). Please note that the maintainers of the BLAST server expect you to use their service reasonably. If you are planning to run hundreds of queries, please switch to the local version in order to avoid blocking the server (in which case, the server may end up blocking your queries).

Recipe 12: Accessing, Downloading, and Reading Web Pages

INSTEAD OF ACCESSING REMOTE web pages through a browser (e.g., Internet Explorer, Firefox, Safari, etc.), you can do the same from a program. This can be useful if you want to read data from many pages or want to extract information automatically. Such a program will have to execute the following actions: (1) connect to the server hosting the web page, (2) access the web page, and (3) download the content of the web page.

To this aim, Python provides two modules dealing with URL functionalities: `urllib` and `urllib2`. These two modules differ in some aspects, as explained next.

USING URLLIB

If you just want to access a static (e.g., HTML) web page and download its source code, you can use the `urlopen()` method of `urllib`. It allows you to read a web page in a way similar to the way you would read a file:

```
from urllib import urlopen
url = urlopen('http://www.uniprot.org/uniprot/P01308.fasta')
doc = url.read()

print doc
```

Source: Adapted from code published by A.Via/K.Rother under the Python License.

This code produces the output:

```
>sp|P01308|INS_HUMAN Insulin OS = Homo sapiens GN = INS PE = 1 SV = 1
MALWMRLLPLLALLALWGPDPAAAFVNQHLCGSHLVEALYLVCGERGFFYTPKTRREAEDLQ
VGQVELGGGPGAGSLQPLALEGSLQKRGIVEQCCTSICSLYQLENYCN
```

In this example, the `urllib.urlopen()` function accesses the URL (given as an argument to the function) and returns a handle, which can be read with the `read()` method, stored into a variable, and printed or saved to a file. You will have noticed that the web page (the FASTA sequence of human insulin) looks nicely formatted. This is because the URL provided corresponds to a text file, and not to an HTML file. Therefore, when you print the content of this web page, you won't see any HTML tags. If you try to access and print the content of, for example, www.uniprot.org/uniprot/P01308, you will see that the output looks much less easy to read. In such cases, you have to save the web page content to a variable or a file and then use an HTML parser in order to extract the information you need. One way to do it is by means of regular expressions (see Chapter 9). Another is to use a specialized HTML parser such as the one described in Recipe 13.

USING URLLIB2

You can do the same with `urllib2`:

```
import urllib2
url = urllib2.urlopen('http://www.uniprot.org/ \
uniprot/P01308.fasta')
doc = url.read()
print doc
```

Source: Adapted from code published by A.Via/K.Rother under the Python License.

This code writes:

```
>sp|P01308|INS_HUMAN Insulin OS = Homo sapiens GN = INS PE = 1 SV = 1
MALWMRLLPLLALLALWGPDPAAAFVNQHLCGSHLVEALYLVCGERGFFYTPKTRREAED
LQVGQVELGGGPGAGSLQPLALEGSLQKRGIVEQCCTSICSLYQLENYCN
```

The difference between the `urllib` and the `urllib2` modules is that whereas `urllib.urlopen()` accepts only URLs (i.e., web addresses), `urllib2.urlopen()` also accepts `Request` objects, which make it possible to send data to a server. When is this useful? Using `Request` objects allows you to fill in web forms automatically. For example, if you want to search Uniprot by keyword, you can send the keyword to the Uniprot search page and get back the web page created by your query. `Request` objects are created taking two arguments: a URL and a dictionary of data, which need to be suitably encoded using the `urllib.urlencode()` function to be sent.

Recipe 13: Parsing HTML Files

FOR PARSING HTML DOCUMENTS, Python provides the HTMLParser class that you can find in the HTMLParser module (yes, the same name). The class has a method feed(), which parses an HTML string passed as an argument and calls custom methods to take care of individual tags. To use the HTMLParser class in your programs, you need to create a subclass from it. In the following example, the MyHTMLParser subclass automatically receives the feed() method from its parent class but adds several methods to process the data.

```python
from HTMLParser import HTMLParser
import urllib

class MyHTMLParser(HTMLParser):

    def handle_starttag(self, tag, attrs):
        self.start_tag = tag
        print "Start tag:", self.start_tag

    def handle_endtag(self, tag):
        self.end_tag = tag
        print "End tag :", self.end_tag

    def handle_data(self, data):
        self.data = data.strip()
        if self.data:
            print "Data :", self.data

parser = MyHTMLParser()
url = 'http://www.ncbi.nlm.nih.gov/pubmed/21998472'
page = urllib.urlopen(url)
data = page.read()
parser.feed(data)
```

Source: Adapted from code published by A.Via/K.Rother under the Python License.

In the last paragraph of this example, an instance of `MyHTMLParser` is created and an HTML page is read from the web and passed as an argument to `feed()`. The page is retrieved from a URL using the `urlopen()` method of the `urllib` module (see Recipe 12), but you could read it from a local file. In the latter case, you can write

```
page = open(filename)
data = page.read()
parser.feed(data)
```

In any case, no changes to the `MyHTMLParser` class are required.

When you call the `feed()` method, it automatically calls three methods for handling the HTML elements: `handle_starttag()`, `handle_endtag()`, and `handle_data()`. These HTML handlers, which have been inherited by `MyHTMLParser` from `HTMLParser`, are able to retrieve tags and tag contents, and are subsequently customized in order to print a sentence followed by the retrieved tag (or tag content). The `feed()` method expects methods with such names to be there. The `handle_starttag(self,tag,attrs)` method is called whenever an opening tag (like `<body>`) is encountered in the HTML document. The `tag` argument is the start tag and `attrs` is a list of (name, value) pairs containing the attributes found inside the tag brackets. The `handle_endtag(self,tag)` method is analogously called for closing tags (e.g., `</body>`), and `handle_data(self,data)` handles data between start and end tags (e.g., plain text). Additional methods are available, such as `handle_comment(data)`, which handles comments. Notice that in `handle_data(self,data)`, the `data` string has been stripped in order to obtain an empty string in the case that `data` was made of several blank spaces. This makes it possible to print `data` only if it contains at least one nonblank character.

For example, for the `<title>` tag, the previous program will print

```
Start tag: title
Data : Human genetic variation is associated with Plas... [J
Infect Dis. 2011] - PubMed - NCBI
End tag : title
```

The strength of the `HTMLParser` is that you do not have to worry about recognizing tags and data in the HTML document. The `feed()`

method takes care of that and calls the other methods automatically. The way the methods of `MyHTMLParser` are called may seem unintuitive to you at first, because you do not see any explicit method call in the code. The example program prints the encountered opening and closing tags and data so that you can follow the order in which the methods are called. You can use the code above to craft your own HTML parsers.

A simple way to save to a file the output of the previous program is to redirect it using the ">" symbol in the terminal shell when you run the program:

```
python myParser.py > HtmlOut.txt
```

Recipe 14: Splitting a PDB File into PDB Chain Files

I F YOU WORK WITH macromolecular structures, you may need to split a multichain PDB file into separate files, each with the coordinates of a single polypeptide chain. In this recipe a PDB file (e.g., 2H8L.pdb) is read line by line and written to a series of output files. Whenever a new chain starts, a new file is opened and all the following lines are written to that file until a new chain identifier (ID) is encountered. The output chain file-names will be of the form: '2H8LA.pdb', '2H8LB.pdb', and so on. That is, they have the same name as the original PDB file except for the chain ID (which is concatenated to the name).

Instead of parsing the entire PDB file with the Bio.PDB module (see Chapter 21), the unpack() method of the struct module is used to convert PDB ATOM lines into tuples from which the chain identifier can easily be extracted. This approach generally is useful for parsing PDB-formatted lines. A complete description of the PDB format for each record line type can be found at www.wwpdb.org/docs.html. See also Box 10.1 in Chapter 10.

To split the chains, the script tracks the current chain ID in the chain_old variable. When the chain ID changes, it means that the reading of the coordinates of one chain is over, the current file can be closed, and a new output file for the new chain can be opened. After the last chain ID, the last chain file must be closed after the for loop is exited. Notice that a trick has been used: the variable chain_old is initialized to the "@" value, which is never found in a PDB file (no chain has identifier "@"). This ensures that when the first ATOM line is read, the "if chain != chain_old:" condition is definitely met, and a new output file is opened for the first chain.

```
from struct import unpack
import os.path

filename = '2H8L.pdb'
in_file = open(filename)
pdb_id = filename.split('.')[0]
pdb_format = '6s5s1s4s1s3s1s1s4s1s3s8s8s8s6s6s6s4s2s3s'
chain_old = '@'
for line in in_file:
    if line[0:4] == "ATOM":
        col = unpack(pdb_format, line)
        chain = col[7].strip()
        if chain != chain_old:
            if os.path.exists(pdb_id+chain_old+'.pdb'):
                chain_file.close()
                print "closed:", pdb_id+chain_old+'.pdb'
            chain_file = open(pdb_id+chain+'.pdb','w')
            chain_file.write(line)
            chain_old = chain
        else:
            chain_file.write(line)
chain_file.close()
print "closed:", pdb_id+chain_old+'.pdb'
```

Source: Adapted from code published by A.Via/K.Rother under the Python License.

You can also write this code in the body of a function split_chains(filename), taking the PDB filename as argument. This would be very useful if you need to split several PDB files into chains.

Recipe 15: Finding the Two Closest Cα Atoms in a PDB Structure

T**HE SCRIPT IN THIS** recipe reads a PDB file (2H8L.pdb, in this case), and finds the two residues with the closest pair of Cα atoms. The script first extracts the residue number and the coordinates of the Cα (CA) atoms for all the residues of an input chain. Then it calculates the distance between all pairs of CA atoms belonging to different residues and retains only the two residues with the smallest distance between their CA atoms.

The calc_dist(p1, p2) function calculates the distance between two points p1, p2 represented in the form of tuples as p1 = (x1, y1, z1) and p2 = (x2, y2, z2). The distance is calculated on the basis of the Pythagorean theorem as the square root of the sum of the squared differences of the x-, y-, and z-coordinates of each point. To this aim, the sqrt() function from the math module is used.

The min_dist(arglist) function finds the closest CA atoms in a PDB chain. It takes as input a list of tuples, where each tuple contains the x-, y-, and z-coordinates of a CA atom as the second, third, and fourth elements of the tuple, and the corresponding residue type and number as the first element of the tuple. Then, for each pair of CA atoms, it calls the calc_dist(atom1, atom2) function to calculate their distance.

If the distance of a CA atom pair is smaller than the distance of the previous pair, the pair is retained; otherwise it is skipped. By iterating over all pairs, the last pair retained will be the one separated by the smallest distance.

The get_list_ca_atoms(PDB) function reads the input file and builds a tuple (residue type+number, x, y, z) for each CA atom of chain A. As in Recipes 14 and 16, the unpack() method of the struct module is used to convert PDB ATOM lines into a tuple from which the residue type, residue number, and (*x*, *y*, *z*) coordinates can be easily identified (see Recipe 16 for a more accurate description of the tuple elements). List comprehension (see Chapter 4) is used to strip the content of each column returned by the unpack() method.

```python
from math import sqrt
from struct import unpack

def calc_dist(p1, p2):
    '''returns the distance between two 3D points'''
    tmp = pow(p1[0] - p2[0], 2) + \
          pow(p1[1] - p2[1], 2) + \
          pow(p1[2] - p2[2], 2)
    tmp = sqrt(tmp)
    return tmp

def min_dist(arglist):
    '''
    returns the closest residue pair and their
    CA_CA distance
    '''
    # initialize variables
    maxval = 10000
    residue_pair = ()
    # read arglist starting from the 1st position
    for i in range(len(arglist)):
        # save x,y,z coordinates from the arglist
        # i-element into the atom1 variable
        atom1 = arglist[i][1:]
        # run over all other elements
        for j in range(i + 1, len(arglist)):
            atom2 = arglist[j][1:]
            # calculate the distance
            tmp = calc_dist(atom1, atom2)
            # check if the distance is lower than
            # the previously recorded lowest value
            if tmp < maxval :
                # save the new data
                residue_pair = (arglist[i][0], \
                                arglist[j][0])
                maxval = tmp
    return residue_pair, maxval
```

```python
def get_list_ca_atoms(pdb_file, chain):
    '''
    returns a list of CA atoms, the residues
    they belong to, and their x,y,z coordinates
    from the input PDB file
    '''
    in_file = open(pdb_file)
    CA_list = []
    pdb_format = '6s5s1s4s1s3s1s1s4s1s3s8s8s8s6s6s6s4s2s3s'
    for line in in_file:
        tmp = unpack(pdb_format, line)
        tmp = [i.strip() for i in tmp]
        # only save CA coords belonging to input chain
        if tmp[0] =="ATOM" and tmp[7] == chain and \
           tmp[3] == "CA":
            # create a tuple (aa_number, x, y, x)
            tmp = (tmp[5]+tmp[8], float(tmp[11]), \
            float(tmp[12]), float(tmp[13]))
            # add the tuple to the list
            CA_list.append(tmp)
    in_file.close()
    return CA_list

# obtain the list of CA atoms of Chain A
CA_list = get_list_ca_atoms("2H8L.pdb", "A")
# identify the closest atoms
res_pair, dist = min_dist(CA_list)
print 'The distance between', res_pair, 'is:', dist
```

Source: Adapted from code published by A.Via/K.Rother under the Python License.

Recipe 16: Extracting the Interface between Two PDB Chains

I N A 3D STRUCTURE, the interface between two polypeptide chains can be defined as all pairs of residues, of which one belongs to the first chain and the other to the second chain, at a distance threshold (e.g., distance between Cα atoms < 6.0 Å). This recipe shows how to determine such a set of residue pairs both without and with Biopython.

WITHOUT BIOPYTHON

The PDB structure is parsed by means of the unpack method of the struct module (see also Recipes 14 and 15). This makes it possible to convert, on the basis of the format expressed by the pdb_format variable, ATOM lines from a PDB file into tuples of elements, each containing a piece of information of the atom (name, serial number, etc). Element 3 corresponds to the atom name, Element 7 to the chain ID, and Elements 11, 12, and 13 to the atom (x, y, z) coordinates. The amino acid type and number can be obtained by concatenating Elements 5 and 8.

Two lists corresponding to the two protein chains, e.g., A and B, are then built, each containing tuples in the form (amino_acid, x, y, z), where the amino_acid variable is the concatenation of the amino acid name and the amino acid number and x, y, and z are the spatial coordinates of its CA. Then the distance between all pairs of CA, one from list A and one from list B, is calculated. When the distance is smaller than 6.0 Å, the pair is retained; otherwise it is ignored.

```
import struct
from math import sqrt

def calcDist(p1, p2):
    tmp = pow(p1[0]-p2[0], 2) + pow(p1[1]-p2[1], 2) + \
        pow(p1[2]-p2[2], 2)
    tmp = sqrt(tmp)
    return tmp

def getInterface(filename, chain1, chain2):
    in_file = open(filename)
    pdb_format = '6s5s1s4s1s3s1s1s4s1s3s8s8s8s6s6s6s4s2s3s'
    A, B, result = [], [], []
    for line in in_file:
        if line[0:4] == "ATOM":
            col = struct.unpack(pdb_format, line)
            a_name = col[3].strip()
            chain = col[7].strip()
            amino_numer = col[5].strip() + col[8].strip()
            x = col[11].strip()
            y = col[12].strip()
            z = col[13].strip()
            if a_name == "CA":
                if chain == chain1:
                    A.append((amino_numer, x, y, z))
                if chain == chain2:
                    B.append((amino_numer, x, y, z))
    #calculate pairs of atoms with distance < 6
    for i in range(len(A)):
        for j in range(len(B)):
            v1 = (float(A[i][1]), float(A[i][2]), float(A[i][3]))
            v2 = (float(B[j][1]), float(B[j][2]), float(B[j][3]))
            tmp = calcDist(v1, v2)
            if tmp < 6:
                result.append((A[i][0], B[j][0], tmp))
    return result

print getInterface("2H8L.pdb", "A", "B")
```

Source: Adapted from code published by A.Via/K.Rother under the Python License.

WITH BIOPYTHON

The same task can be accomplished more easily using Biopython. Here, the distance between two atoms is calculated using the "−" operator, as shown in Example 21.1.

```
from Bio import PDB
parser = PDB.PDBParser()
s = parser.get_structure("2H8L","2H8L.pdb")
first_model = s[0]
chain_A = first_model["A"]
chain_B = first_model["B"]
for res1 in chain_A:
    for res2 in chain_B:
        d = res1["CA"]-res2["CA"]
        if d <= 6.0:
            print res1.resname,res1.get_id()[1], res2.resname,\
                res2.get_id()[1], d
```

Source: Adapted from code published by A.Via/K.Rother under the Python License.

Recipe 17: Building Homology Models Using Modeller

M ODELLER [1,2] IS A package designed to build homology models of protein three-dimensional structures. It is written in Python and can be downloaded from http://salilab.org/modeller/download_installation. html. The package consists of a number of Python scripts that you have to modify by adding the name and location of a target-template alignment file, of a target sequence file, and of a template PDB file. Also, you can set modeling parameters and add or remove specific instructions, depending on what you want to do. Once you have downloaded Modeller, you can copy a default Modeller Python script to a working directory where you want to generate your models and then open the script with a text editor and modify it. Here, a simple example of a Modeller script that builds a three-dimensional model from a single template structure (1eq9A) and a target-template alignment (`alignment.ali`) is displayed and discussed:

```
from modeller import *
from modeller.automodel import *
log.verbose()
env = environ()
env.io.atom_files_directory = ['.', '../atom_files']

a = automodel(env,
            alnfile = 'alignment.ali',
            knowns = '1eq9A',
            sequence = 'MyTarget_Seq')
```

```
a.starting_model = 1
a.ending_model = 1

a.make()
```

Source: Adapted from code published by A.Via/K.Rother under the Python License.

In the first two rows, all standard `modeller` classes are imported, as well as the `automodel` class. The `automodel` class is needed to build a "model object." `log.verbose()` makes the program produce more detailed output and `environ()` creates an environment (env) that is required to build models. The `env.io.atom_files_directory` variable is set to a list of directory path(s) where the program will look for the PDB coordinate file(s) of the template(s). The PDB file of the template must be saved in at least one of the directories specified in the `env.io.atom_files_directory` variable. In the example, the first element of this list indicates the current directory (`.`), the second element is a directory (`atom_files`) that is located in the parent directory of the current one (`../`). In the subsequent lines, input files and parameters are set. Modeller needs to know the filename of the target-template alignment file (`alnfile = 'alignment.ali'`), the target sequence filename (`sequence = 'MyTarget_Seq'`), and the template structure filename(s) (`knowns = '1eq9A'`). The `alignment.ali` file must be saved in the directory where you run the script. These files, plus the environment variable (`env`), are used to initialize a "model object" by creating an instance of the `automodel` class. At this stage, no models have been built yet.

The `a.starting_model` and `a.ending_model` attributes of the "model object" are then set to the number of models that will be generated. They define the indices of the first and last models. In the example, only one model is generated, and therefore, both variables are set to 1. Once all parameters are set, `a.make()` is the method that actually does the job.

Templates for building homology models can be identified using, for example, HHpred (http://toolkit.tuebingen.mpg.de/hhpred), which also creates target-template alignments. The HHpred alignment file format (PIR format, file extension `.ali`) is accepted by Modeller. Here is one example of such a file (where the omitted sequences have been replaced by three dots) for this recipe:

```
>P1;MyTarget_Seq
sequence:MyTarget_Seq: 1: : 230: :: : 0.00: 0.00
IIGGTDVEDGKAPYLAGLVYNNSATYCGGEEHV...NSDH*
>P1;1eq9A
structureX:1eq9A: :A: :A:Chymotrypsin; FIRE ANT, serine
   proteinase, hydrolase; HET PMS; 1.70A {Solenopsis invicta}
   SCOP b.47.1.2:Solenopsis invicta:1.70:0.26
IVGGKDAPVGKYPYQVSLRLS-GSHRCGASILD...----*
```

REFERENCES

[1] A. Sali and T.L. Blundell, "Comparative Protein Modelling by Satisfaction of Spatial Restraints," *Journal of Molecular Biology* 234 (1993): 779–815.

[2] N. Eswar, M.A. Marti-Renom, B. Webb, M.S. Madhusudhan, D. Eramian, M. Shen, U. Pieper, and A. Sali, "Comparative Protein Structure Modeling with MODELLER," *Current Protocols in Bioinformatics* Supplement 15 (2006): 5.6.1–5.6.30.

Recipe 18: RNA 3D Homology Modeling with ModeRNA

MODERNA [1] (HTTP://IIMCB.GENESILICO.PL/MODERNA/) IS a Python library for analyzing, manipulating, and modeling RNA 3D structures, similar to Modeller for proteins (see Recipe 17). This recipe uses a few of the more than 30 functions in ModeRNA to remodel a short fragment from a tRNA structure. To get started, you need to download and install ModeRNA from its web page and download the structure of Phenylalanyl-tRNA from yeast (PDBID: 1EHZ) from the Protein Data Bank (http://rcsb.org/). When in doubt, the ModeRNA website provides more detailed documentation.

This is the code to build the model:

```
from moderna import *

ehz = load_model('1ehz.ent', 'A')
clean_structure(ehz)
print get_sequence(ehz)
print get_secstruc(ehz)

m = create_model()
copy_some_residues(ehz['1':'15'], m)
write_model(m, '1ehz_15r.ent')

temp = load_template('1ehz_15r.ent')

ali = load_alignment('''> model sequence
ACUGUGAYUA[UACCU#PG
> template: first 15 bases from 1ehz
GCGGA--UUUALCUCAG''')
```

```
model = create_model(temp, ali)
print get_sequence(model)
```

Source: Adapted from code published by A.Via/K.Rother under the Python License.

When you execute this script, the following output is generated (within 1–2 minutes):

```
GCGGAUUUALCUCAGDDGGGAGAGCRCCAGABU#AAYAP?UGGAG7UC?UGUGTPCG"UCC
    ACAGAAUUCGCACCA
((((((((..((((........))))).((((...........))))).....
    (((((.......))))))))))))))....
ACUGUGAYUA[UACCU#PG
```

In the following, the program is explained line by line.

The `import` statement in the first line imports the main functions of ModeRNA, which are used in the following commands. Then, the tRNA structure is loaded into Python. The `load_model()` command reads chain A from the RNA structure file `'1ehz.ent'` and saves it in the variable `ehz`. The object returned by the `load_model()` function provides a simplified version of the `Bio.PDB` library that can be thus addressed in a shorter way. The `clean_structure()` function removes from the structure water molecules, ions, and some other problematic atoms that would conflict with the modeling, while `get_sequence()` returns the RNA sequence. Note that the printed sequence contains modified bases, indicated by characters other than ACGU; `get_secstruc()` returns base pairs in dot-bracket notation (see Recipe 8).

In the next paragraph of the program, the first 15 residues are extracted from the RNA chain and written to a separate file. This is achieved using three functions: `create_model()`, `copy_some_residues()`, and `write_model()`. The `create_model()` function creates an empty RNA model. Next, the residues numbered from 1 to 15 are copied to that model and finally written to a file by the `write_model()` function.

To build a model, similar to Modeller (see Recipe 17), ModeRNA needs two things: the template structure and a target-template pairwise sequence alignment. As a template, the 15-residue file created in the previous step is used. It is loaded by the `load_template()` function. The target-template alignment is given as a FASTA string to the `load_alignment()` function, but reading the alignment from a file would work as well. In this case, the argument of the `load_alignment()` function will be the alignment filename. Note that the second sequence in the alignment must be identical to the sequence of the template structure (you can use the `get_sequence()` function to verify that). Furthermore, the two sequences in the alignment must have exactly the same length.

Finally, the `create_model()` command starts the automatic homology modeling procedure in ModeRNA: bases are exchanged and modifications are added as specified in the alignment. Where there are gaps, ModeRNA attempts to find a fitting piece of RNA structure in its internal library and inserts it to connect the residues from the template (the A and U at the borders of the gap, in this case). After that you can write the model to a file with `write_model()` or further work on it in Python. As you can see from the output, the sequence of the model is identical to the one of the target from the alignment.

Although the Python commands to build RNA models may look easy, obtaining high quality models is complicated. The template structure needs to be chosen carefully, the alignment needs to be curated manually, and the resulting model may need further refinement. The ModeRNA library allows you to perform modeling of RNA structures step by step, and therefore facilitates work with RNA 3D structures in general.

REFERENCE

[1] M. Rother, K. Rother, T. Puton, and J.M. Bujnicki, "ModeRNA: A Tool for Comparative Modeling of RNA 3D Structure," *Nucleic Acids Research* 39 (2011): 4007–4022.

Recipe 19: Calculating RNA Base Pairs from a 3D Structure

W HEN ANALYZING AN RNA 3D structure, one of the first things getting attention are the base pairs. Key interactions in RNA are not only canonical Watson-Crick pairs but also many noncanonical base pairs. This recipe explains how to calculate the base pairs from Python using PyCogent.

There are several programs that can calculate base pairs from the PDB structure of an RNA molecule: for a long time, RNAView has set an unchallenged standard. The RNAView software (http://ndbserver.rutgers. edu/services/download) identifies all 12 types of noncanonical base pairs defined by both Leontis and Westhof [1, 2]. The program can be compiled on Linux. Afterwards, all base pairs (e.g., the PDB file 1ehz.pdb) can be calculated with the Linux shell command:

```
rnaview 1ehz.pdb
```

RNAView creates a couple of text files as an output. The file base-pairs.out contains a list of base pairing interactions in tabular format:

```
1_72, A:      1 G-C 72      A: +/+ cis      XIX
2_71, A:      2 C-G 71      A: +/+ cis      XIX
3_70, A:      3 G-C 70      A: +/+ cis      XIX
4_69, A:      4 G-C 69      A: +/+ cis      XIX
5_68, A:      5 G-C 68      A: +/+ cis      XIX
6_67, A:      6 A-U 67      A: W/W cis      n/a
7_66, A:      7 A-U 66      A: -/- cis      XX
```

The PyCogent library [3] (see Recipe 1) contains a parser that allows you to parse these files:

```
from cogent.app.rnaview import RnaView
from cogent.parse.rnaview import RnaviewParser

rna_prog = RnaView()
result = rna_prog('1ehz.pdb')
bpairs = result['base_pairs']
errors = result['StdErr'].read()
stdout = result['StdOut'].read()

bp_dict = RnaviewParser(bpairs)
print 'INFORMATION:'
sys.stderr.write(errors)
print stdout
print 'BASE PAIRS:'
for key in bp_dict:
        print key, bp_dict[key]
```

The program consists of two parts. First, the `cogent.app.rnaview` module is used to execute the RNAView command line tool on a PDB file. The `rnaview` module contains a program wrapper `RnaView` (see Chapter 14), which returns a dictionary with separate entries for the standard output (messages on the screen), errors, and the base pair output. In the second step, a parser for the base pair format from the module `cogent.parse.rnaview` is used. The `RnaviewParser` parser produces a dictionary with the residue numbers as keys and various information, including the base pair type, as values.

ALTERNATIVES TO RNAVIEW

A few interesting, more recent alternatives to RNAView exist. The FR3D software (http://rna.bgsu.edu/FR3D) can identify all kinds of base pairs but requires the commercial MATLAB package. Without MATLAB, FR3D can be used via a web interface (http://rna.bgsu.edu/WebFR3D). The website also provides a huge catalog of exemplary base pair structures that count as a gold standard (http://rna.bgsu.edu/FR3D/basepairs). RNAView is a no-cost alternative, which is why we selected it for this recipe. Base pairs can also be calculated online using the MC-Annotate tool [4] (www-lbit.iro.umontreal.ca/mcannotate-simple). Finally, to calculate Watson-Crick base pairs, you can use the ModeRNA software [5] (http://iimcb.genesilico.pl/moderna; see Recipe 18), writing the following fragment of Python code:

```
from moderna import *
struc = load_model('1ehz.pdb', 'A')
for bp in get_base_pairs(struc):
    print bp
```

ModeRNA also calculates noncanonical pairs, but at the time of writing, this feature is in a development stage. Keep in mind that each program calculates base pairs in a slightly different way, and therefore the resulting base pairs are not 100% identical.

REFERENCES

[1] M. Sarver, C.L. Zirbel, J. Stombaugh, A. Mokdad, and N.B. Leontis, "FR3D: Finding Local and Composite Recurrent Structural Motifs in RNA 3D Structures," *Journal of Mathematical Biology* 56 (2008): 215–252.

[2] H. Yang, F. Jossinet, N. Leontis, L. Chen, J. Westbrook, H. Berman, and E. Westhof, "Tools for the Automatic Identification and Classification of RNA Base Pairs," *Nucleic Acids Research* 31 (2003): 3450–3460.

[3] R. Knight, P. Maxwell, A. Birmingham, J. Carnes, J.G. Caporaso, B.C. Easton, M. Eaton, M. Hamady, H. Lindsay, Z. Liu, C. Lozupone, D. McDonald, M. Robeson, R. Sammut, S. Smit, M.J. Wakefield, J. Widmann, S. Wikman, S. Wilson, H. Ying, and G.A. Huttley, "PyCogent: A Toolkit for Making Sense from Sequence," *Genome Biology* 8 (2007): R171.

[4] P. Gendron, S. Lemieux, and F. Major, "Quantitative Analysis of Nucleic Acid Three-Dimensional Structures," *Journal of Molecular Biology* 308 (2001): 919–936.

[5] M. Rother, K. Rother, T. Puton, and J.M. Bujnicki, "ModeRNA: A Tool for Comparative Modeling of RNA 3D Structure," *Nucleic Acids Research* 39 (2011): 4007–4022.

Recipe 20: A Real Case of Structural Superimposition: The Serine Protease Catalytic Triad

S ERINE PROTEASES ARE ENZYMES cleaving peptide bonds in proteins. The catalysis occurs in the active site of the enzyme, which is made up of three residues very close in space: a histidine (His57), an aspartic acid (Asp102), and a serine (Ser195). The residue numbers of the three catalytic residues are Hist57, Asp102, and Ser195 in almost all serine protease structures. Ser195 serves as the nucleophilic amino acid. It has been observed that the active site triad is structurally very well conserved; i.e., the relative position of the residues homologous to His57, Asp102, and Ser195 in three-dimensional (3D) space is very well conserved in even globally different serine proteases. In most cases, the catalytic residues are not close in sequence. Serine proteases are classified in two categories (chymotrypsin/trypsin-like and subtilisin-like) depending on the global fold of the enzymes.

In this example, Biopython is used to superimpose the structures of two completely different serine proteases on the backbone atoms (CA and N) of the amino acids of the catalytic triad (for the use of Biopython to superimpose structures, see also Chapter 21, Example 21.3).

The first structure (PDB 1EQ9) is a chymotrypsin from *Solenopsis invicta*, the South American imported fire ant, and the second (PDB 1FXY) is a chimeric protein obtained by recombining the N-terminal subdomain from the human coagulation factor X with the C-terminal subdomain from a trypsin. By superimposing the two structures onto the His57, Asp102, and Ser195 residues, you can examine whether the catalytic triad is still well conserved between these two enzymes despite their overall differences.

In a structural superimposition one of the two structures (the target) is kept in a fixed position, and the other (the probe) is rotated and translated until the RMSD between the two structures reaches a minimum. A superimposition must be carried out between sets of atoms of equal size. Therefore, before superimposing two structures, you must decide which atoms of each structure you want to superimpose. These could even be all the atoms present in the structure. In this case, you have to check if the two structures have the same number of atoms and if each atom in one structure has a corresponding atom in the other structure. In general, you may obtain very good results by superimposing a small number of atoms. In particular, if you want to study a specific region or domain of a protein, e.g., a binding site, it will be sufficient to superimpose the atoms of the residues belonging to the binding site. In this case, the RMSD between corresponding binding site atoms in the two structures will be calculated and minimized.

In this example, the structure 1FXY (probe) is superimposed onto the structure 1EQ9 (target). The superimposition produces a rotation-translation matrix, which can be used to generate a rotated and translated structure out of 1FXY (1FXY-superimposed.pdb).

At the beginning of the program, the two structures are retrieved from the PDB, i.e. downloaded from the PDB Web repository. Next, they are read into Biopython structure objects (struct_1 and struct_2). Then, the residues of the two catalytic triads are recorded in six variables in the form of residue objects. Notice that the second argument of the get_structure() method is the path to the PDB files downloaded from the PDB. Indeed, the retrieve_pdb_file() method creates directories (eq for 1eq9.ent and fx for 1fxy.ent) where it saves the downloaded file.

For each triad, backbone atom objects are extracted from the residue objects and appended to a list. The first list is called target because it contains the atoms that will be kept fixed during the superimposition (extracted from 1EQ9), the second is called probe and contains atoms that will be moved in the coordinate system during the rotation-translation (extracted from 1FXY). target and probe are used to calculate the rotation-translation matrix that will be applied to all atoms (struct_2_ atoms) of the 1FXY structure (struct_2). Finally, the rotated and translated structure (1FXY-superimposed.pdb) is saved to a file.

```
# Superimpose the catalytic triads of two different serine
# proteases(on CA and N atoms of res H57, D102, and S195 of chain A)

from Bio import PDB

# Retrieve PDB files
pdbl = PDB.PDBList()
pdbl.retrieve_pdb_file("1EQ9")
pdbl.retrieve_pdb_file("1FXY")

# Parse the two structures
from Bio.PDB import PDBParser, Superimposer, PDBIO
parser = PDB.PDBParser()
struct_1 = parser.get_structure("1EQ9", "eq/pdb1eq9.ent")
struct_2 = parser.get_structure("1FXY", "fx/pdb1fxy.ent")

# get the catalytic triads
res_57_struct_1 = struct_1[0]['A'][57]
res_102_struct_1 = struct_1[0]['A'][102]
res_195_struct_1 = struct_1[0]['A'][195]

res_57_struct_2 = struct_2[0]['A'][57]
res_102_struct_2 = struct_2[0]['A'][102]
res_195_struct_2 = struct_2[0]['A'][195]

# Build 2 lists of atoms for calculating a rot.-trans. matrix
# (target and probe).
target = []
backbone_names = ['CA', 'N']
for name in backbone_names:
    target.append(res_57_struct_1[name])
    target.append(res_102_struct_1[name])
    target.append(res_195_struct_1[name])

probe = []
for name in backbone_names:
    probe.append(res_57_struct_2[name])
    probe.append(res_102_struct_2[name])
    probe.append(res_195_struct_2[name])

# Check whether target and probe lists are equal in size.
# This is needed for calculating a rot.-trans. matrix
assert len(target) == len(probe)

# Calculate the rotation-translation matrix.
sup = Superimposer()
```

```
sup.set_atoms(target, probe)

# Apply the matrix. Remember that it can be applied only on
# lists of atoms.
struct_2_atoms = [at for at in struct_2.get_atoms()]
sup.apply(struct_2_atoms)

# Write the rotation-translated structure
out = PDBIO()
out.set_structure(struct_2)
out.save('1FXY-superimposed.pdb')
```

Source: Adapted from code published by A.Via/K.Rother under the Python License.

Now you can open 1EQ9.pdb and 1FXY-superimposed.pdb with PyMOL (see Chapter 17) and verify whether the catalytic residues in the two structures are well conserved (i.e., if they are well superimposed).

Appendix A: Command Overview

A.1 UNIX COMMANDS

A.1.1 Listing Files and Directories

`ls`	Lists files and directories
`ls -a`	Lists hidden files and directories
`mkdir`	Creates a directory
`cd directory`	Changes to named directory
`cd`	Changes to home directory
`cd ~`	Changes to home directory
`cd..`	Changes to parent directory
`pwd`	Displays the path of the current directory

A.1.2 Handling Files and Directories

`cp file1 file2`	Copies file1 and calls it file2
`mv file1 file2`	Moves or renames file1 to file2
`rm file`	Removes a file
`rmdir directory`	Removes a directory
`cat file`	Displays a file
`more file`	Displays a file a page at a time
`head file`	Displays the first few lines of a file
`tail file`	Displays the last few lines of a file
`grep 'keyword' file`	Searches a file for keywords
`wc file`	Counts the number of lines, words, and characters in a file

A.1.3 Redirection

`command > file`	Redirects standard output to a file
`command >> file`	Appends standard output to a file
`command < file`	Redirects standard input from a file
`cat file1 file2 > file0`	Concatenates file1 and file2 to file0
`sort`	Sorts data
`who`	List users currently logged in

A.1.4 File System Security (Access Rights)

`ls -lag`	Lists access rights for all files
`chmod [options] file`	Changes access rights for named file
`command &`	Runs command in background
`^C`	Kills the job running in the foreground
`^Z`	Suspends the job running in the foreground
`bg`	Backgrounds the suspended job
`jobs`	Lists current jobs
`fg%1`	Foreground job number 1
`kill%1`	Kills job number 1
`ps`	Lists current processes
`kill 26152`	Kills process number 26152

A.1.5 `chmod` Options

Symbol	Meaning
u	User
g	Group
o	Other
a	All
r	Read
w	Write (and delete)
x	Execute (and access directory)
+	Add permission
−	Take away permission

A.1.6 General Rules

- If you've made a typo, use Ctrl-U to cancel the whole line.

- UNIX is case-sensitive.

- There are commands that can take options.

- The options change the behavior of the command.

- UNIX uses command-line completion.

- `%command_name -options <file> [Return]`

- `%man <command name> [Enter]`

- `%whatis <command name> [Enter]`

- Ctrl-A sets the cursor at the beginning of the line.

- Ctrl-E sets the cursor at the end of the line.

- You can use up and down arrows to recall commands.

- The command `whereis` tells you where a given program is.

- You can use a text editor to write stuff (e.g., gedit).

A.2 PYTHON COMMANDS

A.2.1 Overview

- Python is an interpreted language.

- The Python interpreter automatically generates bytecode (`.pyc` files).

- It is 100% free software.

Strengths

- It is quick to write, and there is no compilation.

- It is fully object oriented.

- It has many reliable libraries.

- It is an all-round language.

Weakness

- Writing very fast programs is not easy.

A.2.2 The Python Shell

Overview

You can use any Python command from the interactive Python shell (>>>):

```
>>> print 4**2
16
>>> a = 'blue'
>>> print a
blue
>>>
```

Tips

- All Python commands work in the same way in the Python shell and in programs.

- You can leave the command line by Ctrl-D (Linux, Mac) or Ctrl-Z (Windows).

- The Python shell works great as a pocket calculator.

- Writing code blocks with more than two lines in the Python shell gets painful quickly.

- You can define code blocks by writing extra lines:

  ```
  >>> for i in range(3):
  ...       print i,
  ...       012
  >>>
  ```

A.2.3 Python Programs

- All program files should have the extension .py.

- Each line should contain exactly one command.

- Code blocks are marked by indentation. Code blocks should be indented by four spaces or one tab.

- When you are developing on UNIX, the first line in each Python program should be

  ```
  #!/usr/bin/env python
  ```

Code Formatting Conventions

- Use spaces instead of tabs (or use an editor that converts them automatically).

- Keep lines shorter than 80 characters long.

- Separate functions with two blank lines.

- Separate logical chunks of long functions with a single blank line.

- Variables and function names are in lowercase.

The Dogma of Programming

- First, make it work.

- Second, make it nice.

- Third, and only if it is really necessary, make it fast.

A.2.4 Operators
Arithmetic Operators

7 + 4	Addition; results in 11 (works for strings and lists)
7 - 4	Subtraction; results in 3
7 * 4	Multiplication; results in 28
7 / 4	Division of integers; results in 1 (rounded down)
7 / 4.0	Division by float; results in 1.75
7 % 4	Modulo operator, returns the remainder of a division; results in 3
7 ** 2	Raising to a power; results in 49
7.0 // 4.0	Floor division (cuts off after point); results in 1.0

Assignment Operators
Assignment operators create or modify variables.

Variables
Variable are containers for data. Variable names may be composed of letters, underscores, and, after the first position, digits. Lowercase letters are common, but uppercase is also allowed (usually used for constants). Some words such as `print`, `import`, and `for` are forbidden as variable names.

`a = 10`	Assigns the integer value 10 to the variable a
`b = 3.0`	Variable containing a floating-point number
`a3 = 10`	Variable with a digit in its name
`PI = 3.1415`	Variable written with uppercase letters
`invitation = 'Hello World'`	Variable containing text (string)
`invitation = "Hello World"`	Variable containing text with double quotes

Modifying Variables

`+=, -=, *=,`	Recursive operators
`/=, %=, **=`	`x += 1` is equivalent to `x = x+1`

Comparison Operators
All comparison variables result in either `True` or `False`.

`a == b`	a equal to b
`a != b`	a not equal to b
`a < b`	a smaller than b
`a > b`	a bigger than b
`a <= b`	a smaller or equal to b
`a >= b`	a smaller or equal to b
`in, not in`	The `in` and `not in` operators check if the object on their left is contained (or not contained) in the string, list, or dictionary on their right and return the Boolean value `True` or `False`.
`is, is not`	The `is` and `is not` operators check whether the object on their left is identical (or not identical) to the object on their right and return the Boolean value `True` or `False`.
`a and b`	Boolean operator: if both the condition a and b are `True`, it returns `True`; else it returns `False`.
`or`	Boolean operator: if either the condition a or b or both are `True`, it returns `True`; else it returns `False`.
`not a`	Boolean operator: if the condition a is not verified, it returns `True`; else it returns `False`.

A.2.5 Data Structures

Overview of Data Types

Integers	Are numbers without digits after the decimal point
Floats	Are numbers with digits after the decimal point
Strings	Are *immutable ordered* collections of characters and are indicated with single (`'abc'`) or double (`"abc"`) quotation marks
Lists	Are *mutable ordered* collections of objects and are indicated with square brackets (`[a,b,c]`)
Tuples	Are *immutable ordered* collections of objects and are indicated with round brackets (`(a,b,c)`) or by listing the collection of items separated by commas (`x, y, z`)
Dictionaries	Are unordered collections of `key:value` pairs
Sets	Are collections of unique elements
Boolean	Are either `True` or `False`

Type Conversions

`int(value)`	Creates an integer from a float or string
`float(value)`	Creates a float from an `int` or string
`str(value)`	Creates a string from any variable

A.2.6 Strings

String variables are containers for text. Strings can be marked by many kinds of quotes, which are all equivalent.

`s = 'Hello World'`	Assigns the text to a string variable
`s = "Hello World"`	String with double quotes
`s = '''Hello World'''`	Multiline string with triple single quotes
`s = """Hello World"""`	Multiline string with triple double quotes
`s = 'Hello\tWorld\n'`	String with a tab (`\t`) and a newline (`\n`)

Accessing Characters and Substrings

Using square brackets, any character of a string can be accessed by its position. The first character has the index 0. Substrings can be formed by square brackets with two numbers separated by a colon. The position corresponding to the second number is not included in the substring. If you try to use indices bigger than the length of the string, an `IndexError` will be created.

`print s[0]`	Prints the first character
`print s[3]`	Prints the fourth character
`print s[-1]`	Prints the last character
`print s[1:4]`	Second to fourth position; results in `'ell'`
`print s[:5]`	From start to fifth position; results in `'Hello'`
`print s[-5:]`	Fifth position from the end until the end; results in `'World'`

String Functions

A number of functions can be used on every string variable:

`len(s)`	Length of the string; results in `11`
`s.upper()`	Converts to uppercase; results in `'HELLO WORLD'`
`s.lower()`	Converts to lowercase; results in `'hello world'`
`s.strip()`	Removes spaces and tabs from both ends
`s.split(' ')`	Cuts into words; results in `['Hello', 'World']`
`s.find('llo')`	Searches for a substring; returns starting position
`s.replace('World','Moon')`	Replaces text; results in `'Hello Moon'`
`s.startwith('Hello')`	Checks beginning; returns `True` or `False`
`s.endswith('World')`	Checks end; returns `True` or `False`

A.2.7 Lists

A list is a sequence of elements that can be modified. In many cases, all elements of a list will have the same type, but this is not mandatory.

Accessing Elements of Lists

When you use square brackets, any element of a list can be accessed. The first character has the index 0.

`data = [1,2,3,4,5]`	Creates a list
`data[0]`	Accesses the first element
`data[3]`	Accesses the fourth element
`data[-1]`	Accesses the last element
`data[0] = 7`	Reassigns the first element

Creating Lists from Other Lists

You can extract sublists from lists by applying square brackets in the same way as you would extract substrings.

data = [1,2,3,4,5]	Creates a list
data[1:3]	[2,3]
data[0:2]	[1,2]
data[:3]	[1,2,3]
data[-2:]	[4,5]
backup = data[:]	Creates a copy of the list

Modifying Lists

l[i] = x	The ith element of l is replaced by x.
l[i:j] = t	The elements of l from i to j are replaced by t (iterable).
del l[i:j]	This deletes the elements of l from i to j.
l[i:j:k] = t	The elements l[i:j:k] are replaced by the elements of t (t must be a sequence such that len(l[i:j:k]) = len(t)).
del s[i:j:k]	This deletes the elements of l from i to j with step k.
l.append(x)	This is the same as l[len(l):len(l)] = [x]. It appends the element x to the list l.
l.extend(x)	This is the same as l[len(l):len(l)] = x (where x is any iterable object).
l.count(x)	This returns the number of elements x in l.
l.index(x[, i[, j]])	This returns the smaller k such that l[k] = x and i ≤ k ≤ j.
l.insert(i, x)	This is the same as l[i:i] = [x].
l.pop(i)	This cancels the ith element and returns its value. l.pop() is the same as del l[-1]; return l[-1].
l.remove(x)	This is the same as del l[l.index(x)].
l.reverse()	This reverses the elements of l.
l.sort()	This sorts the list l.
l.sort([cmp[, key[, reverse]]])	This sorts the list l. Optional arguments for the control of the comparison can be passed to the sort() method. cmp is a customized function for the comparison of element pairs that must return a negative value, zero, or a positive value depending on if the first element of the pair is lower than, equal to, or greater than the second element.
sorted(l)	This creates a new list made of a simple ascending sort of l (without modifying l).

Functions Working on Lists

`data = [3,2,1,5]`	Example data
`len(data)`	Length of data; returns 4; also works for many other types
`min(data)`	Smallest element of data; returns 1
`max(data)`	Biggest element of data; returns 5
`sum(data)`	Sum of data; returns 11
`range(4)`	Creates a list of numbers; returns `[0,1,2,3]`
`range(1,5)`	Creates a list with start value; returns `[1,2,3,4]`
`range(2,9,2)`	Creates a list with step size; returns `[2,4,6,8]`
`range(5,0,-1)`	Creates a list counting backward from the start value; returns `[5,4,3,2,1]`

A.2.8 Tuples

A tuple is a sequence of elements that cannot be modified. This means that once you have defined it, you cannot change or replace its elements. They are useful to group elements of different types.

```
t = ('bananas','200g',0.55)
```

Notice that brackets are optional; i.e., you can use either `Tuple = (1,2,3)` or `Tuple = 1,2,3`.

A tuple of a single item must be written either `Tuple = (1,)` or `Tuple = 1`.

You can use square brackets to address elements of tuples in the same way as you would address elements of lists.

A.2.9 Dictionaries

Dictionaries are an unordered, associative array. They have a set of `key:value` pairs:

```
prices = {'banana':0.75, 'apple':0.55, 'orange':0.80}
```

In the example, `'banana'` is a key, and `0.75` is a value.

Dictionaries can be used to look up things quickly:

```
prices['banana'] # 0.75
prices['kiwi'] # KeyError
```

Accessing Data in Dictionaries

By applying square brackets with a key inside, you can request the values of a dictionary. Keys can be strings, integers, floats, and tuples.

`prices['banana']`	This returns the value of `'banana'` (`0.75`).
`prices.get('banana')`	This returns the value of `'banana'` but avoids the `KeyError`. If the key does not exist, it returns `None`.
`prices.has_key('apple')`	This checks whether `'apple'` is defined.
`prices.keys()`	This returns a list of all keys.
`prices.values()`	This returns a list of all values.
`prices.items()`	This returns all keys and values as a list of tuples.

Modifying Dictionaries

`prices['kiwi'] = 0.6`	Sets the value of `'kiwi'`
`prices.setdefault('egg',0.9)`	Sets the value of `'egg'` if it is not defined yet

The None *Type*

Variables can contain the value None. You can also use None to indicate that a variable is empty. None is also used automatically when a function does not have a return statement.

```
traffic_light = [None, None, 'green']
```

A.2.10 Control Flow

Code Blocks/Indentation

After any statement ending with a colon (:), all indented commands are treated as a code block and are executed within the loop if if the condition applies. The next unindented command marks the end of the code block.

Loops with for

Loops repeat commands. They require a sequence of items that they iterate, e.g., a string, list, tuple, or dictionary. Lists are useful when you know the number of iterations in advance and when you want to do the same thing to all elements of a list.

```for base in 'AGCT':``` ``` print base```	Prints four lines containing A, G, C, and T
```for number in range(5):``` ``` print number```	Prints five lines with the numbers from 0 to 4
```for elem in [1, 4, 9, 16]:``` ``` print elem```	Prints the four numbers each to a separate line

Counting through Elements of Lists

The enumerate() function associates an integer number starting from zero to each element in a list. This is helpful in loops where an index variable is required.

```
>>> fruits = ['apple','banana','orange']
>>> for i, fruit in enumerate(fruits):
...     print i, fruit
...
0 apple
1 banana
2 orange
```

Merging Two Lists

The zip() function associates the elements of two lists to a single list of tuples. Excess elements are ignored.

```
>>> fruits = ['apple','banana','orange']
>>> prices = [0.55, 0.75, 0.80, 1.23]
>>> for fruit,price in zip(fruits,prices):
...     print fruit, price
...
apple 0.55
banana 0.75
orange 0.8
```

Loops over a Dictionary

You can access the keys of a dictionary in a for loop. However, their order is not guaranteed.

Conditional Statements with if

if statements are used to implement decisions and branching in the program. They must contain an if block and optionally one or many elif and else blocks:

```
if fruit == 'apple':
    price = 0.55
elif fruit == 'banana':
    price = 0.75
elif fruit == 'orange':
    price = 0.80
else:
    print 'we dont have%s'%(fruit)
```

Comparison Operators

An expression with `if` may contain any combination of comparison operators, variables, numbers, and function calls:

- `a == b,  a != b` (equality)

- `a < b,  a > b,  a <= b,  a >= b` (relations)

- `a or b,  a and b,  not a` (Boolean logic)

- `(a or b) and (c or d)` (priority)

- `a in b` (inclusion, when b is a list, tuple, or string)

Boolean Value of Variables

Apart from the comparison operators, the `if` statement also takes the values of variables directly into account. Each variable can be interpreted by Boolean logic. All variables are `True`, except for

```
0, 0.0, '', [], {}, False, None
```

Conditional Loops with `while`

`while` loops require a conditional expression at the beginning. These work in exactly the same way as in `if...elif` statements:

```
>>> i = 0
>>> while i < 5:
...     print i,
...     i = i + 1
...
0 1 2 3 4
```

When to Use `while`

- When there is a loop exit condition

- When you want to start a loop only upon a given condition

- When it may happen that nothing is done at all

- When the number of repeats depends on user input

- When you are searching for a particular element in a list

A.2.11 Program Structures

Functions

Functions are subprograms. They help you to structure your code into logical units. A function may have its own variables. It also has an input (parameters) and output (returned values). In Python, a function is defined by the `def` statement, followed by the function name, the argument(s) in brackets, a colon (:), and an indented code block:

```
def calc_discount(fruit, n):
    '''Returns a lower price of a fruit.'''
    print 'Today we have a special offer for:', fruit
    return 0.75 * n
print calc_discount('banana', 10)
```

Parameters and Return Values

Input for a function is given by arguments. Arguments may have default values. Then they are optional in the function call. *Do not use lists or dictionaries as default values!*

The output of a function is created by the `return` statement. The value given to `return` goes to the program part that called it. More than one value is returned as a tuple. In any case, the `return` statement ends the function execution.

```
def calc_disc(fruit,n = 1):      # A function with an optional
    print fruit                  # parameter
    return n*0.75

def calc_disc(fruit,n = 1):      # A function returning a tuple
    fruit = fruit.upper()
    return fruit, n*0.75
calc_disc('banana')              # Function calls
calc_disc('banana',100)
```

Good Style for Writing Functions

- Each function should have one purpose only.

- The name should be clear and start with a verb.

- The function should have a triple-quoted comment at the beginning (a documentation string).

- The function should return results in only one way.

- Functions should be small (fewer than 100 lines).

A.2.12 Modules

A module is a Python file (the filename ending with .py). Modules can be imported from another Python program. When a module is imported, the code within is automatically executed. To import from a module, you need to give its name (without .py) in the `import` statement. It is helpful to explicitly list the variables and functions required. This helps with debugging.

`import math`	Includes and interprets a module
`from math import sqrt`	Includes one function from a module
`from math import pi`	Includes a variable from a module
`from math import sqrt, pi`	Includes both
`from math import *`	Includes everything from a module (merges namespaces)

Finding Out What Is in a Module

The contents of any module can be examined with `dir()` and `help()`.

`dir(math)`	Shows everything inside the module
`dir()`	Shows everything in the global namespace
`help(math.sqrt)`	Displays the help text of a module or function
`__name__`	Name of a module
`__doc__`	Help text of a module
`__builtins__`	Container with all standard Python functions

Where Python Looks for Modules

When importing modules or packages, Python looks in

- the current directory,

- the `Python2.6/lib/site-packages` folder, and

- everything in the PYTHONPATH environment variable. In Python, it can be accessed with

```
import sys
print sys.path
sys.path.append('my_directory')
import my_package.my_module
```

Packages

For very big programs, you might find it useful to divide the Python code into several directories. There are two things to keep in mind when doing that:

- To import the package from outside, you need to ensure a file __ init__.py (it may be empty) is in the package directory.

- The directory with the package needs to be in the Python search path (see above).

A.2.13 Input and Output

Reading Text from the Keyboard into a Variable

User input can be read from the keyboard with or without a message text:

a = raw_input()	Reads text from the keyboard to a string variable
a = raw_input('please enter a number')	Displays the text, then reads a string from the keyboard

Printing Text

The Python print statement writes text to the console where Python was started. The print command is very versatile and accepts almost any combination of strings, numbers, function calls, and arithmetic operations separated by commas. By default, print generates a newline character at the end.

print 'Hello World'	Displays a text
print 3.4	Displays the number
print 3 + 4	Displays the result of the calculation
print a	Displays the contents of the variable a
print '''line one line two line three'''	Displays text stretching over multiple lines
print 'number', 77	Displays the text, a tab, and the number

`print int(a) * 7`	Displays the result of the multiplication after converting the variable to an integer
`print`	Displays an empty line

String Formatting

Variables and strings can be combined using formatting characters. This works also within a `print` statement. In both cases, the number of values and formatting characters must be equal.

```
s = 'Result:%i'%(number)
print 'Hello%s!'%('Roger')
print '(%6.3f/%6.3f)'%(a,b)
```

The formatting characters include the following:

- `%i`: an integer

- `%4i`: an integer formatted to length 4

- `%6.2f`: a float number with length 6 and 2 after the comma

- `%10s`: a right-oriented string with length 10

Reading and Writing Files

Text files can be accessed using the `open()` function. It returns an open file whose contents can be extracted as a string or strings that can be written. If you try to open a file that does not exist, an `IOError` will be created.

`f = open('my_file.txt')` `text = f.read()`	Reads a text file and its contents into a string variable
`f = open('my_file.txt','w')` `f.write(text)`	Creates a new text file and writes text from a string variable into it
`f = open('my_file.txt','a')` `f.write(text)`	Appends text to an already existing file
`f.close()`	Closes a file after usage; closing in Python is good style but not always mandatory
`lines = f.readlines()`	Reads all lines from a text file to a list
`f.writelines(lines)`	Writes a list of lines to a file
`for line in open(name):` `    print line`	Goes through all lines and prints each line

`lines = ['first line\n', 'second`
`line\n']`
`f = open('my_file.txt','w')`
`f.writelines(lines)`
	Creates a list of lines with newline characters at the end and saves it to a text file

Directory Names in Python

When opening files, you can also use full or relative directory names. However, on Windows, you must replace the backslash "\" with a double backslash "\\" (because "\" is also used for "\n" and "\t").

```
f = open('..\\my_file.txt')
f = open('C:\\python\\my_file.txt')
```

The csv *Module*

An open file behaves like a list of strings. Thus, it is possible to read it line by line using the csv module:

```
>>> import csv
>>> reader = csv.reader(open('RSMB_HUMAN.fasta'))
>>> for row in reader: print row
...
['>gi|4507127|ref|NP_003084.1| U1 small nuclear
    ribonucleoprotein C [Homo sapiens]']
['MPKFYCDYCDTYLTHDSPSVRKTHCSGRKHK
    ENVKDYYQKWMEEQAQSLIDKTTAFQQGKIPPTPFSAP']
['PPAGAMIPPPPSLPGPPRPGMMPAPHMGGPP
    MMPMMGPPPPGMMPVGPAPGMRPPMGGHMPMMPGPPMMR']
['PPARPMMVPTRPGMTRPDR']
[]
>>>
```

Similarly, lists and tuples can be written to files using csv:

```
>>> import csv
>>> table = [[1,2,3],[4,5,6]]
>>> writer = csv.writer(open('my_file.csv','w'))
>>> writer.writerow(table)
```

Options of csv File Readers/Writers

Both csv readers and writers can handle a lot of different formats. Options you can change include the following:

- delimiter: the symbol separating columns

- quotechar: the symbol used to quote strings

- lineterminator: the symbol at the end of lines

```
reader = csv.reader(open('my_file.csv'), \
    delimiter = '\t', quotechar = '"')
```

A.2.14 Managing Directories

With the os module, you can change to a different directory:

```
import os
os.chdir('..\\python')
```

You can also get a list with all files:

```
os.listdir('..\\python')
```

The third most important function is for checking whether a file exists:

```
print os.access('my_file.txt', os.F_OK)
```

A.2.15 Getting the Current Time and Date

The time module offers functions for getting the current time and date:

```
import time
s = time.asctime()        # as string
i = time.time()           # as float
```

A.2.16 Accessing Web Pages

You can access the HTML code of web pages and downloadable files from the web in a way similar to the way you read files:

```
import urllib
url = 'http://www.google.com'
page = urllib.urlopen(url)
print page.read()
```

A.2.17 Regular Expressions

Regular expressions allow performing pattern matching on strings and search and replace text in a sophisticated way:

```
import re
text = 'all ways lead to Rome'
```

Searching Whether Text Exists

```
re.search('R...\s', text)
```

Finding All Words

```
re.findall('\s(.o)', text)
```

Replacing

```
re.sub('R[meo]+','London', text)
```

How to Find the Right Pattern for Your Problem

Finding the right Regex requires lots of trial-and-error search. You can test regular expressions online before including them into your program:

```
http://www.regexplanet.com/simple/
```

Characters Used in Regex Patterns

Some of the most commonly used characters in Regular Expression (RE) patterns are as follows:

Character	Action
\d	Match decimal character [0..9]
\w	Match alphanumeric [a..z] or [0..9]
\A	Match at the start of the text
\Z	Match at the end of the text
[ABC]	Match one of characters A,B,C
.	Match any character (wildcard)
^A	Match any character but A
a+	Match one or more of pattern a
a*	Match zero or more of pattern a
a|b	Match either pattern a or b
(a)	re.findall() returns a
\s	Match an empty space

`re` *Module Methods*

Method	Returned Object	Attribute Of	Function
compile()	Regex object		Compiles a RE
match()	Match object	Regex object	Determines if the RE matches at the beginning of the string
search()	Match object	Regex object	Scans through a string, looking for any location where the RE matches
findall()	List	Regex object	Finds all substrings where the RE matches and returns them as a list
finditer()	Iterator object	Regex object	Finds all substrings where the RE matches and returns them as an iterator
span()	Tuple	Match object	Returns a tuple containing the (start, end) positions of the match
start()	Integer	Match object	Returns the starting position of the match
end()	Integer	Match object	Returns the ending position of the match
group()	String	Match object	Returns the string matched by the RE
groups()	Tuple	Match object	Returns a tuple containing the strings for all the subgroups
split(s)	List	Regex object	Splits the string into a list, splitting it wherever the RE matches
sub(r, s)	String	Regex object	Finds all substrings where the RE matches and replaces them with a different string
subn(r, s)	Tuple	Regex object	Does the same thing as sub() but returns the new string and the number of replacements

Regular Expression Compilation Flags

You can customize pattern matching by adding an argument to the `re.compile()` function. For example, if you want to perform a case-insensitive search, you can add `re.IGNORECASE` to the `re.compile()` arguments.

This is a list of the possible options that you can use to customize your search:

Flag	Function
DOTALL, S	Makes a dot (.) match any character, including newlines
IGNORECASE, I	Does case-insensitive matches
LOCALE, L	Does a locale-aware match, i.e., takes into account C libraries for local systems (e.g., other languages)
MULTILINE, M	Does multiline matching, affecting ^ and $: ^ ($) matches at the beginning (end) of each line in a string
VERBOSE, V	Enables verbose REs, which can be organized more cleanly and understandably, i.e., is allowed to write a RE in a more readable way (e.g., by introducing whitespaces or comments)

A.2.18 Debugging

Exceptions

No matter how good your program is, sometimes things will go wrong. However, you can make sure it doesn't blow up in the wrong moment, so that, e.g., important data can be saved. Exceptions are the Python mechanisms that let you know that something went wrong.

```
a = raw_input('enter a number')
try:
    inverse = 1.0/int(a)
except ZeroDivisionError:
    print "zero can't be inverted"
```

except, else, *and* finally

Whenever the according kind of error occurs within a try clause, the except code block will be executed without breaking the program execution.

Typical exceptions include the following:

- ZeroDivisionError: when dividing by zero

- KeyError: a key in a dictionary does not exist

- ValueError: a type conversion failed

- IOError: a file could not be opened

What Exceptions to Catch

Generally, everything that can go wrong even though a program has been written with great diligence is worth catching. For example,

- File operations

- Web operations

- Database operations

- User input, especially numbers

While it is possible to wrap an entire program into a single try... except clause, this often makes little sense, because the code gets less transparent and more difficult to debug.

Interrupting Program Execution

You can interrupt the program execution and start the debugger by inserting these commands at any point:

```
import pdb
pdb.set_trace()
```

The debugger shows a shell that works like the normal Python command line, but with some extra commands:

- n (next): execute next statement

- s (step): execute next statement and descend into functions

- l (list): show source code

- c (continue): continue execution until the next breakpoint

- help: print help message

- q (quit): abort the program

Breakpoints

- The command 'b <line number>' sets a breakpoint at the given line.

- The command 'b' displays all breakpoints set.

A.2.19 Comments

In Python, lines can be commented by the hash symbol (#) or by triple quotes. Python uses the triple quoted comments to generate documentation automatically.

`# comment`	Single commented line
`b = a/2.0 # half distance`	Comment after regular command
`"""This program` `calculates fruit prices. """`	Multiline comment enclosed by triple quotes
`'''This program` `calculates fruit prices.'''`	Multiline comment enclosed by triple single quotes

Appendix B: Python Resources

B.1 PYTHON DOCUMENTATION

For general questions on Python, you will find these web links to be useful.

- **Python Homepage** www.python.org

  This is the official Python website.

- **Python Tutorial** http://docs.python.org/tut/tut.html

  This site provides a long tutorial for learning Python from scratch. Some chapters will be useful to help you thoroughly understand a specific topic. Use this if you don't know anything about a particular topic.

- **The Python Standard Library** http://docs.python.org/lib/lib.html

  The standard library is good for quickly finding a particular piece of information. Use this if you have an idea of what you are looking for but don't remember how it is spelled in Python.

- **Global Module Index** http://docs.python.org/2.7/py-modindex.html

  This is the documentation of all modules that come with Python. Look here if you know which module you want to use but don't know how it works.

- **Python Tutorials** http://awaretek.com/tutorials.html

  This is a large and exhaustive collection of Python tutorials covering most of the language aspects. It's for beginners, advanced users, and experienced developers.

B.2 PYTHON COURSES

- **Instant Python** www.hetland.org/python/instant-python.php

  This is a short and straight crash course in the programming language Python.

- **Think Python** www.greenteapress.com/thinkpython/

  Here you can download *Think Python*, which is an introduction to Python programming for beginners. It is a Free Book, which you are free to copy, distribute, and modify, as long as you attribute the work and don't use it for commercial purposes.

- **Website Programming Applications IV (Python) Information**

  http://bcu.copsewood.net/python/

  Here you can find a 12-week Python course with notes and tutorial exercises.

- **Uta Priss's Python Course** www.upriss.org.uk/python/PythonCourse. html

  Here you will find a set of Python exercises with answers grouped by topics.

- **Python Short Course** www.wag.caltech.edu/home/rpm/python_ course/

  This is a short, nice Python course provided by Richard P. Muller (California Institute of Technology) in .ppt, .pdf, and .html formats.

- **Dive into Python** www.diveintopython.net

  Dive into Python is a free Python book for experienced programmers.

- **Software Carpentry** http://software-carpentry.org

  This is a programming course for scientists. It has been curated since 1998.

B.3 BIOPYTHON DOCUMENTATION

- **Biopython Home Page** http://biopython.org/wiki/Main_Page

  This is the official website of Biopython.

- **Biopython Installation** http://biopython.org/DIST/docs/install/Installation.html

  Here you can find instructions on how to install Biopython on any operating system.

- **Getting Started** http://biopython.org/wiki/Getting_Started

  Here you can find basic information to get started with Biopython.

- **Biopython Documentation** http://biopython.org/wiki/Documentation

  This is a collection of links and references for Biopython beginners and advanced users.

- **Biopython Tutorial and Cookbook** http://biopython.org/DIST/docs/tutorial/Tutorial.html

  This site provides a long tutorial for learning Biopython from scratch.

- **Biopython Cookbook** http://biopython.org/wiki/Category:Cookbook

  Here you can find a (growing) list of Biopython working examples (in alphabetical order) that are not present in the Biopython tutorial.

B.4 OTHER PYTHON LIBRARIES

- **Matplotlib** http://matplotlib.org/

  This powerful library for creating diagrams is part of the Scientific Python library (SciPy).

- **Python Imaging Library (PIL)** www.pythonware.com/products/pil/

  PIL is a versatile library for manipulating images.

- **PyCogent** http://pycogent.org/

  PyCogent is a biological library with many functions similar to those of Biopython. However, its main strengths are handling RNA and phylogenetic analyses.

- **ModeRNA** http://iimcb.genesilico.pl/moderna/

  ModeRNA is a Python library to examine, manipulate, and build 3D models of RNA structures.

B.5 MOLECULAR VISUALIZATION SYSTEMS WRITTEN IN PYTHON

- **PyMOL** www.pymol.org

  PyMOL is an open-source, user-sponsored, molecular visualization system. It can produce high-quality 3D images of molecules.

- **UCSF Chimera** www.cgl.ucsf.edu/chimera/

  UCSF Chimera is a highly extensible program for interactive visualization and analysis of molecular structures and related data. High-quality images and animations can be generated.

B.6 VIDEOS

- **Videos Tagged with Python (from Best Tech Videos)**

  www.bestechvideos.com/tag/python/

  This is a collection of handpicked technical videos tagged with Python dedicated to finding the best educational content for developers, designers, managers, and other people in IT.

- **The pydb Debugger** www.bestechvideos.com/2006/12/16/introducing-the-pydb-debugger/

  This is a live demo showing how the pydb debugger works.

B.7 GENERAL PROGRAMMING

- **Learning to Program** www.freenetpages.co.uk/hp/alan.gauld/

  This is a general guide for absolute beginners to programming. The majority of the course is written in Python even though, in principle, it can be easily applied to any programming language.

Appendix C: Record Samples

C.1 A SINGLE PROTEIN SEQUENCE FILE IN FASTA FORMAT

```
>sp|P03372|ESR1_HUMAN Estrogen receptor OS = Homo sapiens GN =
   ESR1 PE = 1 SV = 2
MTMTLHTKASGMALLHQIQGNELEPLNRPQLKIPLERPLGEVYLDSSKPAVYNYPEGAAY
EFNAAAAANAQVYGQTGLPYGPGSEAAAFGSNGLGGFPPLNSVSPSPLMLLHPPPQLSPF
LQPHGQQVPYYLENEPSGYTVREAGPPAFYRPNSDNRRQGGRERLASTNDKGSMAMESAK
ETRYCAVCNDYASGYHYGVWSCEGCKAFFKRSIQGHNDYMCPATNQCTIDKNRRKSCQAC
RLRKCYEVGMMKGGIRKDRRGGRMLKHKRQRDDGEGRGEVGSAGDMRAANLWPSPLMIKR
SKKNSLALSLTADQMVSALLDAEPPILYSEYDPTRPFSEASMMGLLTNLADRELVHMINW
AKRVPGFVDLTLHDQVHLLECAWLEILMIGLVWRSMEHPGKLLFAPNLLLDRNQGKCVEG
MVEIFDMLLATSSRFRMMNLQGEEFVCLKSIILLNSGVYTFLSSTLKSLEEKDHIHRVLD
KITDTLIHLMAKAGLTLQQQHQRLAQLLLILSHIRHMSNKGMEHLYSMKCKNVVPLYDLL
LEMLDAHRLHAPTSRGGASVEETDQSHLATAGSTSSHSLQKYYITGEAEGFPATV
```

C.2 A SINGLE NUCLEOTIDE SEQUENCE FILE IN FASTA FORMAT

```
>ENSG00000188536|hemoglobin alpha 2
ATGGTGCTGTCTCCTGCCGACAAGACCAACGTCAAGGCCGCCTGGGGTAAGGTCGGCGCGCACGCT
GGCGAGTATGGTGCGGAGGCCCTGGAGAGGATGTTCCTGTCCTTCCCCACCACCAAGACCTACTTC
CCGCACTTCGACCTGAGCCACGGCTCTGCCCAGGTTAAGGGCCACGGCAAGAAGGTGGCCGACGCG
CTGACCAACGCCGTGGCGCACGTGGACGACATGCCCAACGCGCTGTCCGCCCTGAGCGACCTGCAC
GCGCACAAGCTTCGGGTGGACCCGGTCAACTTCAAGCTCCTAAGCCACTGCCTGCTGGTGACCCTG
GCCGCCCACCTCCCCGCCGAGTTCACCCCTGCGGTGCACGCCTCCCTGGACAAGTTCCTGGCTTCT
GTGAGCACCGTGCTGACCTCCAAATACCGTTAA
```

C.3 AN EXAMPLE OF AN RNA SEQUENCE IN FASTA FORMAT

```
>ENSG00000188536|hemoglobin alpha 2
ATGGTGCTGTCTCCTGCCGACAAGACCAACGTCAAGGCCGCCTGGGGTAAGGTCGGCGCGCACGCT
GGCGAGTATGGTGCGGAGGCCCTGGAGAGGATGTTCCTGTCCTTCCCCACCACCAAGACCTACTTC
CCGCACTTCGACCTGAGCCACGGCTCTGCCCAGGTTAAGGGCCACGGCAAGAAGGTGGCCGACGCG
CTGACCAACGCCGTGGCGCACGTGGACGACATGCCCAACGCGCTGTCCGCCCTGAGCGACCTGCAC
GCGCACAAGCTTCGGGTGGACCCGGTCAACTTCAAGCTCCTAAGCCACTGCCTGCTGGTGACCCTG
GCCGCCCACCTCCCCGCCGAGTTCACCCCTGCGGTGCACGCCTCCCTGGACAAGTTCCTGGCTTCT
GTGAGCACCGTGCTGACCTCCAAATACCGTTAA
```

C.4 A MULTIPLE SEQUENCE FILE IN FASTA FORMAT

```
>sp|P03372|ESR1_HUMAN Estrogen receptor OS = Homo sapiens GN =
    ESR1 PE = 1 SV = 2
MTMTLHTKASGMALLHQIQGNELEPLNRPQLKIPLERPLGEVYLDSSKPAVYNYPEGAAY
EFNAAAAANAQVYGQTGLPYGPGSEAAAFGSNGLGGFPPLNSVSPSPLMLLHPPPQLSPF
LQPHGQQVPYYLENEPSGYTVREAGPPAFYRPNSDNRRQGGRERLASTNDKGSMAMESAK
ETRYCAVCNDYASGYHYGVWSCEGCKAFFKRSIQGHNDYMCPATNQCTIDKNRRKSCQAC
RLRKCYEVGMMKGGIRKDRRGGRMLKHKRQRDDGEGRGEVGSAGDMRAANLWPSPLMIKR
SKKNSLALSLTADQMVSALLDAEPPILYSEYDPTRPFSEASMMGLLTNLADRELVHMINW
AKRVPGFVDLTLHDQVHLLECAWLEILMIGLVWRSMEHPGKLLFAPNLLLDRNQGKCVEG
MVEIFDMLLATSSRFRMMNLQGEEFVCLKSIILLNSGVYTFLSSTLKSLEEKDHIHRVLD
KITDTLIHLMAKAGLTLQQQHQRLAQLLLILSHIRHMSNKGMEHLYSMKCKNVVPLYDLL
LEMLDAHRLHAPTSRGGASVEETDQSHLATAGSTSSHSLQKYYITGEAEGFPATV
>sp|P62333|PRS10_HUMAN 26S protease regulatory subunit
    10B OS = Homo sapiens GN = PSMC6 PE = 1 SV = 1
MADPRDKALQDYRKKLLEHKEIDGRLKELREQLKELTKQYEKSENDLKALQSVGQIVGEV
LKQLTEEKFIVKATNGPRYVVGCRRQLDKSKLKPGTRVALDMTTLTIMRYLPREVDPLVY
NMSHEDPGNVSYSEIGGLSEQIRELREVIELPLTNPELFQRVGIIPPKGCLLYGPPGTGK
TLLARAVASQLDCNFLKVVSSSIVDKYIGESARLIREMFNYARDHQPCIIFMDEIDAIGG
RRFSEGTSADREIQRTLMELLNQMDGFDTLHRVKMIMATNRPDTLDPALLRPGRLDRKIH
IDLPNEQARLDILKIHAGPITKHGEIDYEAIVKLSDGFNGADLRNVCTEAGMFAIRADHD
FVVQEDFMKAVRKVADSKKLESKLDYKPV
>sp|P62509|ERR3_MOUSE Estrogen-related receptor gamma OS = Mus
    musculus GN = Esrrg PE = 1 SV = 1
MDSVELCLPESFSLHYEEELLCRMSNKDRHIDSSCSSFIKTEPSSPASLTDSVNHHSPGG
SSDASGSYSSTMNGHQNGLDSPPLYPSAPILGGSGPVRKLYDDCSSTIVEDPQTKCEYML
NSMPKRLCLVCGDIASGYHYGVASCEACKAFFKRTIQGNIEYSCPATNECEITKRRRKSC
QACRFMKCLKVGMLKEGVRLDRVRGGRQKYKRRIDAENSPYLNPQLVQPAKKPYNKIVSH
LLVAEPEKIYAMPDPTVPDSDIKALTTLCDLADRELVVIIGWAKHIPGFSTLSLADQMSL
LQSAWMEILILGVVYRSLSFEDELVYADDYIMDEDQSKLAGLLDLNNAILQLVKKYKSMK
LEKEEFVTLKAIALANSDSMHIEDVEAVQKLQDVLHEALQDYEAGQHMEDPRRAGKMLMT
LPLLRQTSTKAVQHFYNIKLEGKVPMHKLFLEMLEAKV
```

C.5 A GENBANK ENTRY

```
LOCUS         AY810830                705 bp    mRNA      linear    HTC 22-JUN-2006
DEFINITION    Schistosoma japonicum SJCHGC07869 protein mRNA, partial cds.
ACCESSION     AY810830
VERSION       AY810830.1  GI:60600350
KEYWORDS      HTC.
SOURCE        Schistosoma japonicum
  ORGANISM    Schistosoma japonicum
              Eukaryota; Metazoa; Platyhelminthes; Trematoda; Digenea;
              Strigeidida; Schistosomatoidea; Schistosomatidae; Schistosoma.
REFERENCE     1  (bases 1 to 705)
  AUTHORS     Liu,F., Lu,J., Hu,W., Wang,S.Y., Cui,S.J., Chi,M., Yan,Q.,
              Wang,X.R., Song,H.D., Xu,X.N., Wang,J.J., Zhang,X.L., Zhang,X.,
              Wang,Z.Q., Xue,C.L., Brindley,P.J., McManus,D.P., Yang,P.Y.,
              Feng,Z., Chen,Z. and Han,Z.G.
  TITLE       New perspectives on host-parasite interplay by comparative
              transcriptomic and proteomic analyses of Schistosoma japonicum
  JOURNAL     PLoS Pathog. 2 (4), E29 (2006)
   PUBMED     16617374
REFERENCE     2  (bases 1 to 705)
  AUTHORS     Liu,F., Lu,J., Hu,W., Wang,S.-Y., Cui,S.-J., Chi,M., Yan,Q.,
              Wang,X.-R., Song,H.-D., Xu,X.-N., Wang,J.-J., Zhang,X.-L.,
              Wang,Z.-Q., Xue,C.-L., Brindley,P.J., McManus,D.P., Yang,P.-Y.,
              Feng,Z., Chen,Z. and Han,Z.-G.
  TITLE       Direct Submission
  JOURNAL     Submitted (07-MAR-2005) Chinese National Human Genome Center at
              Shanghai, 351 Guo Shoujing Road, Shanghai 201203, China
FEATURES             Location/Qualifiers
     source          1..705
                     /organism="Schistosoma japonicum"
                     /mol_type="mRNA"
                     /db_xref="taxon:6182"
                     /clone="SJCHGC07869"
     CDS             <1..545
                     /note="similar to insulin receptor precursor"
                     /codon_start=3
                     /product="SJCHGC07869 protein"
                     /protein_id="AAX26719.2"
                     /db_xref="GI:76155430"
                     /translation="HVESDKVPVASIHATLNGPGSIRITWSNPVKPNGLIIHYLLRYR
                     PRNHDQSYTDSNHSSSDVSLPWLTKCISMSHWSADHSEHALTSSSYIAINQKEVSRSK
                     RGYNANSSTTDGGISIKDLSPGSYEFQILAVSLAGNGEWSPTVIFNIPFYTDHNGTIN
                     RMFIELLLFTVCVPCMPHHV"
ORIGIN
        1 ctcatgttga atctgataaa gttcctgtag catctattca tgcaacattg aatggtccgg
       61 gaagtatccg tattacgtgg tctaatccag tcaaacctaa tggtttaatt atacattatt
      121 tattgcggta tagaccaagg aatcatgatc agagttatac agatagtaac cattcgtctt
      181 cagatgtgtc gctgccatgg ttgacaaaat gtatttcgat gagtcattgg tcggctgacc
      241 attctgaaca cgcattgact tcaagttcat atatagctat taatcaaaaa gaagtatcac
      301 gaagtaaacg tggttataat gctaatagta gtactactga tggcggaatc tcaattaaag
      361 atttatcacc aggtagctat gaatttcaaa ttttagccgt ttctcttgct ggtaacggag
      421 aatggagtcc aaccgtaata ttcaatattc cattctatac agaccataat ggcacaataa
      481 accgtatgtt tatagaactc ttattattta cagtttgtgt cccatgtatg ccgcatcacg
      541 tgtaatgttt tgattaagga gattcaaatt ttatacgttc tctcataagt gatctttact
      601 tttaattgtg tgctctaaga atatacgcat tttcggttca atagattcta aaacaatgca
      661 attatgagtt agatttcatt aatgcatatg taagctaatt ttcta
```

C.6 AN EXAMPLE OF A PDB FILE HEADER (PARTIAL)

```
HEADER    ADENYLATE KINASE                        28-JUL-95   3AKY
TITLE     STABILITY, ACTIVITY AND STRUCTURE OF ADENYLATE KINASE
TITLE    2 MUTANTS
COMPND    MOL_ID: 1;
COMPND   2 MOLECULE: ADENYLATE KINASE;
COMPND   3 CHAIN: A;
COMPND   4 SYNONYM: ATP\:AMP PHOSPHOTRANSFERASE, MYOKINASE;
COMPND   5 EC: 2.7.4.3;
COMPND   6 ENGINEERED: YES;
COMPND   7 MUTATION: YES
SOURCE    MOL_ID: 1;
SOURCE   2 ORGANISM_SCIENTIFIC: SACCHAROMYCES CEREVISIAE;
SOURCE   3 ORGANISM_COMMON: BAKER'S YEAST;
SOURCE   4 ORGANISM_TAXID: 4932;
SOURCE   5 EXPRESSION_SYSTEM: ESCHERICHIA COLI;
SOURCE   6 EXPRESSION_SYSTEM_TAXID: 562;
SOURCE   7 EXPRESSION_SYSTEM_PLASMID: PUAKY
KEYWDS    ATP:AMP PHOSPHOTRANSFERASE, MYOKINASE, ADENYLATE KINASE
EXPDTA    X-RAY DIFFRACTION
AUTHOR    U.ABELE,G.E.SCHULZ
REVDAT   2   24-FEB-09 3AKY    1              VERSN
REVDAT   1   14-NOV-95 3AKY    0
JRNL        AUTH    P.SPUERGIN,U.ABELE,G.E.SCHULZ
JRNL        TITL    STABILITY, ACTIVITY AND STRUCTURE OF ADENYLATE
JRNL        TITL 2 KINASE MUTANTS.
JRNL        REF     EUR.J.BIOCHEM.                 V. 231    405 1995
JRNL        REFN                     ISSN 0014-2956
JRNL        PMID    7635152
JRNL        DOI     10.1111/J.1432-1033.1995.TB20713.X
REMARK   1
REMARK   1 REFERENCE 1
REMARK   1  AUTH    U.ABELE,G.E.SCHULZ
REMARK   1  TITL    HIGH-RESOLUTION STRUCTURES OF ADENYLATE KINASE
REMARK   1  TITL 2 FROM YEAST LIGATED WITH INHIBITOR AP5A, SHOWING
REMARK   1  TITL 3 THE PATHWAY OF PHOSPHORYL TRANSFER
REMARK   1  REF     PROTEIN SCI.                   V.   4   1262 1995
REMARK   1  REFN                     ISSN 0961-8368
```

C.7 AN EXAMPLE OF PDB FILE ATOMIC COORDINATE LINES (PARTIAL)

```
ATOM      1  N   GLU A   3      14.566  13.214  -5.148  1.00124.25           N
ATOM      2  CA  GLU A   3      14.723  13.413  -3.720  1.00104.91           C
ATOM      3  C   GLU A   3      15.364  14.778  -3.535  1.00 90.84           C
ATOM      4  O   GLU A   3      16.534  14.887  -3.163  1.00 89.79           O
ATOM      5  CB  GLU A   3      15.618  12.318  -3.132  1.00110.77           C
ATOM      6  CG  GLU A   3      15.697  11.047  -3.981  1.00106.03           C
ATOM      7  CD  GLU A   3      14.442  10.186  -3.891  1.00 98.15           C
ATOM      8  OE1 GLU A   3      13.330  10.748  -3.800  1.00102.08           O
ATOM      9  OE2 GLU A   3      14.568   8.944  -3.933  1.00 89.07           O
ATOM     10  H   GLU A   3      15.307  12.837  -5.660  1.00  0.00           H
ATOM     11  N   SER A   4      14.641  15.804  -3.965  1.00 75.08           N
ATOM     12  CA  SER A   4      15.109  17.173  -3.848  1.00 62.63           C
ATOM     13  C   SER A   4      13.930  18.076  -3.553  1.00 54.45           C
ATOM     14  O   SER A   4      12.858  17.909  -4.136  1.00 57.43           O
ATOM     15  CB  SER A   4      15.780  17.610  -5.145  1.00 68.59           C
ATOM     16  OG  SER A   4      16.975  16.884  -5.352  1.00 81.78           O
ATOM     17  H   SER A   4      13.751  15.640  -4.335  1.00  0.00           H
ATOM     18  HG  SER A   4      16.747  15.954  -5.444  1.00  0.00           H
ATOM     19  N   ILE A   5      14.130  19.027  -2.645  1.00 40.73           N
ATOM     20  CA  ILE A   5      13.056  19.926  -2.242  1.00 29.28           C
ATOM     21  C   ILE A   5      13.594  21.322  -1.912  1.00 22.93           C
ATOM     22  O   ILE A   5      14.740  21.479  -1.485  1.00 19.39           O
ATOM     23  CB  ILE A   5      12.288  19.365  -1.004  1.00 25.75           C
ATOM     24  CG1 ILE A   5      11.029  20.188  -0.731  1.00 40.48           C
ATOM     25  CG2 ILE A   5      13.186  19.406   0.225  1.00 19.34           C
ATOM     26  CD1 ILE A   5       9.779  19.701  -1.453  1.00 43.17           C
ATOM     27  H   ILE A   5      15.021  19.138  -2.251  1.00  0.00           H
ATOM     28  N   ARG A   6      12.811  22.334  -2.260  1.00 16.17           N
ATOM     29  CA  ARG A   6      13.027  23.673  -1.755  1.00 17.99           C
ATOM     30  C   ARG A   6      11.709  24.062  -1.136  1.00 18.10           C
ATOM     31  O   ARG A   6      10.677  24.081  -1.809  1.00 16.57           O
ATOM     32  CB  ARG A   6      13.388  24.629  -2.895  1.00 19.64           C
ATOM     33  CG  ARG A   6      14.657  24.227  -3.637  1.00 18.34           C
ATOM     34  CD  ARG A   6      15.031  25.230  -4.717  1.00 37.27           C
ATOM     35  NE  ARG A   6      13.950  25.410  -5.682  1.00 59.35           N
ATOM     36  CZ  ARG A   6      13.635  24.529  -6.626  1.00 61.85           C
```

C.8 AN EXAMPLE OF THE SEQRES LINES OF A PDB FILE (FROM FILE 1TDL)

```
SEQRES 1 A 223 ILE VAL GLY GLY TYR THR CYS GLY ALA ASN THR VAL PRO
SEQRES 2 A 223 TYR GLN VAL SER LEU ASN SER GLY TYR HIS PHE CYS GLY
SEQRES 3 A 223 GLY SER LEU ILE ASN SER GLN TRP VAL VAL SER ALA ALA
SEQRES 4 A 223 HIS CYS TYR LYS SER GLY ILE GLN VAL ARG LEU GLY GLU
SEQRES 5 A 223 ASP ASN ILE ASN VAL VAL GLU GLY ASN GLU GLN PHE ILE
SEQRES 6 A 223 SER ALA SER LYS SER ILE VAL HIS PRO SER TYR ASN SER
SEQRES 7 A 223 ASN THR LEU ASN ASN ASP ILE MET LEU ILE LYS LEU LYS
SEQRES 8 A 223 SER ALA ALA SER LEU ASN SER ARG VAL ALA SER ILE SER
SEQRES 9 A 223 LEU PRO THR SER CYS ALA SER ALA GLY THR GLN CYS LEU
SEQRES 10 A 223 ILE SER GLY TRP GLY ASN THR LYS SER SER GLY THR SER
SEQRES 11 A 223 TYR PRO ASP VAL LEU LYS CYS LEU LYS ALA PRO ILE LEU
SEQRES 12 A 223 SER ASP SER SER CYS LYS SER ALA TYR PRO GLY GLN ILE
SEQRES 13 A 223 THR SER ASN MET PHE CYS ALA GLY TYR LEU GLU GLY GLY
SEQRES 14 A 223 LYS ASP SER CYS GLN GLY ASP SER GLY GLY PRO VAL VAL
```

```
SEQRES 15 A 223 CYS SER GLY LYS LEU GLN GLY ILE VAL SER TRP GLY SER
SEQRES 16 A 223 GLY CYS ALA GLN LYS ASN LYS PRO GLY VAL TYR THR LYS
SEQRES 17 A 223 VAL CYS ASN TYR VAL SER TRP ILE LYS GLN THR ILE ALA
SEQRES 18 A 223 SER ASN
```

C.9 AN EXAMPLE OF THE CUFFCOMPARE OUTPUT FOR THREE SAMPLES (Q1, Q2, AND Q3)

Because of the length of each line, the first field of each new line is in bold, and different lines are highlighted using indentation.

Medullo-Diff_00000001 XLOC_000001 Lypla1|uc007afh.1
  q1:NSC.P419.228|uc007afh.1 |100|35.109496|34.188903
  |36.030089|397.404732|2433 q2:NSC.
  P429.18|uc007afh.1|100 |15.885823|15.240240
  |16.531407|171.011325 |2433 q3:NSC.
  P437.15|uc007afh.1|100 |18.338541|17.704857|18.97222
  4|181.643949|2433
Medullo-Diff_00000002 XLOC_000002 Tcea1|uc007afi.2
  q1:NSC.P419.228|uc007afi.2|
  18|1.653393|1.409591|1.897195 |18.587029|2671
  - q3:NSC.P437.108|uc007afi.2|100 |4.624079|4.258801|4
  .989356|45.379750|2671
Medullo-Diff_00000003 XLOC_000002 Tcea1|uc011wht.1
  q1:NSC.P419.228|uc011wht.1|100 |9.011253|8.560848
  |9.461657|101.302266|2668 q2:NSC.
  P429.116|uc011wht.1|100 |6.889020|6.503460|7.27458
  0|73.238938 |2668q3:NSC.P437.108 |uc011wht.1|90
  |4.170527 |3.817430|4.523625|40.928694|2668
Medullo-Diff_00000004 XLOC_000003 Tcea1|uc007afi.2
  q1:NSC.P419.231|NSC.P419.231.1
  |100|31.891396|30.892103 |32.890690|379.023601
  |1568q2:NSC.P429.121|NSC.P429.121.1 |100|27.991543
  |27.007869|28.975218 |313.481210|1532 -
Medullo-Diff_00000005 XLOC_000002 Tcea1|uc007afi.2
  q1:NSC.P419.236 |NSC.P419.236.1
  |100|1.164739|0.868895 |1.460583|13.879881|- - -
Medullo-Diff_00000006 XLOC_000004 Atp6v1h|uc007afn.1
  q1:NSC.P419.55|uc007afn.1 |100|39.526818 |38.58510
  2|40.468533|455.599775|1976 q2:NSC.
  P429.43|uc007afn.1 |100|25.034945 |24.182398|25.887
  493|271.738343|1976 q3:NSC.P437.37|uc007afn.1
  |100|20.848047 |20.043989|21.652104|205.866771|1976

C.10 SOLVENT ACCESSIBILITY OF AMINO ACIDS IN KNOWN PROTEIN STRUCTURES

Data are derived from Domenico Bordo and Patrick Argos, "Suggestions for 'Safe' Residue Substitutions in Site-Directed Mutagenesis," *Journal of Molecular Biology* 217 (1991): 721–729, and converted into relative values. The data for this table were calculated from 55 proteins (5,624 residues) in the PDB database.

Amino Acid			Solvent Exposed Are		
Name	3-letter	1-letter	> 30 Å² (exposed)	< 10 Å² (buried)	10–30 Å²
Alanine	Ala	A	48%	35%	17%
Arginine	Arg	R	84%	5%	11%
Aspartic Acid	Asp	D	81%	9%	10%
Asparagine	Asn	N	82%	10%	8%
Cysteine	Cys	C	32%	54%	14%
Glutamic Acid	Glu	E	93%	4%	3%
Glutamine	Gln	Q	81%	10%	9%
Glycine	Gly	G	51%	36%	13%
Histidine	His	H	66%	19%	15%
Isoleucine	Ile	I	39%	47%	14%
Leucine	Leu	L	41%	49%	10%
Lysine	Lys	K	93%	2%	5%
Methionine	Met	M	44%	20%	36%
Phenylalanine	Phe	F	42%	42%	16%
Proline	Pro	P	78%	13%	9%
Serine	Ser	S	70%	20%	10%
Threonine	Thr	T	71%	16%	13%
Tryptophan	Trp	W	49%	44%	7%
Tyrosine	Try	Y	67%	20%	13%
Valine	Val	V	40%	50%	10%

Appendix D: Handling Directories and Programs with UNIX

L EARNING GOAL: You can navigate through directories and run programs using a UNIX shell.

D.1 IN THIS CHAPTER YOU WILL LEARN

- What a computer "shell" is

- How to use UNIX/Linux shell commands

- How to set UNIX/Linux environment variables

- How to run programs from the command line

D.2 STORY: THE UNIX COMMAND SHELL: ANOTHER WAY TO TALK TO YOUR COMPUTER

To directly communicate with the operating system (OS) of your computer, you can use a so-called *shell*. The shell is a *program* that provides an interface to the OS and, as such, provides access to the capabilities of your computer on a more basic level than does a graphical desktop. In particular, through *shell programming languages*, you can send commands to the computer and execute programs of the operating system. Shells have capabilities such as displaying the contents of directories or the status of a program during its execution. However, their main aim is to invoke ("launch") other programs. OS shells are generally one of two types: textual or graphical. A command-line shell supplies a textual interface to the

OS. Conversely, a graphical shell (or graphical desktop) provides the user with a graphical interface, which allows the user to interact with the computer using images rather than typing commands. In a graphical desktop, you can navigate through directories (in a file browser), examine what happens on your computer (by different dialogs), and start programs (by clicking icons). The important thing to know is that a graphical desktop is not monolithic but delegates most of the work by starting other programs. On a smartphone where you can install your own applications, this concept is even more visible. In a command-line shell (or simply shell), you do the same by typing commands instead of clicking icons. A brief historical background on OS and user interfaces is provided in Box D.1.

The relative advantages of textual shells and graphical interfaces are often debated. Certain operations can be performed much faster from a shell than from a graphical interface, but the latter is simpler to use in many aspects. The best choice is often determined by the way in which you use your computer. The UNIX/Linux command shell is referred to in many parts of this book, in particular for installing programs and in Chapter 14. This appendix explains the very basics of the UNIX shell on Linux or Mac OS X.

D.2.1 Problem Description: Managing Sequence Files

Sequences and the related annotations are often recorded in text files (see Box D.2 for a definition of text files and a selection of text editors). Text files are stored in directories (folders) on your computer. To manage a collection of sequence data, you often need to rearrange the files and sort them into your own system of directories. What do you have to do if you want to see what files are there, create and delete folders, copy and rename files, and move them around?

BOX D.1 A BIT OF HISTORY

Different OSs support different shell types and interfaces. Nowadays, the two major classes of operating systems are UNIX and Windows. Linux and Mac OS X are UNIX-like systems, i.e., operating systems based on UNIX, which was developed at the Bell Laboratories in the late 1960s. Linux is an open-source version of UNIX specifically developed for microcomputers (i.e., home computers and laptops). Mac OS X uses Darwin, an OS based on BSD (Berkeley Software Distribution), the version of UNIX developed at the University of California, Berkeley, in the 1970s and early 1980s.

BOX D.2   TEXT FILES AND TEXT EDITORS

A text file is a kind of file that contains plain text; i.e., it contains no format-ting (e.g., italics) and no metadata (e.g., page layout). Text files are struc-tured as a sequence of text lines and are platform independent, which means that they can be read on UNIX, Windows, Macintosh, etc. without differences in the formatting. In particular, they are readable by computer programs and are the types of files used to write programming source code. To create, read, and write text files, you can use programs called "text editors," which are often provided by the OS. Alternatively, you can down-load them from the Internet. Some are very simple to use (e.g., *pico, nano, gedit*), and some others are more complicated (*emacs, vi*). Text editors usu-ally have a simple visual help listing the available commands (in the form of text or icons), including file saving and exiting. You can choose your text editor on the basis of your personal taste. In this book we will refer to *gedit* (http://projects.gnome.org/gedit/). A screenshot of the gedit graphical user interface is reported in Figure D.1. It displays the content of the P62805. seq file (i.e., the Uniprot sequence P62805 in FASTA format; you can retrieve it at http://www.uniprot.org/uniprot/P62805.fasta), which is men-tioned later in the chapter.

You can control each of these operations by typing specific instruc-tions (UNIX commands) in the command shell of your computer. In this appendix, you will learn how to use UNIX commands to handle files and directories with sequence data. The same commands will also help you to use programs that have no graphical interface, to install programs that

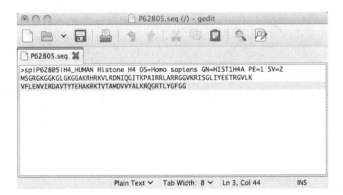

FIGURE D.1   The gedit graphical user interface. *Note:* The gedit interface displays the content of the P62805 Uniprot sequence in FASTA format (www.uniprot.org/uniprot/P62805.fasta).

have no .exe file available, to compile C programs, to log in to remote computers, to have more control of your system, to perform operations that repeat often, to better know how your computer works, etc.

D.2.2 Example UNIX Session

```
mkdir sequences
cp P62805.seq sequences/P62805-b.seq
rm P62805.seq
cd sequences
ls
P62805-b.seq
```

D.3 WHAT DO THE COMMANDS MEAN?

The P62805.seq file can be retrieved at www.uniprot.org/uniprot/P62805.fasta and saved in your home directory.

The commands displayed in the example in Section D.2.2 must be typed at the prompt in a terminal. To do this, you first have to start a terminal session.

Q & A: WHAT EXACTLY IS A PROMPT?

The location where things you enter on the keyboard will appear is called a *prompt*. It is often marked by a cursor (a thin _ or | symbol that is sometimes blinking). The UNIX prompt is commonly marked by a symbol such as % (Mac OS), $, or > (Linux). Sometimes the prompt also contains the name of the user or the current directory followed by one of these symbols. In the Python shell (see Chapter 1), there is a prompt as well: the >>>.

D.3.1 Starting a Command-Line Shell

To use the command-line shell, you first need to open a terminal window on the screen displaying a prompt where you can enter commands (see Figure D.2). The terminal window is also called *text console, shell terminal, shell console, text terminal, terminal window*, etc. In this book, we call it *terminal*, but sometimes we simply refer to it as *shell*. The procedure to start a terminal differs from one OS to another. On Linux you usually click on the "Terminal" icon from the Applications > Accessories menu or press Ctrl-T. On Mac OS X you click on the "Terminal" icon in the Application > Utilities directory. On Windows, you can open a terminal by Start → Execute → cmd. However, the Window command syntax is mostly different from the UNIX/Linux one and won't be

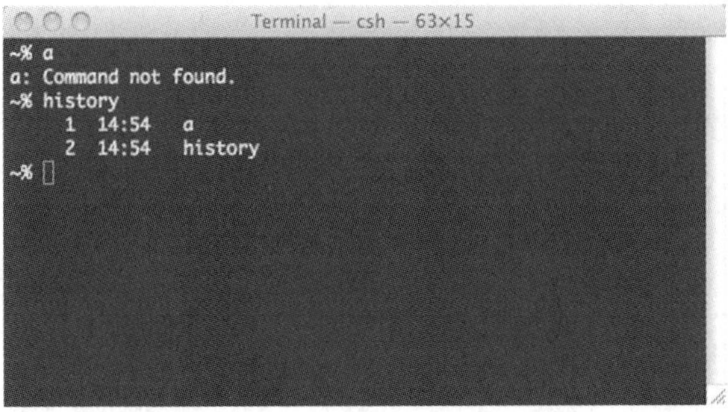

FIGURE D.2   The text console. *Note:* The text console shows the result of typing a nonexistent command (a) and an existent command (history).

described here. See, for example, http://ss64.com/nt/ for a complete list of Windows CMD commands.

D.3.2 Using the Command-Line Shell

Now you have a terminal on your computer screen that displays a prompt at the top left (see Figure D.2), and you are likely wondering about what you need to do in order to send a command to the computer. To answer this question, you need some additional information:

- Like a graphical file browser, a command shell is always located in a particular directory. All files on your computer are stored in a directory structure; i.e., the file system is arranged in a hierarchical structure, like a phylogenetic tree (see Figure D.3). The directory on top of the hierarchy is called the *root* directory. Any new terminal is started in your home directory. This means that when you start a terminal, the actions you perform by typing commands at the prompt will have an effect on your home directory files and subdirectories, unless you move to a different directory or you explicitly specify that a command must act on a file or directory located elsewhere. The directory where a terminal is when you are typing a command is also called *current working directory*.

- To send a command to your computer, you first have to write a UNIX *command* immediately after the prompt and second press the Return or Enter key. No action will be performed unless you do both steps. A command consists of the command name (to be written in the first position near the prompt) and zero or more arguments, separated by a blank

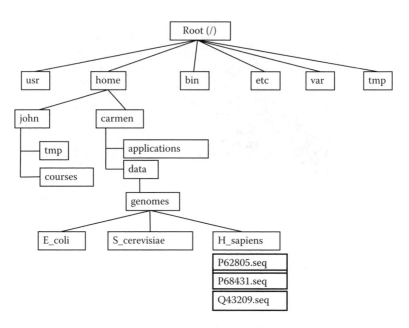

FIGURE D.3   The hierarchical structure of directories.

space. Furthermore, commands may have "switches" (or "options"), which change the behavior of a command. They must follow the command name and are usually preceded by a "-". (See also Box D.3.)

If you type

```
ls -l
```

BOX D.3   HISTORY AND AUTOCOMPLETION IN UNIX SHELLS

The up and down arrow keys show you the most recently used commands in the console. In UNIX shells, the [Tab] key tries to autocomplete names of programs and files. In other words, by typing the beginning of a command name, a filename, or a directory name and pressing the [Tab] key, the rest of the name will be completed automatically. If the shell finds more than one name beginning with the letter(s) you have typed, it will beep, prompting you to type more letters before pressing the [Tab] key again. In some versions or settings, a list of the available names beginning with the letters you typed will appear on the screen. Open a terminal on your computer and type

```
his[Tab]
```

and see what happens.

and then press the Return key, you will see that a long (i.e., detailed) list of the content (in terms of files and subdirectories) of the current directory is displayed on the screen. Try also

```
ls
```

What are the differences?

On the other hand, if you now type "a" and then press the Return key, you will see that you receive an error message saying command not found (see Figure D.2). Let's analyze these events step by step:

- Before you type "ls –l", "ls", or "a", and every time there is an empty line after the prompt, the terminal is ready to receive the input.

- When you type "ls –l", "ls", or "a", you do not get any computer reaction; you can enter as many characters as you want before actually sending them to the computer.

- When you press Return, you receive a feedback from the computer. Pressing the Return button causes the characters (i.e., your input command) to be sent to the system.

- The feedback consists of a list of filenames and/or directory names or an error message that says a: command not found. The latter means that the computer does not recognize "a" as a command name. This implies that *you are supposed to know command names*. For example, now you know that ls is a command name and that –l is a possible command option ("switch").

What happens behind the scenes when you type an existing command name such as ls at the prompt of your terminal window and then you press the Return key? The execution of a program on your computer is started. The output of this program will be what you expect from the command you gave to the computer.

D.3.3 Common UNIX Commands

In the UNIX session in Section D.2.2, the most common UNIX commands occur. Their purpose explained line by line is as follows:

```
mkdir sequences
```

creates a directory with the name sequences. mkdir, which is the UNIX command; sequences is its argument.

```
cp P62805.seq sequences/P62805-b.seq
```

copies the file P62805.seq, which was previously downloaded to the current directory from http://www.uniprot.org/uniprot/P62805.fasta, into the sequences subdirectory with a different name (P62805-b.seq instead of P62805.seq). The cp command requires two arguments: first the file to be copied and second the directory or filename where you want to copy it.

```
rm P62805.seq
```

removes the file P62805.seq from the current working directory. There is no undeleting in UNIX, so the file is lost forever!

```
cd sequences
```

moves the terminal to the sequences subdirectory (cd = change directory).

```
ls
```

displays the content of the sequences directory.

Q & A: When I Start a New Terminal, in Which Directory Am I?

When you first log in or start a new terminal, your current working directory is your *home* directory. Your home directory has the same name as your username, for example, john, and it is where your personal files and subdirectories are saved.

Q & A: Where Can I Find the Description of a Command and Its Options?

There is a command manual where you can find the description of each command and of the option(s) a particular command can take. The manual pages for a <command name> can be accessed by typing

```
man <command name> [Enter]
```

Manual pages can be exited by typing q.

The command

```
whatis <command name> [Enter]
```

returns a brief description of <command name>.

You will also find a detailed description of all UNIX/Linux commands and options on the Internet. The most common ones are listed in Appendix A, "Command Overview."

Q & A: What if I Make a Typo While Typing a Command?

If you notice you have made a typo, you can press Ctrl-U to cancel the whole line. Alternatively, you can simply use the Delete key on your keyboard to remove all characters. Pressing Enter works sometimes, although there is a risk that you execute another command unintentionally. Just never try pressing Enter when your command starts with rm. (It deletes files.)

Q & A: How Many Commands Do I Need to Know?

There are hundreds of UNIX commands. For example, Siever and colleagues[*] report 687 commands in Linux. Notice that even in the same operating system, the number and types of commands may differ depending on the distribution you use and the software packages that have been installed. Fortunately, the most common commands (e.g., to list the content of a directory or to delete a file) remain unchanged between distributions.

When you know between 10 and 20 commands, you will be able to navigate through your system swiftly: move from one directory to another; list the contents of a directory; verify in which location you are; create, delete, copy, and rename files and directories; and, of course, start programs. With another 20 commands, you can control what happens on your computer very directly: install programs, add user accounts, and monitor and terminate running programs. When you are able to fluently use 100 or more UNIX commands, you are probably an experienced system administrator.

D.3.4 More UNIX Commands
Deleting Files

```
rm P62805.seq
```

---

[*] E. Siever, A. Weber, S. Figgins, R. Love, and A. Robbins, *Linux in a Nutshell*, 5th ed. (Sebastopol, CA: O'Reilly Media, 2005).

has the effect of permanently removing the file `P62805.seq`. After the process `rm P62805.seq` has finished running, the shell returns the UNIX prompt ($, %, >, etc.), indicating that it is waiting for further commands.

Listing Files and Directories

```
ls
```

This command is used for listing files and subdirectories present in the directory where you typed the command. Suppose you are `Carmen` (see Figure D.3), then:

```
ls
applications data
```

A file or a directory in the directory tree is univocally identified by its *path*, which is a sort of unique "address" for that file or directory. So, the full path of the `H_sapiens` directory in Figure D.3 is

```
/home/carmen/data/genomes/H_sapiens/
```

And if Carmen wishes to list the contents of this directory, she can type

```
ls /home/carmen/data/genomes/H_sapiens/
```

However, if she typed this command from her home directory (`/home/carmen/`), she would be able to directly list the directories that are "below"; i.e.,

```
ls data/genomes/H_sapiens/
```

Shortcuts exist for some special directories:

- `/`: root directory (at the top of the hierarchy)
- `.`: current directory
- `..`: parent directory
- `~`: home directory

Try the following commands in your terminal and see what happens:

```
ls /
ls .
ls ..
ls ~
```

Q & A: I Find Writing These Long Directory Names Annoying. Is This Really Necessary?

No. UNIX terminals can autocomplete names for you. Try writing the first character of a directory and press [TAB]. The terminal will try to guess the name and complete it. Only if there are several alternatives, you won't see anything. Then you need to supply more characters and press [TAB] again or press [TAB] twice to see a list of alternatives.

Changing Directories

You can move from one directory to another using the command cd (change directory) followed by the path of the directory you want to move to. Try the following commands (replace /home/carmen/ with the path of your home directory):

```
cd ..
ls
cd /
ls
cd /home/carmen/
ls
```

If you type, from any directory,

```
cd
```

or

```
cd ~
```

you will be moved to your home directory.

Notice that if you are in Carmen's home directory and want to move to John's, either you can specify the full path

```
cd /home/john/
```

or you can go first up to the parent directory (.. /) and then down to `john`, i.e.,

```
cd ../john
```

If you want to go *two* directories up, type

```
cd ../../
```

and so on.

Copying Files

```
cp file1 file2
```

makes a copy of *file1* and calls it *file2*. You end up with two files with the same content.

Moving (or Renaming) Files and Directories

```
mv file1 file2
```

moves a file from one place to another. This has the effect of moving (rather than copying) the file, so you end up with one file instead of two. It can also be used to rename a file by moving the file to the same directory but giving it a different name. Moving can also be applied to entire directories.

Creating Directories

```
mkdir <directory name>
```

creates a subdirectory in your current working directory.

Removing Directories

```
rmdir <directory name>
```

removes a directory. You have to empty the directory before deleting it.

Visualizing the Path of the Current Directory

```
pwd
```

enables you to find out in which directory you are in relation to the root directory.

Search a File for Keywords

```
grep <keyword> myfile.txt
```

searches files for specified keywords or patterns. The output is printed to the screen and consists of a list of the lines in `filename.txt` that contain `<keyword>`. Notice that unless you specify the `-i` option, the `grep` command is case-sensitive. In other words, "Science" and "science" will be considered different keywords.

Displaying the Command History List
If you type

```
history
```

you will get a list of the last commands entered, each accompanied by some additional information such as the time you sent the command. This means that everything you type in the terminal is recorded by the terminal (but this information is kept confidential). The `history` command accesses your command history and prints it to the screen.

Redirecting the Output of a Command
The > symbol can be used to redirect the output of a command. For example,

```
grep HUMAN filename.txt > output.txt
```

writes a list of the lines in `filename.txt` that contain the keyword `HUMAN` to the `output.txt` file.

More UNIX/Linux commands and utilities are provided in Appendix A.

D.3.5 UNIX Variables

A UNIX variable is a symbolic name that your system associates to a piece of information, which represents the "value" of the variable. The variable *name* can be used every time you want to recall the variable *value*. They work similarly to string variables in Python. Examples of variables (name/value pairs) are USER (your login name), OSTYPE (the type of operating system you are using), HOST (the name of the computer you are using), HOME (the path of your home directory), PRINTER (the default printer to send print jobs), and PATH (the directories the shell searches to find a command). The value of UNIX variables can be set using specific commands and procedures that will be described next.

UNIX variables can be of two types: *environment variables* and *shell variables*. The latter apply only to the current instance of the shell terminal.

An example of a shell variable is the history variable, the value of which corresponds to the number of shell commands that will be saved in the history list (which can be displayed by typing history at the prompt). The default value of this variable is 100. If now you type

```
history
```

the last 100 commands will appear on the screen (unless you have typed fewer commands so far). If you set the history variable to 200 in a given terminal, the variable will keep the default value (100) in a different terminal.

Environment variables, unlike shell variables, apply to all active terminal sessions. You can change them as well, and the new value will be active in all new terminal sessions you open after that. So, when you change environment variables, they won't work in old terminal windows apart from the one in which you changed them. Environment variables are indicated with uppercase names, and shell variables are indicated with lowercase names.

Why are UNIX variables useful? They are basically used to configure programs. For example, if you want to run a local BLAST search, you can type (provided you have already downloaded and installed the BLAST package; see Recipe 11)

```
blastp -query input_file -db database -out output_file
```

Where does the shell look for the blastp program? One time-consuming option is that the shell searches every directory in your computer. Otherwise, you can write the program name preceded by the complete path to its directory or move to the directory where the program is whenever you want to run it (e.g., /Applications/blast/blastp). This requires you to know the location of every program you run (including UNIX commands). A better approach is that the shell knows where to look for programs. It accesses a variable called PATH that contains a list of all directories where programs can be found. If you want to run a program the path of which is not present in the PATH list, you have to modify the value of the PATH variable by adding the missing path. When the system returns a message saying "command: Command not found," this indicates that either that command doesn't exist at all or it is simply in a directory the path of which is not recorded in your PATH variable. This is why when installing BLAST, you need to add the location of the blastp program to the PATH environment variable.

Q & A: What Is the Difference between a Program and a Command in UNIX?

There is none. For each command in the UNIX shell, there is a program somewhere that is executed when you type its name. The shell does not know any commands by itself, but every time you type something, it sees whether it can find a program that matches your input and executes it.

Display, Set, and Unset UNIX Variables

The shell command to display environment variables is printenv or simply env. Open a terminal and type

```
printenv
```

Shell variables can be both set and displayed using the set command. They can be unset by using the unset command. For example, to change the number of shell commands saved in the history list, you can type

```
set history = 200
```

Notice that this sets the history variable only for the lifetime of the current terminal session.

If you want to display the value associated to an environment or shell variable, you have to use the command echo followed by $<variable name>. For example, if you want to display the path of your home directory, type

```
echo $HOME
```

Notice that many environment variables are set by the system. The commands to set and unset an environment variable are setenv and unsetenv, respectively, followed by two arguments, <variable name> and <new value>, separated by a space. For example, if you are using the *tcshell* (which is a specific kind of shell) and want to switch to the *bash* shell (which is another kind of shell), you can type

```
setenv SHELL /bin/bash
```

What if you want to permanently change variable values? This can be done by setting variable values in special files, which are called *initialization files*. When you log in to a UNIX/Linux host, the OS always reads the initialization files present in your home directory. These files have special names and contain instructions to set up your working environment. Initialization files for the C shell and tcshell are .login and .cshrc, and for the bash shell they are .bash_login and .bashrc. Note that all these filenames begin with a dot. At login the shell automatically first reads the .cshrc (or .bashrc) file followed by .login (or .bash_login).

However, .login (.bash_login) is read only at the login, whereas .cshrc (.bashrc) is read each time a shell terminal is started. It is a good rule to set environment variables in the .login (.bash_login) file and the shell variables in the .cshrc (.bashrc) file. The setting instruction syntax depends on the shell you are using.

As an example, we show how to set the history variable in the .cshrc file. First you have to open the file with a text editor (see Box D.2). *Gedit* (http://projects.gnome.org/gedit) is an easy, user-friendly text editor.

```
gedit ~/.cshrc
```

Add the following line to the file:

```
set history = 200
```

Save the file and exit. If you now want the shell to read the .cshrc file, either you start a new terminal or you force the shell to reread the file in the current terminal. This can be done with the command source:

```
source .cshrc
```

If you want to display the new history value, you can now type

```
echo $history
```

D.4 EXAMPLES

Example D.1 Finding All Chicken Proteins

As a result of BLAST, you have a set of 2,000 protein sequences downloaded from GenBank/Uniprot as one big FASTA file (called sequences.fasta). They are from many different organisms, but you are interested in chicken proteins only. You need to prepare a subset of these sequences to start evaluation. You know that all proteins contain a species name in Latin.

```
> gi 1234567 | gallus gallus | lysozyme
THISISAPRQTEINSEKWENCE...
```

Solution:

```
grep -A 2 gallus sequences.fasta > chicken.fasta
```

What does this command do? grep searches keywords in files. It is case-sensitive, so it does matter whether you search "gallus" or "GALLUS". The -A 2 switch collects not only the line with the keyword but also a second line immediately after that (the one with the sequence). Finally, the > symbol stores the output in the chicken.fasta text file. See also Box 9.2, Chapter 9 for details on the grep command.

Example D.2 Running BLAST from the Command Line

You can run BLAST either on the Internet or locally (see Recipe 11). Running BLAST locally has many advantages. For example, you can align a query sequence to your own database of protein or gene sequences, or you can align pairs of sequences several times and, e.g., only if a condition is fulfilled. To do so, you have to download the

BLAST+ package first, install it, and run it from the shell command line by specifying options and arguments. All you need to accomplish this task can be found in the "Introduction to BLAST" document provided by NCBI, which can be found at www.ncbi.nlm.nih. gov/books/NBK1762/.

D.5 TESTING YOURSELF

Exercise D.1 Using the `cd` and `ls` Commands

Open a console and go to the Desktop folder to see what files are there. Then go back to your home directory.

Exercise D.2 Output Redirection

Create a text file with the names of all files on the Desktop using a single command on the terminal.

Exercise D.3 Using `grep`

Find out whether any file in the folder `/etc/` contains your account name.

Exercise D.4 More Advanced Actions

Create a working directory in your home directory, e.g., `Exercises`. Download from the Internet ten sequences of your favorite genes or proteins in FASTA format and copy them, one after the other, to a text file (`my_sequences.fasta`) in the `Exercises` directory. Using `grep` and output redirection, put only the header lines of these sequences in a different file.

Exercise D.5 Running BLAST

Copy a single sequence record (header + sequence) from `my_sequences.fasta` (Exercise D.4) to a different file (`query_sequence.seq`). Format `my_sequences.fasta` in order to be able to use it as a BLAST database. Use BLAST to align `query_sequence.seq` to the formatted `my_sequences.fasta` file.

Hint: See Recipe 11.

Index